Recording Science in the Digital Era

From Paper to Electronic Notebooks and Other Digital Tools

Recording Science in the Digital Era

From Paper to Electronic Notebooks and Other Digital Tools

Cerys Willoughby
University of Southampton, UK
Email: Cerys.Willoughby@soton.ac.uk

ROYAL SOCIETY
OF CHEMISTRY

Print ISBN: 978-1-78801-420-5
PDF ISBN: 978-1-78801-633-9
EPUB ISBN: 978-1-78801-844-9

A catalogue record for this book is available from the British Library

The Royal Society of Chemistry is a charity, registered in England and Wales, Number 207890, and a company incorporated in England by Royal Charter (Registered No. RC000524), registered office: Burlington House, Piccadilly, London W1J 0BA, UK, Telephone: +44 (0) 20 7437 8656.

Visit our website at www.rsc.org/books

Printed in the United Kingdom by CPI Group (UK) Ltd, Croydon, CR0 4YY, UK

Preface

Technology has had an enormous impact on science, providing new ways to capture vast amounts of data in digital form, providing mechanisms to control instruments and automate research, facilitating complex calculations and simulations, and offering ways to ask and answer new questions in science. Despite computers and software becoming essential day-to-day tools for scientists, the majority are still capturing the records of their research using paper notebooks. Electronic Laboratory Notebooks (ELNs) and other electronic notebooks challenge the traditional paper notebook. Such digital tools can (and should) make life easier for researchers. For example, they can perform complex calculations, connect to instruments to automatically collect data, connect to repositories to access previous research, provide facilities that enable collaboration and witness approval, provide access to content from any location, and so on. However, such tools can also have drawbacks, for example, incompatibility between systems, legacy data formats, issues around data security, and concerns about intellectual property protection. Just as importantly, such tools are often very complex to use in comparison to a paper notebook – what could be easier to use than pen and paper? The march of technology is inevitable and the need for researchers to work with ELNs and other digital systems is increasingly likely, for example, to meet the requirements of organisations, funders, and regulatory bodies; or to collaborate with partners around the world. Members of the scientific community need to understand what these tools can do and how they can make things better for their research, but they also need to understand the issues that exist in this area and how functionality may develop in the future.

Although there is a relatively long history of ELN development, the marketplace is still very much open. There are a few dominant players,

Recording Science in the Digital Era: From Paper to Electronic Notebooks and Other Digital Tools
By Cerys Willoughby

Published by the Royal Society of Chemistry, www.rsc.org

particularly targeted at corporate laboratories, but the lack of adoption of such tools outside of these environments means there is a vast opportunity for developers to create new tools with capabilities that would be of immense value to researchers. A key aspect of developing digital tools is to ensure their usability, but in order to do this, the needs and working practices of researchers must be understood. Many different groups are interested in ELNs, particularly for how they might help in meeting new requirements for data management and sharing, but the reality appears to be that many, particularly in academic environments, do not have a good understanding of what the benefits of ELNs are and the choices available.

The aim of this book is to document this area of growing importance for all those involved in science. The book aims to not only highlight the pros and cons of moving from paper to digital recording of scientific research, but also to consider how digital tools can be designed and improved to better support users of these systems. An improved design and functionality can not only encourage more participation, but also ensure that these systems capture and enable access to high quality records that can best facilitate good science into the future. There are limited materials available discussing ELNs and other tools for capturing and managing the scientific record. Existing materials typically describe the design or use of ELNs and the results of pilot studies, but there is a gap discussing the impact of these tools on the quality and content of the scientific record. There is also a limited amount of discussion about how researchers record and organise their records, and how a 'user-centred' focus can improve the design of tools for scientific researchers, including ELNs. This book brings together research into the motivations for recording scientific research, how scientific researchers (and chemists in particular) record their experiments and other research, the current benefits and problems with ELNs and other digital tools, and additional needs in this digital world that can inform both the functionality and interface design of tools to support researchers into the future. The book provides examples of different kinds of electronic notebooks and ELNs for both field and laboratory scientists, and provides some practical advice on implementing an ELN and making the transition from paper to digital record-keeping.

In order for high quality data to be effectively used and re-used by the scientific community, it must be both discoverable and trustworthy. This requires high quality context in the form of metadata to be available with the data, and for good practices to be following in curation and management of the data and records. This book also provides an overview of digital curation, data management, and preservation, as well as discussing some existing tools and how tools such as ELNs can be better designed to capture metadata 'at source'. The book also looks at issues in Open Science and how digital tools influence the dissemination of data and results, and how they can help to improve data sharing and public engagement in science. The book also considers how new technologies may change how the scientific record is used and captured in the future. Although a primary topic of this book is Electronic Laboratory Notebooks, and many of the examples are from chemistry, examples of digital tools and their applications within science are provided from across numerous disciplines, including both field-based and laboratory-based sciences. The same issues are faced across all disciplines, and although the specific digital tools used to mitigate them may be different, in the majority of cases, they are attempting to solve the same problems.

I have a long list of people who I would like to thank for their help, both in the production of this book, and for the research behind the examples included in the book. First, I would like to express my thanks to Colin Bird for his very helpful comments and advice on the chapters, along with his endless support and encouragement. I thank Jeremy Frey for his guidance during my research and for helping provide the opportunity to write this book. I would like to thank Sami Kanza for her helpful advice and comments on the chapters of this book. I would like to thank Connor Sheppard and Helen Armes from the Royal Society of Chemistry for their help and advice with the book. I would like to thank Colin Bird, Tom Clabon, Laura Cowen, Jeremy Frey, Steve Hughes, Sami Kanza, Katja McLaughlin, David Rankine, Mark Thomas, James Thompson, Tess Vaughan, and Phil Willoughby for generously allowing me to use their likenesses in images used in the book.

I would also like to thank the various people who have contributed to my studies and user research. From Southampton University, Thomas Logothetis, Colin Flowers, Simon Coles, and students and researchers from the School of Chemistry. I have been fortunate enough to visit a number of locations around the world to engage in user research, and wherever I have gone, I have received a great welcome and hospitality. Thanks in particular go to Brynn Hibbert and his team, Adrian Plummer and Danmar Gloria at UNSW, Sydney; Mat Todd and his team, Paul Ylioja, Swapnil Anam, and Murray Robertson at the University of Sydney; Karl Mueller and his team, Kelly Murphy, Nikki LaFemina, and Eric Poweleit at Penn State University; and Mark Rutland, his family, and colleagues from KTH. Apologies to anyone I have forgotten, my memory is more fallible than most!

Cerys Willoughby

Contents

Recording Science in the Digital Era: From Paper to Electronic Notebooks and Other Digital Tools
By Cerys Willoughby

Published by the Royal Society of Chemistry, www.rsc.org

CHAPTER 1

Introduction

1.1 Background

The paper notebook is, and has long been, the trusty companion of the majority of those who engage in scientific research. Some notebooks are beautifully illustrated personal records of a lifetime of research, within which we can trace the path of scientific discovery from the seed of an idea to fully fledged theories and proven results. Some notebooks are trophies of work, achievement and involvement, which when flicked through bring back nostalgic memories of breakthroughs and hard work. Other notebooks reside upon shelves, never visited, but kept 'just in case' they are needed, to investigate why an experiment might have failed or to hold the numbers explaining why a data point is different than usual. Some have a formal structure born from years of education, where the 'correct way to write a notebook' was drummed into the researcher, and others are freer flowing and more narrative. Some of the notebooks contain detailed representations of the memories and knowledge of the author who wrote them, while others are simply a reflection of the minimum information that the researcher was taught to record.

In this digital era, there is pressure to abandon paper in favour of the benefits of digital tools. For the majority of scientists, computers are essential everyday tools that enable their research. The majority of researchers are currently working with a mixture of paper and digital materials. Recording of scientific activities is still often completed within a paper notebook or using paper-based forms, even though the majority of instruments are now capturing data in digital form. Once the data is captured, it is usually processed and analysed using computer software, creating a disconnect between the paper record and the digital results. Although researchers are comfortable with the use of computers for everyday tasks,

Recording Science in the Digital Era: From Paper to Electronic Notebooks and Other Digital Tools
By Cerys Willoughby

Published by the Royal Society of Chemistry, www.rsc.org

there are barriers to the adoption of tools designed to replace the paper notebook such as Electronic Laboratory Notebooks (ELNs). This is particularly true for complex software where anxieties about ease of use can lead to scientists shunning the tools.[1] There are also questions about what impact such tools may have on the scientific record. Although it may seem likely that such tools would have a positive effect by potentially standardising the record, improving consistency, and incorporating digital data more easily, there is also a possibility that the technology may change the information that is recorded, or even change researcher behaviour in a negative way.

In order to design digital tools that support scientists carrying out scientific activities, it is necessary to understand how scientists work with both paper and digital tools, and to examine the impacts different interface designs may have upon what they record. In this book, we will examine note-taking and data management behaviour of a range of scientific researchers and examine the requirements for digital tools to facilitate the creation of a quality scientific record that can be used to understand and reproduce scientific research. We will look at how digital tools have developed in science, and in particular those that facilitate the capture and management of the scientific record. We will also take a look at techniques that can be used to derive requirements and improve the usability of digital tools for scientists. A number of studies are presented that explore how different techniques for eliciting information from the memories of researchers can be used to capture information of value to the scientific record, with both positive and negative consequences. We will also examine the role of digital tools in collaboration and Open Science, and how digital tools might be enhanced in the future to support researchers in performing and recording their scientific research.

1.2 The Scientific Record

The purpose of science is to generate knowledge about the world and beyond through systematic accumulation of information from observation, measurement, and experimentation. Scientific progress and the refinement of knowledge is entirely dependent upon the research that has gone before. It is not enough for scientists to just produce information and knowledge, that information and knowledge must also be captured and shared so that the wider scientific community can access it and build upon it. The effective recording and communication of scientific ideas, theories, observations, measurements, and results is therefore fundamental to science. To build upon the research of others it is necessary to trust the hypotheses, results, and conclusions that are derived from these scientific activities. The foundation of the scientific method is reproducibility. In order to verify that theories and observations of other researchers are correct, it is necessary to be able to repeat those experiments and to get the same results.

Although the first recording of scientific endeavour is a matter for debate, scientists have been recording and sharing their observations and ideas for millennia.[2] Since the 17th century, the 'scientific record' has been recorded primarily through the publication of academic journals and also conference presentations and proceedings.[3] However, the published record is not the complete scientific record. The scientific paper and paper notebook represent two ends of a process of 'inscription' and both artefacts can be considered to be part of the scientific record.[4,5] The paper notebook records the day-to-day scientific activities of researchers and often also records the planning stages of the research, whereas the journal article typically only represents the end of the process where the results and findings from the research are presented formally. Other information artefacts produced during the scientific process include data from instruments and 'intermediate' documents produced during research activities, such as research reports, presentations, drafts of papers, and meeting minutes, as shown in Figure 1.1.

The notes captured in the paper notebook represent traces of the memories, thoughts, and activities of the author, and the notebook itself accumulates these traces. The notes are typically captured in chronological order, and therefore reading through the notes enables the reader to go backwards and forwards through time to explore how the research has progressed or how ideas originated. The notes captured in the notebook are not just an aid to memory for the author. For scientific activities that are funded by industry or for commercial purposes, the contents of notebooks can be essential documents as a legal record to protect Intellectual Property (IP), and the content of notebooks and their associated data can be brought into sharp public attention during IP disputes or when fraud is suspected.[6–9] When the notebook does come to the forefront, it is vitally important that the content of the notebooks is 'good', for the protection of intellectual property and 'defending the research', but also for providing a record of the thoughts of the experimenter and detailed procedures of

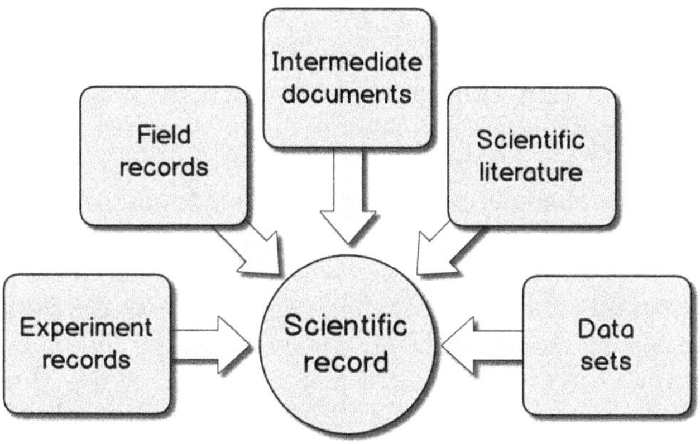

Figure 1.1 Elements contributing to the scientific record.

the experiments.[10] Paper notebooks are by nature blank and provide no guidance on what information should be recorded within them. Consistency in recording comes from standard guidelines for recording experiments used both in academia and formalised by regulatory organisations, for example the *Good Laboratory Practice Handbook*.[10,11] Industrial organisations are more likely to have specific requirements about recording than academic environments, where instead researchers are more likely to learn through trial and error. The record must be usable for many years hence, and must not only allow the experiments and conditions within them to be replicated by others, but must also be sufficient for the scientist producing it to be able to understand it throughout the duration of their research.[12] Even though technology has changed, and often the way that data is created has changed, many academic researchers are still keeping their paper notebooks in exactly the same way they were when they were taught decades ago.[13]

Paper notebooks have a somewhat confused role. On the one hand, they are extremely personal, individual accounts of scientific endeavour, but on the other, they are written to be read by others. Notebooks construct an 'institutional memory' for the immediate research group, university, or other organisation, fulfilling functions as organisational documents that must fill expectations beyond those of the individual scientist who created them.[12,14] In reality, however, most notebooks are shared with very few people and it can often come as a shock when researchers are asked to surrender their notebooks to their institution when they move on.

Notebooks that are not in current use can typically be found residing unused on shelves 'just in case', or even locked away in an archive.[14,15] This is not always the case, and some notebooks may be passed over to a colleague or a new researcher to inform their research. Even if notebooks are readily available and shared, there are still significant problems involved in extracting information of value from these records. Content in the notebook may be illegible or difficult to read, and the records may be incomplete. Notebooks are difficult to search and may require considerable specialist knowledge to interpret. They can be impossible to understand without significant extra context, and very often the only person that can decipher the content of the notebook is the person who actually authored the notebook.[14] The frustration of trying to understand the content of someone else's notebook is neatly summed up by Hoffman (p. 280):[16]

> Yet almost everyone who examines research notebooks shares the experience of how incommensurable, fragmentary, crude, puzzling, and thus hard to conceive such material is. Apparently, research notebooks do not at all permit a direct look into the researcher's mind or – through his or her eyes – at the course of events at the research site. Only when contextualized by additional sources and supplementary assumptions does the content of research notebooks become intelligible.

As a consequence of these problems, knowledge gained from previous scientific activity can be effectively lost to current and future generations of researchers. A particular consequence of this loss of knowledge is an unnecessary duplication of unproductive effort. This is particularly true for failed experiments. The lessons learnt from failed activities are extremely valuable for understanding the causes of failure, improving research protocols, and avoiding dead-ends in research, but negative results are rarely published. The knowledge from these failed experiments and activities remain trapped within the notebooks recording them, leaving others to blindly repeat the same mistakes again.

1.3 The Impact of Technology

Technology pervades our lives in numerous ways, many of which we are probably not even aware of. When we think about technology, we may picture the computer sat on our desks that we use for work or study, but technology is everywhere and intimately involved in our everyday interactions with businesses and individuals. The smart phone technology we carry in our pockets is incredibly sophisticated and enables us to perform countless tasks with a simple touch of the screen. The smartphone does not just facilitate communication, but also enables us to play games, keep up with the news, checking the weather, capture photographs, make videos, access our banks, search for information, and pay our bills. Our smartphones contain numerous tiny sensors that can determine our location, our movement, and our environment, feeding data to applications that can help us find our way and monitor our fitness. The pervasive nature of technology extends beyond the obvious and the visible. Everything is instrumented; we have smart cars, smart buildings, and smart cities. The impact of technology is not just about how much faster or easier tasks can be done, but how they change our behaviour and what they enable us to achieve.

Technology is undoubtedly having a major impact on scientific research and teaching, shaping the way that science is performed, results captured, and findings disseminated. Computers have gone from being devices primarily for calculation to now providing a vast array of functionality that enables scientists to not only work more efficiently, but also changing the way they work and enabling new scientific questions to be answered. Both inside and outside of the lab, most instruments are now digital and many are computer controlled. A large proportion of data from scientific activities is now 'born digital'. The ability of instruments to produce vast quantities of data rapidly leads to enormous challenges in terms of data management and ensuring the preservation of data over long periods of time. Computer software rather than paper and calculators also now provide the main tools used by scientists for capturing, analysing, and annotating their data. Some disciplines of science perform research entirely *in silico* or using data-intensive methods by taking pre-existing data in digital form or new data from instruments and using computational methods such as simulations and modelling to model processes or find patterns and derive new hypotheses.

Along with the increasing dominance of technology and digitisation of data, has come a drive towards more openness in science. Digital data is much easier to share, and Internet technologies, in particular, provide new ways to store and access scientific information, for example, ranging from email, social-networking, and cloud-storage to scientific directories, digital-format scholarly journals, and scientific data repositories. Organisations and institutions are becoming more global and more collaborative. Whether research and results are being shared within a single organisation or group, or to the wider scientific community, there is a need to ensure that scientific records and their associated data are effectively managed, preserved, and made accessible to others.

1.4 Current Issues in Scientific Record-keeping

Progress can only be made and research built upon if the research and underlying data are made available. The scientific journal and conference as the primary methods of communication of new scientific ideas and discoveries have rarely provided access to the underlying data or context. Published articles have historically placed greater emphasis on the conclusions made from the data than the actual data itself, and almost never describe the results or lessons learned from failed research.[17] Historically, data was collected and used for a specific purpose and subsequently discarded or at least ignored. The idea of reusing and repurposing of data were rarely considered and certainly not considered to be useful to anyone outside the originator's domain. In fact, despite the undoubted value of sharing data and scientific work, researchers have long been reluctant to share their data, particularly where work is funded by industry and where others would gain a competitive advantage from access to the data. The priority is to publish or to submit a patent first, and then maybe the data will be shared.

Much of scientific activity and progress depends on quality data. In order to make use of data and build upon the work of other researchers, it is necessary to have trust in the reliability of the data and the work of other researchers. In the last few years, there have been numerous high-profile cases of scientific misconduct. These range from poor scientific practice to outright fraud. Many of these cases were not originally detected during the scientific peer-review process that is meant to act as the gate-keeper of trust in scientific endeavour and communication. Poor scientific record-keeping and data management can easily lead to unintended mistakes and discrepancies between published and actual data through transcription errors, mixed-up data, and missing information. Deliberate manipulation, falsification, and entire fabrication of results is harder to find without access to the raw data, and this lack of access has allowed unscrupulous scientists to get away with fraud for much longer than they should. Trust in science extends to the relationship between science and society and not only does scientific fraud reflect badly on science, but in many cases the frauds have directly and negatively impacted on people's lives. Consequently, mistrust in science has spawned anti-scientific conspiracy theories, for example, one of the most famous cases is the now debunked link between the Measles, Mumps, and Rubella (MMR) vaccine and autism.[18]

Technology provides an opportunity to support transparency in research and the assessment of data quality. Tools such as electronic notebooks, and ELNs in particular, can help to standardise the captured record and can also facilitate the sharing of both data and scientific practice with the wider community. The sharing of data is a tricky problem. Many researchers are still reluctant to share their data, often fearing that the data will be misused or that they will be beaten to publication by competitors, and consequently fear unauthorised access to the data if they store it online or even behind a firewall. However, there is now increasing pressure on researchers to share their data when they publish the results. The principles of Open Science take this a step further by encouraging the sharing and reuse of all of the materials of scientific activities including notebooks, raw data, and other materials, such as intermediate data and code, to enable the reproduction of research and rapid access to research. Technology has added some additional problems. The increasing speed and capabilities of technology have enabled the generation of vast quantities of scientific data from instruments and sensors, leading to the concepts of 'big data' and the 'data deluge'. All this data needs to be stored and managed, and mechanisms are needed to be able to adequately process and interpret it so that it can be used in an appropriate timescale. More and more scientific data is being stored on the web, whether informally by researchers needing a safe place to put their work, or in huge disciplinary repositories, explicitly designed with sharing in mind. The storage and sharing of data brings with it many concerns about the protection of intellectual property and privacy issues. Who owns the data when it is located in the cloud or a repository? How can we ensure we get credit for data we share? How do we stop it being stolen? Does the data contain sensitive or private information? In the days when the data was captured on paper these concerns were more easily solved by simply locking up the data in a filing cabinet or safe. Although paper has vulnerabilities, digital data has unique problems of remote access, degradation, and obsolescence. In order to be useful, data needs to be discovered and meaningful, and effort is required to ensure that data is well described so that it can be found and understood.

1.5 Design of Interfaces for Record-keeping and Digital Curation

The purpose of digital tools is to make our lives easier, but technology can be complicated. Although many scientists now have roles that cross the boundaries between scientist and computer scientist, the majority of scientists want to be able to focus on their science without needing to become a computer programmer or database administrator. Tools designed for any audience needs to fulfil two basic requirements: the tool must provide appropriate functionality to enable the users to achieve their goals, and second, those users must be able to use the tool effectively and efficiently. Sadly, a great deal of software that is developed fails on both counts. The functionality often reflects the goals and assumptions of the developer rather than the needs of users, and the design of software is often far from intuitive for users preventing them from easily achieving their goals.

Usability is essential to the adoption and continued use of software tools. When software is difficult to use, users will avoid using it and ultimately reject it. This can be expensive for both the implementing organisation and the developer of the software. Usability is often a neglected aspect of software design, frequently not considered until late in the development cycle or even not considered until customers complain. Ideally, one or more usability practitioners should be involved from the very beginning of any software project in the concept stages, and then be there to work with both developers and users throughout the software development life cycle. However, usability practitioners are often only involved at a relatively late stage in a project, leaving them to find the existing problems rather than to help avoid them in the first place. Usability is most effective when it is included as part of the whole development process. Adhering to user-centred design practices helps to define user requirements, provides assistance with design, and validates the design through testing and working with users.

Although ELNs have been around for decades, their adoption within certain environments is relatively low.[1] Often, electronic notebooks, especially ELNs, are perceived as complicated, difficult to use, and unreliable. A paper notebook on the other hand is perceived as easy to use and flexible. If researchers feel that a digital tool is making their life more difficult then they will resent it and eventually ignore it, but if the tool facilitates their work and is easy to use, then it will be perceived as an essential tool for their research. The relatively low rates of adoption of ELNs within academic environments highlight that it is not sufficient just to provide tools such as ELNs and expect researchers to use them. In order for any software to be used and to be useful, it must be usable and meet the needs of the researchers that it is trying to support.

For science, digital tools play multiple roles in facilitating research and communication, making high quality data available, and enabling its reuse. In order for data to be effectively used and re-used by the community, it must be both discoverable and trustworthy. This requires high quality context for the data ideally added by the researcher creating the data, but getting researchers to do this is often a very difficult task. To this end, tools must be developed to facilitate capture of the research record and the associated context, but we also need to understand how such tools might influence the quality of the research record. With this in mind, this book strives to put the user at the centre of the discussion about ELNs and other digital tools. In order to be able to create such tools, we need to understand how researchers work, what information they currently capture, how they capture this information, and how the design of digital tools might influence what they record. We also need to consider the needs of the wider scientific community and how the design of such digital tools might facilitate the capture of the additional context required to make the research record discoverable and reusable. We would therefore also benefit from an understanding of what metadata researchers currently use and how digital tools might influence the capture of metadata.

1.6 Organisation of the Book

In this book, we will investigate the various issues raised so far in more detail. In Chapter 2, we will take a brief look at the history of note-taking in science and the motivations behind taking notes. We will also take an in-depth look at the role of memory in determining what we remember and record, and how different techniques from psychology might be used in the design of digital tools to encourage researchers to capture notes that are accurate, complete, and meaningful to others.

In Chapter 3, we will observe the current note-taking and data management behaviours of a range of scientific researchers. We will take a look at the record-keeping process and how researchers learn to keep their records. We will examine the guidance for note-taking given to different scientific disciplines and how notes from different disciplines differ in their structure and organisation. We will take a look at the importance of images in record-keeping and examine strategies for finding content within the notes and making links between notes in paper notebooks and other resources, such as digital data files or photographs. Finally, we will explore record-keeping behaviours, note-taking as ritual, and the private nature of notebooks.

In Chapter 4, we will make a brief run through the history of the digitisation of science and in particular those technologies that support the capture and storage of data in digital form and digital tools for capturing the research record. These technologies include the digitisation and automation of instruments and the evolution of databases for managing scientific data. We will take a brief look at the development and role of Laboratory Information Systems (LIMS) in managing the laboratory and laboratory records. We will explore the problems of managing paper-based and digital materials together and the potential solution provided by electronic notebooks. We will examine some examples of electronic notebooks for use in both the field and the laboratory. We will finally take a brief look at working *in silico* and the role of digital platforms for repeatable processing and analysis of data.

In Chapter 5, we will take an in-depth look at managing and preserving digital research data. We will discuss the issues and challenges around research reproducibility and how data management practices and curation can help. We will discuss the role of metadata, standards, identifiers, and provenance in data curation and their importance for discoverability and reuse. We will also examine the vulnerability of digital data and the steps that need to be taken to ensure that data is preserved for future access and reuse.

In Chapter 6, we return to electronic notebooks, and ELNs in particular, for a more in-depth look at their features and benefits over paper notebooks. We will examine a number of different examples, taking a look at how the similarities and differences in their capabilities change with their target audience. We will also look at some experiences of using an ELN from different disciplines and how an ELN can facilitate and support scientific research in different ways. We will discuss some of the potential barriers to the adoption of ELNs and explore why some scientists are reluctant to use them, despite their benefits.

Making the move from paper notebooks to ELNs for recording scientific research can be a daunting prospect. In Chapter 7, we will look in detail at planning the implementation of an ELN system, including consideration of factors that can influence the success of the project. For example, creating a business case, putting together a project team, overcoming resistance from staff, and managing the project. We also examine how to determine requirements for the project, including identifying the potential benefits of an ELN and how capabilities can be matched to the needs of the organisation. We will also discuss factors that might influence the choice of ELN provider and other considerations to take into account before selecting an ELN. Finally, we will consider the next steps after selecting an ELN such as planning the deployment, piloting the ELN, testing strategies, and training.

In Chapter 8, we will take an in-depth look at a variety of user design techniques, including interaction design, that are used to help make software both useful and usable for the intended audience. The chapter uses a variety of examples based upon the record-keeping needs and behaviours of real researchers and presents potential ELN capabilities based upon these needs. We examine techniques for gathering requirements, tools for modelling user goals, and designing potential capabilities including conceptual models, roles, personas, scenarios, storyboarding, and prototyping. This chapter also includes some guidelines and best practices for designing interfaces for desktop, mobile, and web software.

In Chapter 9, we will explore how different interface designs for digital tools might impact on capturing the scientific record. Many ELNs make use of templates to capture the record of experiments, whilst others enable a more free-form approach to recording. In this chapter, we will take a look at the potential problems with template use and how different template designs might impact the information that is recorded in both positive and negative ways. We also take a look at the role of mobile applications in bridging the gap between paper notebooks and ELNs, and how other kinds of interfaces may help to better support users in capturing their research in electronic form.

In Chapter 10, we will examine the use of metadata in science including the issues and attitudes of researchers towards creating metadata. We will look at the roles of both machine-generated and user-defined metadata and the importance of capturing metadata 'at source'. We will examine some of the ways in which digital tools can help support data curation, including automatic metadata creation and smart assistance. We will also look at how different interface designs might encourage curation and the production of user-defined metadata by researchers. We will examine existing digital tools for metadata design and creation, and consider how usability in tool design can support users with such tools. We will also take a brief look at how collaborative metadata creation can help generate metadata of use to a wider audience.

In Chapter 11, we will discuss the various 'open' initiatives that facilitate Open Science. We will also look at how Open Science can facilitate both collaboration across the scientific community and public engagement with science. We will also discuss how open access to data can provide opportunities to encourage mutually beneficial collaboration between volunteers and scientists through citizen science. We will explore the barriers to Open Science, in particular around data sharing, and look at some of the ways these barriers can be overcome. Finally, we will take a look at the use of digital tools in Open Science, including those for sharing data, replicating research, and communicating scientific knowledge.

In Chapter 12, we investigate a range of newer technologies that enable researchers to interact with their scientific data in new ways and potentially to simplify or improve the way the scientific record is captured. We take a look at technologies such as Virtual Reality, Augmented Reality, Artificial Intelligence, the Internet of Things, and wearables, and examine how they are currently used in a variety of different scientific applications. In this chapter, we will also look at a variety of 'natural interfaces' that might provide easier and more intuitive ways to interact with technology than the standard mouse and keyboard of the familiar desktop computer. We will also explore a variety of uses of natural interfaces within applications for science. Finally, we will take a look at how the technologies might be brought together to provide smart environments and intelligent assistants to facilitate the everyday activities of scientists and how they may help to provide alternative ways of capturing the scientific record.

Finally, in Chapter 13, we will take a look back at the various issues discussed in the book and attempt to bring together the various threads to consider the way ahead for recording science in the digital era.

References

1. S. Kanza, C. Willoughby, N. Gibbins, R. Whitby, J. G. Frey, J. Erjavec, K. Zupančič, M. Hren and K. Kovač, *J. Cheminf.*, 2017, **9**, 31.
2. C. L. Bird, C. Willoughby and J. G. Frey, *Chem. Sci. Rev.*, 2013, **42**(20), 8157.
3. C. A. Lynch in *The Fourth Paradigm: Data-Intensive Scientific Discovery*, ed. T. Hey, S. Tansley and K. Tolle, Microsoft, Redmond, 2009, p. 177.
4. B. Latour and S. Woolgar, *Laboratory Life: The Construction of Scientific Facts*, Princeton University Press, Princeton, 1986.
5. F. L. Holmes, *Investigative Pathways: Patterns and Stages in the Careers of Experimental Scientists*, Yale University Press, New Haven, 2004.
6. W. J. Levelt, P. Drenth and E. Noort, *The Fraudulent Research Practices of Social Psychologist Diederik Stapel*, Tilburg University, Tilburg, 2012.
7. D. Ince, *Significance*, 2011, **8**, 113.
8. L. Harding, 2005, Available at https://www.theguardian.com/science/2005/feb/19/science.sciencenews (accessed Jan 2019).
9. J. W. McAllister, *Philos. Sci.*, 2012, **79**(5), 871.
10. A. Eisenberg, *J. Chem. Educ.*, 1982, **59**(12), 1045.
11. WHO, *Good Laboratory Practice (GLP): Quality Practices for Regulated Non-Clinical Research and Development*, World Health Organisation, Geneva, 2nd edn, 2009.
12. K. Shankar, *J. Doc.*, 2009, **65**(1), 151.
13. S. Carpenter, 2012, Available at https://www.sciencemag.org/careers/2012/09/must-paper-trail-be-paper (accessed Jan 2019).
14. M. H. Elliot, 2004, Available at https://www.scientificcomputing.com/articles/2004/08/its-not-about-paper? (accessed Apr 2016).
15. A. Tabard, W. E. Mackay and E. Eastmond in *Proceedings of the ACM 2008 Conference on Computer Supported Cooperative Work – CSCW '08*, ACM Press, New York, 2008, p. 569.
16. C. Hoffman, *Sci. Context*, 2013, **26**(2), 279.
17. C. L. Bird and J. G. Frey, *Chem. Sci. Rev.*, 2013, **42**(16), 6754.
18. D. Flaherty, *Ann. Pharmacother.*, 2011, **45**(10), 1302.

CHAPTER 2

Motivations for Record-keeping

2.1 Introduction

As observed by Bruno Latour and Steve Woolgar, scientists spend a great deal of their time creating literature of various sorts.[1] Scientists meticulously record their actions and activities in notebooks, but they also spend their time creating tables of data, carefully labelling materials, generating indexes and inventories, scribbling on whiteboards, producing slides for talks, and of course, authoring the all-important scholarly literature. Latour and Woolgar describe these activities as an apparent 'mania for inscriptions' (p. 48) and the role of the laboratory is to turn physical materials into these inscriptions. They observe how the materials of the laboratory, such as animals and chemicals, are transformed into sheets of figures and graphs ready for publication. Some of the material creates its own inscriptions by activating instruments that record it, effectively transforming matter into written documents. The final document is considered the most significant part: the journal article, upon which funding and credibility in science depends. Soon the journal article becomes a part of the published literature and begins the process of inscription all over again as the scientists build upon the inscriptions of others. Whether or not you agree with Latour and Woolgar's assessment of the behaviour of scientists in the lab, it is true that researchers spend vast amounts of time collecting and processing information to generate knowledge for dissemination. Strangely, Latour and Woolgar's study seems to mostly ignore what are perhaps the most important inscriptions of all: the capture of the daily research record.

A majority of the research into scientific writing and record-keeping has focused on the production of formal writing and scholarly publications, and much less attention has been paid to the day-to-day recording of

Recording Science in the Digital Era: From Paper to Electronic Notebooks and Other Digital Tools
By Cerys Willoughby

Published by the Royal Society of Chemistry, www.rsc.org

activities and the organisation of related materials.[2] And yet these activities are fundamental to the capture and dissemination of scientific knowledge. Without the capture of our thought processes and our observations, it is impossible to share our ideas and findings with others. The who, what, when, where, how, and why of our scientific studies would be lost. The day-to-day notes that we typically capture in paper notebooks are of fundamental importance to the scientific record. Information that we fail to record or that is recorded inaccurately impacts on the reliability and completeness of the scientific record itself. In this chapter, we will take a brief look at the history of record-keeping and the reasons why we take notes. We will also take an in-depth look at the role of memory in capturing the scientific record. In future chapters, we will explore how this knowledge of memory can be used to influence the design of interfaces to encourage the effective capture of the scientific record for ourselves and other scientists.

2.2 Early Record-keeping

Record-keeping has a long history, and probably dates back to the very beginning of writing when records of commercial transactions or the storage of grain were recorded to act as an accurate memory over long periods of time.[3] Some of the earliest evidence of scientific note-taking and records come from the Babylonians, who made sophisticated astronomical observations and recorded the details of these observations on clay tablets.[4] In classical times, note-taking as a prelude to formal writing was a common practice, used, for example, by the prolific note-taker and author Pliny the Elder, whilst 200 years earlier, Archimedes had made recordings of his mathematical discoveries and mechanical inventions on parchments made from animal skins.[5] Even earlier than this, Hippocrates and his followers were prolific producers of writings, including various textbooks, lectures, research, notes, essays, and even the medical histories of patients that are still of value today. Hippocrates and his followers' writings provide a record of experiences and observations that have been used for teaching and providing an insight into historical practice for more than 2000 years.

Although we tend to think of note-taking as being scribbles made in paper notebooks, a variety of materials such as clay tablets, animal skin parchment, vellum, papyrus, and paper sheets were used before the invention of the paper notebook in around 1600. Notebooks consisting of bound blank pages were given a variety of names based upon different intended purposes including writing-books, day-books, ephemerides, diaries, memorandums, waste-books, journals, ledgers, common-place books, and later the 'pocket book'.[6] In historical note-taking, notes were often considered temporary and were taken on erasable writing surfaces, and later discarded completely or revised and copied onto more durable surfaces for more permanent storage. Erasable surfaces have been used since antiquity and were still in use until relatively recently. These erasable surfaces included wax tablets, slate blackboards, sand trays with fine material, and chemically treated paper that could be wiped clean. The modern equivalents to the erasable surface would be whiteboards, post-it notes, and some note-taking software, where the contents are either discarded after use or copied into a different tool for keeping and refining.

Paper notebooks were used to record noteworthy information more permanently. Typically, information from books or lectures were captured and were usually arranged by topic so that related ideas could be captured together. All literate school boys would be taught how to create these 'common-place' books, which acted as a convenient collection place for knowledge, where quotations from books or arguments would be organised under headings or subjects ('Heads' or 'Titles') for easy retrieval.[6] Scholars developed impressive techniques for condensing, managing, and retrieving references to the scores of literary sources they recorded in these common-place books that held a lifetime's reading and analysis. These common-place books were often large and designed to be left in the library or study, whilst the chronological journals and the pocket-books, used by Da Vinci, were portable and ideal for taking notes on the go.

In the 17th century, scientists were moving from simply copying and organising content to making and recording their observations of the world within notebooks. Recording these observations served as an aid for the author's memory, but these scientists also recognised that their research was unlikely to be completed within their own lifetimes, and that the records would be of value to others in the future who might continue the work. The Renaissance and the Scientific Revolution that followed encouraged the capture of detailed scientific notes and records. Naturalists such as Linneaus, Dampier, and Banks were exploring the globe and capturing observations of the natural world, whilst other scholars such as Gilbert, Galileo, Boyle, and Newton were documenting the revolutions of astronomy, mathematics, physics, and chemistry through experimentation and invention. The notebooks of such scholars and scientists are objects of historical interest, providing an invaluable record of the workings and evolutionary path of science and a fascinating insight into minds of their authors. These notebooks were typically highly detailed, beautifully illustrated, and their authors meticulous in their recordings, capturing not just their activities, but also their thoughts and their ideas.

2.3 Why Do We Take Notes?

Although historically the communication of science has been in the form of letters or formal publications, for hundreds of years, the primary mechanism for recording science has been the paper notebook. It is within these notebooks that our research takes form and develops, and we capture the details of the process we followed, the observations we made, and the measurements we captured. But why do we take notes? When I asked this question to around 60 researches and staff in chemistry from the KTH

Royal Institute of Technology in Sweden, the answers given revolved around a number of common themes. The primary response was, unsurprisingly, that we take notes 'to remember'. Without taking notes we forget what we have done and how we did it. A related theme to this was about using notes for 'troubleshooting'. We need to remember what we have done so that we can work out what has gone wrong or to analyse our research. The record of previous experiments can be compared with the current experiment to understand why problems or differences might have occurred. The majority of researchers I have talked to about note-taking have also spoken about this use of their notes to help track down problems with their current research. Why did that experiment work, but this one failed? How did the change I made in this experiment impact the results compared with the last experiment?

A related theme discussed by the KTH scientists was reproducibility. Not only do they use their notes to compare results between different experiments, but also to remember the experiment in order to repeat it. Notes are frequently used to act as a reminder of a particular procedure that has previously been performed or to build upon the results of a previous experiment by adapting from the original to a new design. The notes are memory of what worked and what did not. In my experience, the majority of laboratory scientists make use of their own records at least as much as the published literature when planning new research, and even more so when they are building upon previously performed experiments. Researchers, in fact, often find it very hard to directly repeat experiments from the literature because too much contextual information is missing, and standard practices differ in unknown ways between different institutions. Repeating your own experiment, successful or not, is much easier than trying to repeat someone else's, in part because of tacit knowledge, but also because the procedures recorded in the notebook are usually much more detailed. Tacit knowledge is our expert knowledge and skill that we develop by observation and active practice, and is often difficult to share because we may not be consciously aware of it or it may be 'intuitive' and therefore difficult to describe to someone without the same experience. As a result, subtleties, learned through practice and essential for success in performing a task, are often lost when the procedure is written down.

It is no surprise that notebooks serve as an 'aide-memoire', but these are no temporary rough scribbled notes, needed only in the moment. For their value to be realised to the author, they must provide sufficient and reliable information to enable the author to track down and understand problems in their results and to repeat their own processes. A poorly kept record will provide none of those things. Hopefully, most researchers learn quickly when their own notes are inadequate, for their own benefit at least! An individual researcher may be able to piece together enough information from a poor record and their own memory combined, but that is unlikely to be sufficient after a few months and if anyone else needs to try to follow the procedure.

Another major theme for taking notes discussed with the scientists from KTH was around legal requirements that it was necessary to adhere to, and concerns about Intellectual Property (IP) protection. The record does not just supplement their own memories, but it also provides an essential record of what was done, when, and by whom. In the discussions, it was a common opinion that being able to provide 'proof' of having done an experiment or having made a discovery made keeping notes a necessity. IP is a hot topic in science. Research is competitive, whether that is because the research itself is commercially valuable or because of the need to publish first in order to gain prestige and win funding. Disciplines such as biotechnology, electronics, materials science, and novel chemical processes have been spun into multi-billion dollar industries by patenting techniques, technologies, materials, and even organisms.[7] For the majority of scientists, well-kept notebooks are essential documentation to provide the proof necessary for engaging in the patent process and for fighting IP disputes. However, IP is not the only legal issue that concerns record-keeping in science. Pharmaceutical companies are highly regulated and drug testing results need to be backed up with detailed provenance trails, whilst the structure and content of patient records in healthcare and related research are often legally mandated because of their potential to be brought into lawsuits and insurance disputes.[8,9] Funding bodies, institutions, and organisations often have their own requirements and best practices that must be adhered to, and notebooks and other documentation may be regularly audited to assess compliance. The capture of detailed and accurate notes is important to ensure that commercial gains can be made from discovery, but equally important is the protection gained from ensuring that adequate evidence is available to support your claims if the research comes under legal scrutiny.

There are also more personal elements for keeping notes highlighted by the discussions with the KTH scientists. For example, the researchers make use of their notes as a background to discussions and producing reports, organising the data to find hidden patterns, and highlighting important information. Very few of the participants mentioned the use of notes for communication and sharing with others, although using the notes for instruction and 'intellectual continuity' were mentioned. The lack of 'sharing' as a reason for keeping notes emphasises the private nature of the notebooks, which is an element of note-taking that we will come back to in Chapter 3. We had to ask the researchers explicitly 'who uses notes and why?' to elicit thoughts about the role of notes for people other than the authors themselves. Beyond 'me', the potential users of the notes appeared to be limited to others within the same team or organisation: supervisors, colleagues, and 'the next generation'. The 'why' of the notes for others include the same personal reasons as before, but also opened up more collaborative activities, such as enabling others to reproduce the work, knowing what does not need to be repeated, improving procedures, sharing data and procedures, and for the production of manuscripts. To a large extent, these responses reflect the notebook as not just a tool for personal memory, but also as an 'institutional memory' where others can use the knowledge of individuals to learn and build upon. The 'proof' element also came into play with some of the more comical responses, with the notebook acting as proof that you work where you say you do and as a way to impress your supervisor! It was interesting that the notebook was recognised as an asset for the owning organisation or project, but not for a wider audience. In many organisations, notebooks are passed from one researcher to another in the team. For example, notebooks may be passed over to a new member of the team when staff move on, someone new takes over a project, or a new researcher joins the team. The notes then become a foundation for continuing the project work and as a learning tool for new staff.

Reproducibility is a significant issue in science. Not only do researchers need to be able to reproduce their own experiments to grow their research, but others have to be able to reproduce them in order to verify that published results are valid and trustworthy. As stated by Poulter, 'The integrity of science as a discipline rests on the ability of scientists to reproduce the claims of others'.[10] Despite the importance of reproducibility, it is relatively rare for studies to be independently replicated before publication. An exception to this is the *Organic Syntheses* journal that asks for significantly more details about experimental procedures than other journals and experiments are carefully replicated in the lab by one of the journal's Board of Editors. This may seem extreme, but there have been a large number of studies that indicate a vast quantity of scientific results are unable to be reproduced, contain errors, or include a lack of data to convincingly back up their claims. These problems of reproducibility extend across many scientific disciplines and also into areas of human study such as psychology and social science. Everybody makes mistakes, but common causes of discrepancies between the data and the claimed findings can come from poor record-keeping, poor management of data leading to mix-ups or losses, errors in calculations, inexperienced use of statistics, and poor study design. Even more concerning than the accidental mistakes that get published, are examples of unethical malpractice where researchers have deliberately manipulated, falsified, or outright fabricated results. There have been a number of high-profile retractions from journals of published scientific studies when other researchers have been unable to reproduce the results and become suspicious or noticed inconsistencies in the data. Just a few examples include: the retraction of four papers from a study on personalised therapy for cancer treatment at Duke University, where researchers had deliberately manipulated data and misused statistical analyses;[11] the 'Hwanggate' scandal, where four papers were retracted after it was discovered that large proportions of the data had been fabricated in research on patient-specific embryonic stem cell lines, as well as unethical practices around the attainment of donated eggs;[12] Jan Hendrik Schön, a physicist working on nanotechnologies for Bell

Laboratories, falsified or fabricated data for dozens of papers, deleting the original files making verification impossible;[13] and dozens of papers have been retracted from a prominent social psychologist Diederik Stapel, who admitted spending years manipulating data and fabricating research.[14]

Sometimes scientists are accused of fabricating or manipulating data to skew results in the direction they desire. In cases where scientists are unfairly accused, the unpublished scientific record becomes an essential tool for transparency and self-protection, as demonstrated by the so-called 'Climategate' controversy. When a server was hacked at the Climatic Research Unit (CRU) at the University of East Anglia (UEA), thousands of emails and computer files were retrieved and distributed to various sites on the Internet ahead of a climate change summit. Climate change sceptics selectively chose content from the stolen documents to suggest that climate change theory was a conspiracy and that the climatologists at the CRU had fabricated or manipulated data to exaggerate global warming. Several investigations found that there was no evidence of wrong-doing by the scientists at the CRU, but a significant outcome was the recommendation that more effort was needed to gain trust with the public by making available all the supporting data, including computer code so that findings could be replicated and verified independently.[15] We will discuss more about the role of the scientific record in reproducibility of science and openness in science in future chapters.

The use of note-taking as 'proof' of activities and of recording exactly what happened can be thought of as 'showing your working'. Much like school maths exams, if you record only the answer and not how you got to it, the reader is both less informed and suspicious. If the path to the answer cannot be tracked, then it is very much more difficult to solve problems and provide evidence to back up claims. An important extension of showing your working is the idea that the record is not just recording activities and ideas, but also that the act of note-taking actually helps to develop ideas by enabling the author to organise and focus their thoughts.[16] The time required to write the narrative of the experiments or observations provides the opportunity to reflect and assess the success or failure of the activities and how they fit into the research as a whole.[17] Even reading through old notes years later can spark old memories and lead to new inspirations.[18] If the time is not taken to properly record the work and reflect on the activities then opportunities for new ideas and solutions to problems may be missed.

Note-taking practices and the records that are kept will inevitably vary depending on the nature of the organisation, the scientific discipline, and the nature of the individual. For industrial and commercial purposes, note-taking needs to be more formal to meet the needs of IP protection or regulations, and as a result may have a more consistent structure or pattern that needs to be followed. Whereas records in academia may be looser and more ad hoc in structure, reflecting the development and transformation of ideas and different paths in the research. In any case, the notes captured in our notebooks, and their associated unpublished 'inscriptions', are at least as an important part of the scientific record as formal publications.

2.4 The Role of Memory in Capturing the Scientific Record

Just as knowledge and experience of scientific activities is of no use if it is locked inside the researcher's head, the scientific record is of little benefit if it is inaccurate. The information that is captured in our notes must be accurate, meaningful, and complete. Note-taking is an act of memory, allowing the author of the record to forget the specific details of an experiment or their observations in the field, but to access the memory again at a later time. The notes also act as an 'external memory' or 'institutional memory' for the benefit of others who wish to replicate or understand the recorded activities. In some situations, the steps of an activity or important observations are only recorded after the event, often because it is difficult to make notes at the time they are needed. For example, for many laboratory scientists, it is not possible to take a paper notebook into a dust-free laboratory; for field scientists, it is not practical to record notes whilst working in the field at night; or for all scientists, the complexity of the work or time-constraints mean only the most significant details can be recorded at the time or to save effort.[2,19,20] Under these circumstances, it is easy to lose important information because it is forgotten before it has had a chance to be recorded into the 'memory' of the notebook. Our notes prompt the recall of memories, but the notes we record are intrinsically linked to what we remember, both in terms of our memory of events and also what we remember to record. The accuracy of the record is only as good as the information that is entered into to it. If the record excludes information because the author did not remember it, did not think it was important, or it was obvious to them (but unknown to a later reader) as so-called 'tacit knowledge', then the record may cease to be usable in the future when the original memory has been lost.

Digital tools can be used as note-taking and record-keeping devices, becoming the 'external memory' for the author and for others. The information that a researcher enters into such a system is affected by the knowledge in their own memories and by the interface that is provided for them. Such tools can act as a prompt for users, encouraging the capture of some information by explicitly asking for it, but it is also possible to inadvertently cause a loss of information by structuring interfaces in a particular way that hinder recall of important information. In order to design interfaces that have a positive impact on the users of these systems and that do not cause unexpected problems, it is important to have an understanding of the way that knowledge is stored in the mind, our memory processes, and techniques that can be used to stimulate recall.

The knowledge that each of us has is vital to our ability to be able to make sense of the world around us and to function within it. This knowledge not only relates to our understanding of our environment, but also to who we are, and how we achieve our goals and carry out plans. Knowledge is stored in our memory and we access this information whenever we perceive our environment, plan an action, remember the past, or imagine the future. Traditionally, memory and knowledge structures have been investigated with lab-based psychology experiments. More recently, our understanding of memory has been supplemented by studies of patients with brain damage and memory disorders, and the use of brain imaging techniques on both normal and brain-damaged patients. There has also been a drive to increase the number of experiments that are 'ecologically valid', in that they apply to how memory is used and operates in everyday life. An example of ecologically valid studies of memory are on eyewitness memory. Eyewitness memory is the memory an individual has for a dramatic event, often for witnessing a crime or disaster. Because eyewitness testimony is commonly used in the judicial system, the importance of whether the memory is accurate and reliable is highly significant. Research into eyewitness memory has led to findings about how memory for events can be changed and distorted by the influence of other factors. The research has also led to the creation of a variety of techniques for promoting the accurate recall of memories, and which we will discuss a little later in this chapter.

Our knowledge is stored in long-term memory, as opposed to short-term memory that we use for holding small amounts of information for brief periods of time. For example, when we read a measurement from an instrument it is stored in short-term memory for a few seconds before we write it down for later recall. Long-term memory, in contrast, is where information is retained for much longer periods of time, and potentially indefinitely. Information stored in long-term memory is traditionally divided into two main types, implicit and explicit memory. Explicit memory is conscious memory that can be retrieved and 'known', including memory for semantic (general knowledge) information and 'episodic' memory (memory for events). Implicit memory, on the other hand, includes procedural and perceptual memory, information that influences our behaviour, but that is not consciously recalled. For example, perceptual memory involves recognition of things we see and hear, whilst procedural memory is the learning of repetitive skills such as reading, riding a bike, and playing the piano. However, the evidence that these types of memory are actually separate systems with different ways of representing knowledge is controversial. Although many studies apparently show evidence in favour of the traditional separation of implicit and explicit memory, other groups argue that these can be explained with a single memory system.[21–24] These disagreements are characteristic of the difficulties in explaining the internal functions and systems of the brain from human behaviour and cognitive problems alone. Despite these problems, findings from psychological research studies are useful for providing clues as to how we might be able to better design interfaces for digital tools to support researchers and to identify how aspects of human behaviour might cause problems for the capture of experiment records or interactions with our tools.

Studies have shown that the knowledge we have has an impact on how we understand, interpret, memorise, and recall information. The way that the knowledge is represented, acquired, and stored in memory is important for understanding why these effects might occur. Knowledge structures are also important in understanding how the different types of information that are stored in memory might be connected and related. Understanding these knowledge structures is also useful for designing information structures in computing such as metadata, semantics, taxonomies, and schemas. These information structures are useful for capturing and describing scientific activities and data in a way that it is meaningful for the humans that must make use of them. The knowledge structures in memory determine how effectively humans can both process and remember information. If we present information in a way that is incongruent with these knowledge structures or is unexpected, then we can create confusion and undesirable effects. In the context of technology, this can lead to interfaces that fail to generate the expected response from the user, are confusing, and generate avoidance behaviour. For digital tools for capturing the research record, this could mean a failure for the user to engage with the system, cause users to provide inaccurate or inappropriate information, or to fail to record important information. All these effects can lead to missing information and a poor-quality scientific record that is not fit for purpose.

Semantic memory is memory for general knowledge about the world such as knowledge of objects, facts, word meanings, and people, which do not relate to specific instances of time or place. Semantic memory is thought to consist of 'concepts' about things in the world. A concept is a representation in memory of a specific entity or class of entities together with their properties. For example, the concept 'duck' may have the properties 'quacks', 'swims', and 'has webbed feet'. At a higher level (or superset) is the concept 'bird' that has the properties 'lays eggs', 'has feathers', and 'can fly'. The duck is a child concept (or subset) of bird, along with parrot, ostrich, and so on. At a higher level still, is the concept 'animal' with properties 'moves', 'breathes', 'has skin', and 'eats', as shown in Figure 2.1. Evidence for such a hierarchical organisation of concepts in the brain comes from a condition called semantic dementia. In sufferers of semantic dementia, the meaning of words break down systematically from precise referential meanings, through semantically related meanings, and finally to a total loss of meaning. Less familiar meanings break down first, for example, patients forget the names of less familiar animals before more common ones.[25] What these symptoms show is that there is a continuum of meaning in semantic memory at different levels of abstractness. At the most extreme level, information is disconnected from context, and at the

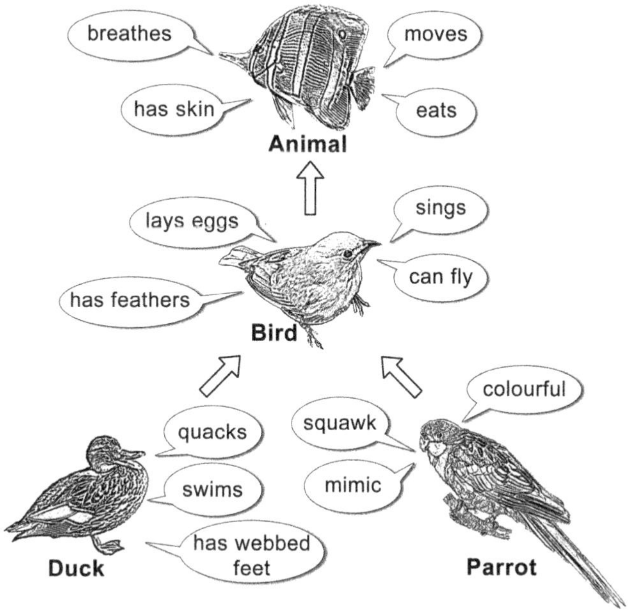

Figure 2.1 Example of representations of the concepts 'duck', 'parrot', 'bird', and 'animal' in memory.

other, it is embedded in specific contexts relating to personal experience. As higher-knowledge structures start to break down, the context-free levels of knowledge are compromised, and the meaning of words and objects presented outside of a physical context can no longer be identified.[26]

A theory known as the spreading-activation theory suggests that semantic memory is a network of concept nodes organised by semantic similarity, such that the more properties two concepts have in common the more connections exist between them.[27] This is similar to the structure of an ontology used to represent relationships between concepts in computing. For example, Figure 2.2 shows an example of an ontology based upon the concepts for birds discussed previously. When a concept is processed because it is seen, read about, heard about, or thought about the node is activated. The activation acts like a signal that spreads out from the node along the paths of the network, activating the semantically most closely related nodes along the network, making it quicker to retrieve information about more closely related concepts than distantly related ones. The spreading activation term is also used in computing to describe a mechanism for information retrieval using semantic networks that functions in the same way, leading to suggestions that using spreading activation techniques on ontologies could be used to support users of personal and task-based information management systems.[28,29]

The structure of memory as a semantic network can also help to explain why certain information is more likely to be retrieved from memory than

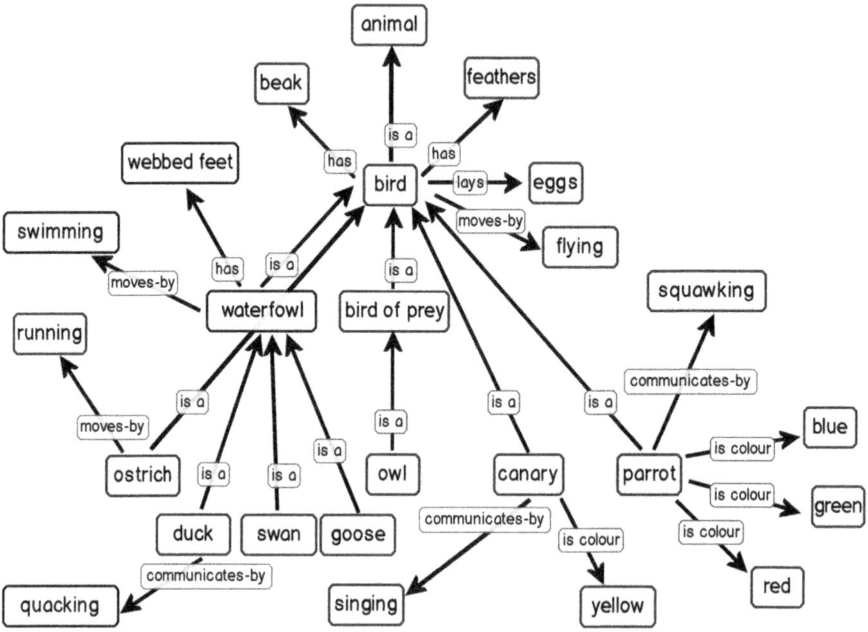

Figure 2.2 Example of how concepts about birds might be represented as an ontology.

others. For example, if a semantic network contains nodes for vehicles with related nodes: bus, fire-engine, car, truck, and so on, and there is also a node for red, fire-engine will have a close connection to red, along with other nodes such as: orange, apples, sunsets, cherries, and so on. If 'vehicle' is use as a cue for information retrieval, then the spreading activation from the vehicle node will activate fire-engine, car, truck, bus, and so on. Activation of each vehicle related node will retrieve others that are semantically related, such as 'have fuel', 'have wheels', and so on. If 'red' is used as a cue, the activation that spreads to fire-engine will be unlikely to retrieve apples, cherries, and sunsets, because they share few semantic similarities. Figure 2.3 shows the potential spreading activation routes from 'red' and 'fire-engine' to 'vehicle' and then other vehicles and the properties of vehicles, but there are a number of other potential routes. For example, if the spreading activation route was to go from 'red' to 'apple' then other red fruits such as 'strawberry' and 'cherry' are more likely to be activated than 'bus'.[27]

There is evidence from some brain-damaged patients and neuroimaging studies that suggest we store perceptual and functional information about objects in different parts of the brain. Living things are usually associated with perceptual information, such as what does it look like, taste like, smell like, or feel like? Man-made objects tend to be associated with functional information, what is it for, how do you use it? There are many brain

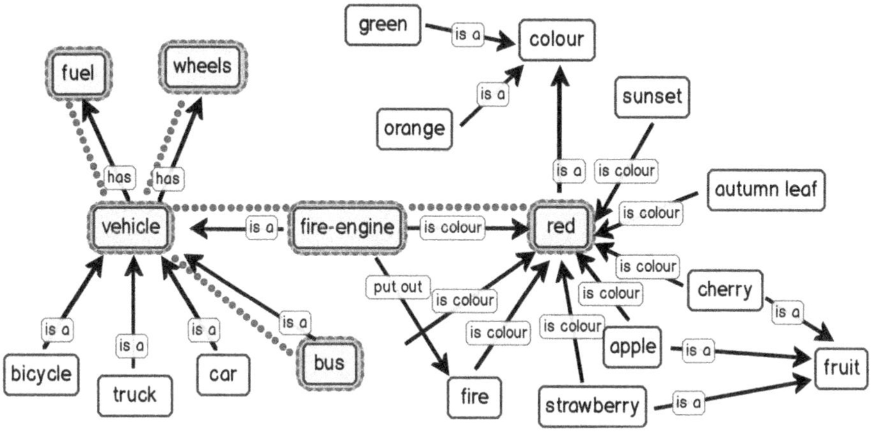

Figure 2.3 Example of spreading activation theory. The dotted line shows how concepts may be activated along the path from 'red' to 'fire-engine' and then 'vehicle' and related concepts such as 'bus' and 'has wheels' and 'has fuel'.

damage cases where patients have a problem recognising living things, with a smaller number of patients that are unable to recognise non-living things.[30]

Different authors use different names for these knowledge structures (such as schemas, categories, frames, and mental models) and have different theories to explain their workings, but they are all roughly talking about the same conceptual model of memory.[31] In order to make sense of the world and to interpret new information, it is necessary to be able to relate the new information to existing information. Our mental representation of previous experience and knowledge enable us to have expectations and predictions about the world. These knowledge structures also have a large impact on how we understand, process, and retrieve information. If we are able to associate information with what we already know through the presence of context, then we are more likely to be able to make sense of it and to remember it. If information fails to be understood then it is less likely to be remembered and less likely to be remembered correctly.[32,33]

Our need for context and understanding has potential impacts in a number of areas related to record-keeping. If a researcher is performing a task without a full understanding of the context of the task, they may not only struggle with the task itself, but then also struggle to remember what they did because of the difficulty of generating their own retrieval strategy based on conceptual understanding. If researchers do not understand particular concepts, they may rationalise these and make erroneous interpretations. These could in turn distort their memories for what they did in the experiment, what they thought happened, and consequently what they then record. The misinterpretations and erroneous recall can have a knock-on effect for another researcher trying to make sense of the record or to duplicate the research. Although the design of digital tools for recording the research record on the surface may not be able to help solve the misunderstandings, the design could encourage researchers to record information about their understanding of the concepts, so that it is easier for mistakes to be discovered by a supervisor or colleague. Digital tools could also be able to make use of data mining and other techniques to provide checking and validation to highlight inconsistencies and discrepancies in recorded methods and results in order to highlight problems.

Studies have demonstrated that our expectations and previous knowledge can interfere with what we remember. Information considered as 'obvious' in terms of our knowledge may be omitted from what we record about our research, but 'unobvious' or unexpected information would be more likely to be highlighted. We also tailor the information we retrieve and how we present it based on who our perceived audience is and when we are asked to take a different perspective when recalling information.[33–37] Social context also appears to shape the way that memories are recalled and retold. For example, people tend to use exaggeration when they recall information if they are trying to entertain an audience, but tend to simplify information when the intention is to inform.[36] The findings of such studies suggest that deliberately changing our perspectives to think about a different audience at the time we record our activities could elicit different information and therefore result in different information being recorded. The information recalled could be more or less accurate depending upon the perspective adopted, or the specific instructions given. If we ask researchers to record information for a particular audience, a corresponding change would be seen in the information recorded based upon the perceived knowledge of the specified audience. For example, if a researcher was asked to write up an experiment for their supervisor the record may be less detailed than if the record is written for the benefit of a colleague or student who needs to repeat the experiment. Researchers may choose to omit details because they are making assumptions about the shared knowledge with the audience, or because they assume that a 'gist' perspective is adequate instead of a fully accurate account. Recording the research for a colleague undertaking similar research is likely to be less detailed than if an account was provided to enable someone from an entirely different research team to understand or repeat the work. The content of the recording may also depend upon the outcome of the research, for example, a researcher may make the assumption that an accurate record is not required because an experiment failed and may even chose not to record a failed experiment. In some cases, what may be recorded is what should have happened, rather than what did happen. For example, students may claim their experiment worked because that is the expectation of their teachers; and thereby fail to record pertinent observations or engage in reasoning about why the results were different than expected. Students may write up an experiment as though it worked, even though it did not, because they do not view the failure as part of the learning experience. Important information about why it the experiment has failed is then lost. If the student tries to repeat the experiment in the future there will be no guidance in the record on how to avoid the problem and there is a good chance that the same mistakes will be made.

That the audience of the communication affects the way memory is communicated is potentially very important in the context of digital tools. The perception of who the audience is, what knowledge they are expected to have, and the perceived function of the record may be very important when recording information that will be used by others. Humans are not very good at judging what knowledge others have and tend to wrongly assume others have the same knowledge as themselves.[38,39] This is another reason why ensuring the capture of as much detail and context as possible in the scientific record is important. Digital tools can support users in this endeavour by ensuring the recording of accurate and complete information, but also by providing ways of bringing together information from different sources to improve the efficiency of recording. For example, many researchers repeat the same or similar experiments a number of times, and it is inefficient to record the same protocol repeatedly. Researchers may choose not to write down the procedure each time they do the experiment or miss 'obvious' steps out from their record. Digital tools can help by enabling researchers to duplicate the protocols and make changes to the appropriate sections to describe differences. Digital tools can also link to common protocols so that the whole process does not need to be written out each time, but it is still available for the author or other researchers at a later time.

The activities we perform usually have a recognisable sequence or procedure, where one thing has to happen before another. For example, when making a cup of tea, you have to heat the water and add the tea bag to the mug before you pour on the water, and then later remove the tea bag. A special kind of schema in memory, known as a script, provides a knowledge structure in the brain for handling these kinds of ordered events and procedures. Scripts deal with events and the consequences of events, and contain the information about the order of events that are usually expected to happen in a given situation. For example, when you go to a restaurant you know that reading the menu comes before ordering the food or paying the bill. Scripts may be important for the planning and execution of everyday activities in order to fulfil particular goals. Scripts also enable us to understand other activities, such as when we observe someone else engaging in an activity or reading the account of a given activity. Scripts are potentially relevant to scientific activities, because the majority of research activities include step-by-step procedures that have to be followed or protocols of some other kind where actions happen in a specific order. For example, performing an analysis of data, carrying out a field survey, or undertaking an experiment all usually require a number of steps to be followed in the correct order. Even without the individual detailed steps we consider as part of the experiment such as mixing chemicals or taking a blood sample, there are other everyday activities to complete such as coming to the lab, putting on a lab coat, fetching equipment, measuring chemicals, making observations, cleaning up, and leaving the lab, with many other steps along the way. Studies have shown that there is a surprising amount of agreement between the 'basic' action language we all use to describe familiar activities, such as visiting a dentist and eating in a restaurant, with very few unique actions described; and we also tend remember the actions in the stereotypical temporal order.[40,41] When presented with descriptions of activities, subjects in studies found it more difficult to recall the correct order if actions were presented out of sequence and also falsely remember actions that were not actually carried out, but that fitted with their expectations of what should happen. Subjects also tend to forget information that is not perceived as relevant to the script unless it was particularly interesting or unusual. For example, in a script for eating a meal in a restaurant, that the waiter poured wine might be forgotten, but the fact that the waiter was naked would be well remembered.

When recording repetitive research, the influence of scripts in memory may have a number of potential impacts. Actions that are meant to be taken may be recorded, even if the researcher forgot to do them. In the chemistry lab, for example, many experiments fail for apparently unknown reasons, but on closer inspection, the cause is often found to be that a step in the procedure has been missed out. Observations that at the time are not perceived as unusual or unimportant may not be recorded, although unusual or unexpected observations are more likely to be recorded. Digital tools have the potential to assist with these problems by prompting the information that is recorded, asking the user to confirm actions or observations, and specifically asking if anything unexpected happened.

Episodic memory is memory for personal events and experiences including details such as what, when, where, and with whom. Episodic memory is what allows us to 'travel back in time' and re-experience something that happened in the past.[42] Our memories are not a literal account of events and what we remember. Although we may have very vivid memories for significant or personally meaningful experiences, more mundane memories are representative and more general.[43] Some researchers believe that memories are linked to other systems in the brain related to the self and goal attainment, and that memories may be stored or remembered depending on whether they relate to factors relating to our ability to attain our goals.[43,44] An event such as carrying out an experiment in a lab may be better or more accurately recalled if the activity is related to some goal of the individual. For example, an experiment may be better remembered by students if they plan to continue study in the subject or work in that field, or by researchers if it is significant to the core of their research. Memory may also be affected based on the perceived success or failure of the experiment. If the experiment failed and the researchers did not see the results they were expecting, this may change what information they recall or what they choose to report. It may be that the experiment is less well remembered or less well recorded, when in fact, accurately recording the experiment may help with capturing the reasons behind the failure or ideas about how to solve the problem.

In the past, it was believed that the amount of practice or rehearsal of an activity was the primary factor that influenced whether information passed into long-term memory, and therefore whether the information would be remembered. The reality is much more complicated than that and there are no definitive laws of memory.[45] Prior knowledge, conditions at encoding (when the memory was generated), influences after encoding, and conditions at retrieval, all can have an effect on what is remembered later.

If I ask you to describe your last experiment to me, your response will use 'free recall'. Free recall is believed to allow information to be recalled in the form closest to the structure that it is actually held in memory. Studies have found that under free recall conditions, memories are recalled based on their distinctive detail or temporal order, suggesting that these memories are organised both temporally and thematically.[46] Making use of these temporal and distinctive detail elements of memory may help improve recall of what happened during research and to ensure that the important elements are captured in the record. For the purposes of tools for record-keeping, we can ask for the information in temporal order (what did you do?) and ask for details of what happened to capture information about the procedure and elicit script-related actions. Subsequently, questions can be asked related to the significance of those events. For example, what were the interesting observations, what worked or did not work as expected, how did it meet expectations, or how did it meet the intended goals?

Recollection of an event depends upon the information stored in memory about the event, but can also be affected by the information available at the point of retrieval. Without the right cues, you might have trouble remembering something, but given the right cue you are much more likely to be able to recall the information you want to remember. Cues act as a prompt for memory recall. For example, with the spreading activation theory we discussed previously, cues to cause the activation of the 'fire-engine' concept could have been 'red' and 'vehicle'. Those same cues might also bring to mind 'bus' and 'sports car'. On the other hand, the cues 'red' and 'fruit' may bring to mind 'strawberry', 'apple', and 'cherry'. The questions 'who', 'what', 'where', and 'when' can make effective cues, although they are not all equally as effective; the cueing order also makes a difference, with the question 'what' on its own apparently most effective, followed by 'where', 'who', and 'when'.[47] Cues can only be effective if the appropriate semantic links exist in memory, but an inappropriate cue is unlikely to retrieve relevant information or memories. An inappropriate cue may instead cue the recall or retrieval of irrelevant information and may actually inhibit the recall of relevant information.

Various studies have shown that cues that are 'self-generated' are more effective for recall than those generated by someone else. These cues may be individual words to describe a set of information, or through the use of drawing diagrams or plans that can be used to recall specific details of a related event.[48–50] This finding suggests that self-generating metadata to describe scientific research may help to improve memory and recall for what was seen and done for a specific experiment. Including drawings of observations or a flow chart of the procedure are also likely to help improve memory and recall, and also contribute additional context to the record. In designing digital tools for record-keeping, the use of free recall and cued recall may be useful. To a large extent, the blank paper notebook can be considered a tool for eliciting free recall of memory because no structure is imposed, whereas a form or table with prompts or questions to complete represents an example of cued recall.

Memory errors in the real world can have real consequences. Eyewitness testimony is one of the main areas of memory and recall research outside of the psychology lab. Mistakes in eyewitness testimony can have significant consequences, and so various techniques have been developed, based on cognitive studies and research studies such as those detailed previously, to help improve memory retrieval and reliability. The primary outcome of such research has been the cognitive interview, designed to improve recall and help minimise memory errors that might lead a police investigation in the wrong direction or lead to innocent people being found guilty. Eyewitness testimony can be subject to a number of different memory errors in addition to forgetting. Not only can witnesses have incomplete memory, but their memory can be misled and influenced by their own biases, and even the questions they are asked. For example, when subjects in a classic study by Loftus and Palmer[51] were asked to estimate the speed of two cars crashing in a video, the estimated speed was significantly higher when the word 'smashed' was used in the question compared to less suggestive words such as 'contacted' and 'bumped'. Not only did this affect the study participants original answer but also their longer-term memory of the event. The use of the word 'smashed' led to significantly higher estimates of speed and to many more subjects claiming to have seen broken glass a week later, when in fact there was none.

Other studies have shown that eyewitness memory can be influenced by post-event information and even be affected by information prior to observing an event, with interference from earlier memories where prior experiences or knowledge are thematically similar. Memories created by misleading information may be very similar to those produced by the actual event, making it difficult for the witness to distinguish between them.[52] For memories of scientific research, memories of previous or subsequently conducted research could interfere with memory for a specific activity, and therefore cause inaccurate information to be recorded or to make it difficult to recall the details. For example, if the same or a similar experiment is repeated a number of times, it may be difficult to remember the exact procedure or results that came from one specific experiment. Steps may be repeated or missed out because of confusion between memory of the current activity and previous activities. This is particularly likely to be a problem where researchers wait until a later time to record their research. If a number of similar activities have been carried out before the research is recorded, then it is unlikely that the record for each individual activity will be remembered precisely without some interference.

As already mentioned, our previous experiences and knowledge influence what we understand and remember. Confirmation bias and social bias are commonly seen in eyewitness testimony where witnesses remember and report what fits with their knowledge and perceptions of crimes and criminals, rather than what actually happened. Information that matches our expectations is easily remembered over time, but information that is inconsistent with expectations or perceived as irrelevant is more rapidly forgotten.[53,54] There is also evidence that information that is inconsistent with expectations can become modified to fit the expectations, and as a result these changes will be remembered as though they really happened.[55] Such biases could lead to researchers recording what they think should have happened, rather than what really did happen.

Despite the importance of complete and accurate eyewitness testimony, the traditional police interview has been found to be relatively poor at generating correct information compared with the cognitive interview, which generates significantly more information and introduces less incorrect information.[56–58] Techniques used in the cognitive interview include context reinstatement, 'report everything', recall from a variety of perspectives, and recall in a different temporal order. In context reinstatement, the witness is encouraged to mentally reconstruct the situation from the event, what they could see, who was there, how they felt, and so on. Asking the witness to report everything they can remember, even if the memory is only partial or incomplete, follows this. Next, the witnesses are asked to adopt the perspective of other people at the scene, what might those people have seen or felt? The final stage encourages the witness to recall events in a different temporal order, for example, running the event backwards or starting from a point in the middle that has relevance for the witness. Similar techniques based on the cognitive interview are used in survey and questionnaire designs to elicit effective recall from students and members of the public.[59–63] Designing tools that run a full cognitive interview protocol for recording research would obviously be excessive, but the methods used are based on the memory retrieval theories described in the previous sections. Elements of the modified cognitive interview such as the reinstatement of context through the use of diagrams, a reminder to report everything, changing the perspective, and recalling information in the correct temporal order may all assist recall in the field and laboratory, and therefore ultimately help improve the captured record.

Using templates or other structured interfaces provides the opportunity to make use of the various techniques for eliciting information from users described in the previous section. These techniques encourage the recall of information that will be of use to the author and others who seek to make use of the information in the future. Templates are frequently used within Electronic Laboratory Notebooks (ELNs) to give a formal structure to the records. These templates can also provide cues that might act as reminders to increase the completeness of information recorded and encourage authors to enter more information. Templates could, for example, include sections to encourage researchers to record important information that they might otherwise forget to record, or to capture information that will be helpful when they come to look up the experiment later. However, there is evidence that the inflexibility and formality of templates might restrict the content that the author enters and that important information may not be recorded if the template does not specifically ask for it.[64,65] In theory, the researcher has ultimate authority over what they record and can therefore choose to improvise if the system does not allow them to capture everything they want, but in reality, the researcher may not realise that important information is missing because the design of the system is not providing any cue or other reminder for it.[66] Using templates might lead to the loss of important experimental information or details of the personal experience of conducting the experiment, such as observations or knowledge that was learnt as part of the experiment process.

In Chapters 9 and 10, we will explore how using some of these techniques to assist with memory recall can help to change the information that is recorded in the scientific record and provide insights into the effects that different interface designs may have on the information that is captured. We will consider how we ensure that the right information is captured, and in particular, we will look at the potential positive and negative effects of templates that are widely used within tools designed to capture the scientific record.

2.5 Conclusion

Record-keeping has a long history in science. Although our research notebooks may not be as sophisticated and beautiful as those of some of the eminent scientists and thinkers from history, they still serve the same purposes of providing memory and protection for our ideas and work. In Chapter 3, we will take a look at the record-keeping process and how researchers learn to record their research and take notes. We will explore how the records are structured and organised, and how researchers manage the modern issues of keeping track of digital files and other records within their paper notebooks. We also examine record-keeping behaviour and the private nature of notebooks.

References

1. B. Latour and S. Woolgar, *Laboratory Life: The Construction of Scientific Facts*, Princeton University Press, Princeton, 1986.
2. K. Shankar, *Arch. Sci.*, 2006, **4**, 367.
3. A. Blair, *Too Much to Know: Managing Scholarly Information Before the Modern Age*, Yale University Press, New Haven, 2010.
4. A. Aaboe, *Philos. Trans. R. Soc. A.*, 1974, **276**(1257), 21.
5. R. S. Gates, M. J. McLean and W. A. Osborn, *J. Res. Natl. Inst. Stand. Technol.*, 2015, **120**, 293.
6. R. Yeo, *Notebooks, English Virtuosi, and Early Modern Science*, The University of Chicago Press, Chicago, 2014.
7. F. L. Macrina, *Scientific Integrity: Text and Cases in Responsible Conduct of Research*, ASM Press, Washington DC, 4th edn, 2014.
8. J. Downing, P. Murray-Rust, A. P. Tonge, P. Morgan, H. S. Rzepa, F. Cotterill, N. Day and M. J. Harvey, *J. Chem. Inf. Model.*, 2008, **48**(8), 1571.
9. K. Shankar, *J. Doc.*, 2009, **65**(1), 151.
10. C. D. Poulter, *J. Org. Chem.*, 2009, **74**(17), 6415.
11. D. Ince, *Significance*, 2011, **8**, 113.
12. R. Saunders and J. Savulescu, *J. Med. Ethics*, 2008, **34**, 214.
13. R. F. Service, *Science*, 2002, **298**(5591), 30.
14. E. Callaway, *Nature*, 2011, **479**, 15.
15. *Nature*, 2010, **468**, 345, Available at https://www.nature.com/articles/468345a (accessed Jan 2019).
16. A. Eisenberg, *J. Chem. Educ.*, 1982, **59**, 1045.
17. M. R. Canfield in *Field Notes on Science and Nature*, ed. M. R. Canfield, Harvard University Press, Cambridge, 2011, pp. 1–18.
18. B. Heinrich in *Field Notes on Science and Nature*, ed. M. R. Canfield, Harvard University Press, Cambridge, 2011, pp. 33–48.
19. L. Arnstein, R. Franza, G. Borriello and S. Consolvo, *IEEE Pervasive Computing*, 2002, **1**(3), 13.
20. P. M. Scholl and K. Van Laerhoven in *Proceedings of the 2014 ACM International Joint Conference on Pervasive and Ubiquitous Computing Adjunct Publication (UbiComp '14 Adjunct)*, ACM, New York, 2014.
21. B. H. Schott, R. N. Henson, A. Richardson-Klavehn, C. Becker, V. Thoma, H. Heinze and E. Düzel, *Proc. Natl. Acad. Sci. U. S. A.*, 2005, **102**(4), 1257.
22. H. Spiers, E. A. Maguire and N. Burgess, *Neurocase*, 2001, **7**, 357.
23. L. M. Reder, H. Park and P. D. Kieffaber, *Psychol. Bull.*, 2009, **135**(1), 23.
24. C. J. Berry, R. N. Henson and D. R. Shanks, *J. Mem. Lang.*, 2006, **55**, 515.
25. J. Hodges and K. Patterson, *Lancet Neurol.*, 2007, **6**, 1004.
26. E. Funnell, *Cogn. Neuropsychol.*, 2001, **84**(4), 323.
27. A. M. Collins and E. F. Loftus, *Psychol. Rev.*, 1975, **82**(6), 407.
28. F. Crestani, *Artif. Intell. Rev.*, 1997, **11**(6), 453.
29. A. Katifori, C. Vassilakis and A. Dix, *Cogn. Syst. Res.*, 2010, **11**(1), 25.

30. J. F. Marques, N. Canessa, S. Siri, E. Catricalà and S. Cappa, *Brain Res.*, 2008, **1194**, 90.
31. C. Kohls and K. Schieter in *Proceedings of the 15th Conference on Pattern Languages of Programs (PLoP '08)*, ACM, New York, 2008.
32. E. T. Bergman and H. L. Roediger, *Mem. Cognit.*, 1999, **27**(6), 937.
33. J. Bransford and M. Johnson, *Verbal Learn. Verbal Behav.*, 1972, **11**, 717.
34. B. Tversky and E. J. Marsh, *Cogn. Psychol.*, 2000, **40**(1), 1.
35. E. J Marsh and B. Tversky, *Appl. Cogn. Psychol.*, 2004, **18**, 491.
36. N. M. Dudukovic, E. J. Marsh and B. Tversky, *Appl. Cogn. Psychol.*, 2004, **18**, 125.
37. R. C. Anderson and J. W. Pichert, *Verbal Learn. Verbal Behav.*, 1978, **17**, 1.
38. R. S. Nickerson, *Psychol. Bull.*, 1999, **125**(6), 737.
39. R. Bromme, R. Rambow and M. Nückles, *J. Exp. Psychol. Appl.*, 2001, 7, 317.
40. L. W. Barsalou and D. Sewell, *J. Mem. Lang.*, 1985, **24**, 646.
41. G. H. Bower, J. B. Black and T. J. Turner, *Cogn. Psychol.*, 1979, **11**, 177.
42. E. Tulving, *Annu. Rev. Psychol.*, 2002, **53**, 1.
43. M. A. Conway, *Neuropsychologia*, 2009, **47**, 2305.
44. M. A. Conway and C. W. Pleydell-Pearce, *Psychol. Rev.*, 2000, **107**(2), 261.
45. H. L. Roediger, *Annu. Rev. Psychol.*, 2008, **59**, 225.
46. S. J. Anderson and M. A. Conway, *J. Exp. Psychol. Learn. Memory Cogn.*, 1993, **19**(5), 1178.
47. W. A. Wagenaar, *Cogn. Psychol.*, 1986, **18**, 225.
48. F. Sharifian, *Issues Edu. Res.*, 2001, **11**(1), 15.
49. C. J. Dando, R. Wilcock, R. Milne and L. Henry, *Appl. Cogn. Psychol.*, 2009, **23**, 698.
50. W. F. Brewer and J. C. Treyens, *Cogn. Psychol.*, 1981, **13**, 207.
51. E. Loftus and J. Palmer, *J. Verbal. Learn. Verbal. Behav.*, 1974, **13**, 585.
52. B. P. Allen and D. S. Lindsay, *Appl. Cogn. Psychol.*, 1998, **12**, 277.
53. M. R. Tuckey and N. Brewer, *Appl. Cogn. Psychol.*, 2003, **17**, 785.
54. C. Stangor and D. McMillan, *Psychol. Bull.*, 1992, **111**(1), 42.
55. M. R. Tuckey and N. Brewer, *J. Exp. Psychol. Appl.*, 2003, **9**, 101.
56. A. Memon, C. A. Meissner and J. Fraser, *Psychol. Public Policy Law*, 2010, **16**, 340.
57. R. P. Fisher, R. E. Geiselman and M. Amador, *J. Appl. Psychol.*, 1989, **74**(5), 722.
58. R. E. Geiselman, R. P. Fisher, D. P. MacKinnon and H. L. Holland, *J. Appl. Psychol.*, 1985, **70**, 401.
59. L. M. Desimone and K. C. Le Floch, *Educ. Eval. Policy Anal.*, 2004, **26**(1), 1.
60. N. Schwarz, *Appl. Cogn. Psychol.*, 2007, **21**(2), 277.
61. B. C. Holleman and J. M. Murre, *Appl. Cogn. Psychol.*, 2008, **22**(5), 709.
62. R. Tourangeau, L. J. Rips and K. Rasinski, *The Psychology of Survey Response*, Cambridge University Press, Cambridge, 2000.
63. T. B. Jabine, M. L. Straf, J. M. Tanur and R. Tourangeau, *Cognitive Aspects of Survey Methodology: Building a Bridge Between Disciplines*, National Academy Press, Washington DC, 1984.

64. K. McGlade, C. Cargo, D. Fogarty, M. Boohan and M. McMullin, *Clin. Teach.*, 2012, **9**(2), 112.
65. D. Swinglehurst, T. Greenhalgh and C. Roberts, *BMJ Open*, 2012, **2**, e001754.
66. K. Shankar, *J. Am. Soc. Inf. Sci. Technol.*, 2012, **58**(10), 1457.

CHAPTER 3

Keeping and Managing Records

3.1 Introduction

In the previous chapter, we looked at the reasons why we might take notes and how our memories, and the design of interfaces for capturing records, might influence what we record. The nature of what is recorded is influenced by both the motivations of scientists to record their work and practices learnt through formal teaching and practical experience. As we discussed in Chapter 2, the main reason for record-keeping for the majority of scientists is for the record to act as an 'external memory'. The external memory functions primarily for the author of the record, but for some teams, the record may also be shared with colleagues and new team members. However, the scientific record has a much broader role for the wider scientific community, in particular enabling the research to be replicated. If the record of the research is insufficient, then neither the original author, nor anyone else, will be able to repeat the research. Information that is not recorded leads to missing context for the research and uncertainty about the who, what, when, how, and why of the research. If the research cannot be repeated, it cannot reliably be built upon; an essential foundation of the scientific process. Ensuring the accuracy and completeness of the scientific record is a critical but complex element of designing digital tools for record-keeping, and we will talk more in depth about the importance of reproducibility in future chapters. In order to design effective digital tools for record-keeping, we also need to understand current record-keeping behaviour. Although the recording needs and behaviours of different kinds of scientists are not substantially different at a high-level, in this chapter, we will consider the activities and needs of a variety of scientists in order to understand their similarities and differences.

Recording Science in the Digital Era: From Paper to Electronic Notebooks and Other Digital Tools
By Cerys Willoughby

Published by the Royal Society of Chemistry, www.rsc.org

The nature of the data, the structure of the records, and the terminology used differs between different scientific domains. Scientists make notes in a wide range of locations and environments, both within the laboratory and out in the field. Some scientists conduct their research in both of these environments, for example, archaeologists, biologists, environmental scientists, and earth scientists often collect materials in the field that may be brought back for later analysis in the lab. Within the laboratory, activities may take place within different locations; different parts of the lab, in different rooms, often moving between the lab and the office, or where different instruments are located in different rooms; and even different buildings.[1–3] Where and when notes are taken may impact on how detailed the notes are and what activities they cover. The structure of notes may also differ depending on the purpose of the notes. For example, a laboratory or formal field notebook is likely to differ in structure compared with an informal notebook used for personal 'notes to self'.[4] As we shall see, the nature of the science and the kinds of activities the researchers are carrying out influences the content and structure of the notebook.

Paper notebooks are very flexible and can be used for different purposes simultaneously.[5] As a part of the record, the notebook can contain any information that the author considers to be useful: thoughts, descriptions, and observations relating to the research activity. But also, the notebook may contain more personal information including contact details, web addresses, reminders, meeting minutes, and 'to-dos'. The information included in the notebook may be in any format the author chooses, for example, text, symbolic abbreviations, bulleted lists, sketches and diagrams, charts, flow charts, maps and plans, tabular data, and calculations.[6] Different disciplines have their own contrasting balance of these kinds of elements. In the field, the notebook may act as a journal of daily movements, a record of locations, weather conditions and observations, a catalogue of items collected or observed, or a narrative of the exploration of new terrain or other discoveries. The field notebook may also include maps, photographs, and drawings. In the lab, the notebook may discuss theories and plans, describe the recipe for an experiment, and present a record of observations and measurements, including tabular data, graphs, and sketches that can be used to assess the success of the work.

3.2 The Record-keeping Process

The majority of researchers go through the same or similar basic processes in the construction of their records as illustrated by the examples provided in Table 3.1. The table describes the high-level processes used to create the scientific record for a number of different scientific disciplines. For the majority of laboratory scientists, there is a process of designing the activity where the protocol of the experiment or activity is recorded. During the experiment or other practical activity, there is active recording of the steps taken, observations made, and data acquired. After the experiment or

Table 3.1 The high-level processes typically recorded in the daily scientific record.

Discipline	High-level processes and descriptions
Synthetic biology[1]	**Design:** Sketching and designing a structure, defining the procedure (materials and protocols to use)
	Construction: Mixing, heating, cooling, data recording, changes to protocol
	Analysis: Visual image (gels), quantitative, statistical
Cell biology[3]	**Preparation:** Creation of working document (experiment protocol and observation sheet)
	Execution: Combination, incubation, dispensing, separation, detection (recording and data capture), storage, and retrieval
	Documentation: Formal record creation, transcription from working document, pasting printed results, adding references
Life sciences[7]	**Experiment:** Combination, dispensing, separation, incubation, detection, labelling, and identification
Chemistry	**Planning:** What to do, materials and safety
	Experiment: Alter method to what did do, observations. Yields. Deviations from plan
	Analysis: Processing, record analysis of results
Environmental chemistry	**Planning:** Literature, reports, talking to collaborators; Figure out protocol, materials, and tool requirements
	Experiment: Preparation, washing, melting, milling, washing, leaching, filtering, and drying
	Analysis: Instrument capture, processing, report generation
Synthetic chemistry	**Planning:** Literature, reaction scheme, materials and safety, record planned procedure
	Experiment: Record actual procedure, measurements, times, and observations
	Analysis: Processing, record analysis of results
Field biology	**Planning:** Locations, materials, collection protocols
	Collection: Record locations, specimen details, measurements, behaviours, weather conditions, diagrams, photographs
	Analysis: Sample processing, observations, record analysis of results
Geology	**Planning:** Locations, previous studies
	Fieldwork: Locations, sections, logs, mapping, sample collection, diagrams, photographs
	Analysis: Thin section analysis, geophysical modelling, geological mapping, record analysis of results
Archaeology	**Planning:** Surveys, locations, collection and recording protocols
	Fieldwork: Locations, maps, context sheets, artefact collection, sections, conditions, soil sample collection, diagrams, photographs
	Analysis: Environmental analysis, artefact analysis, results, formal reports

activity is completed, there is typically an analysis and documentation stage, where the results of the research are recorded and analysed. For field scientists, there is a similar process, although the design or planning stage may be more or less well documented compared with the fieldwork stage itself.

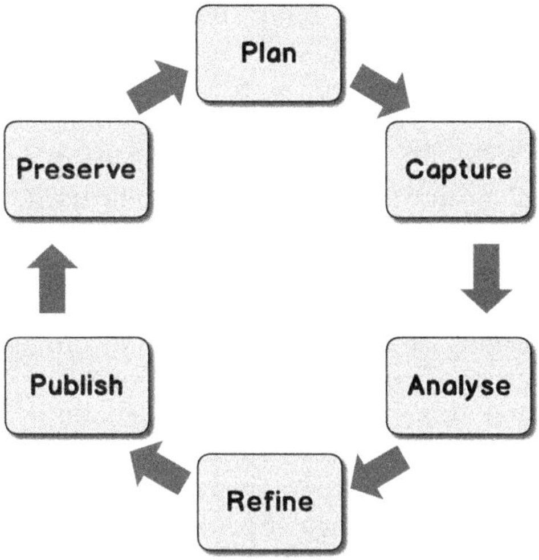

Figure 3.1 The record-keeping process.

The results of the analysis stage of research leads to the production of presentations, quarterly reports, and papers; and the various elements of the research are finally preserved or archived in some form. Figure 3.1 shows a visualisation of this record-keeping process as a circular flow because the documents and data preserved from completed research contribute to the design of future research. The results of previously conducted research are not only utilised by the original researcher and their team, but also by the wider scientific community. This can be seen in practice from interviews with chemistry researchers, who make use of online search and repositories to locate literature and data deposited by other chemists in order to develop the protocols for their own experiments. The published results are also used for comparison with the researcher's own results to verify the success of their experiments and to help locate problems if they occur. The activities that are performed by researchers on a day-to-day basis are focused at the beginning of the cycle, with preservation rarely mentioned as a consideration for most researchers when thinking about their research. Although, as we shall discuss in Chapter 5, this is a situation that is changing.

Planning research involves pulling information from one's own research, but also from other locations, primarily from the published literature, such as journal articles or previous reports, but also including data from repositories, and other sources on the Web. Materials required for planning research may include both digital and paper-based materials. Paper-based materials may include maps, surveys, and photographs, as well as notebooks from previous research projects. Historically, much planning for research would have taken place at the library as the main source of both primary and secondary sources, but because of the proliferation of materials online, planning is now more commonly performed at a computer in the office. Some researchers also communicate with collaborators at the planning stage either locally or remotely. Once information has been gathered and evaluated it is used to put together protocols for the research, such as collection protocols for samples in the field or the plan for an experiment in the laboratory and the materials to be used.

Capture of information is a primary activity associated with the 'enactment' part of the research. In the field, this enactment may be the recording of the research environment or subject of the research, the collection of samples, and the making of measurements. The recording may be a narrative, the capture of data entries, or the generation of digital files through the use of equipment or sensors. Within the laboratory, recording takes place during the enactment of the experiment, where protocols are followed and materials manipulated. Observations, measurements, and the details of instrument or equipment settings or configurations are captured. On the fly changes to the protocols may be made as the experiment progresses depending on the observations made, and these changes are recorded. Data may also be captured through the use of a variety of instruments and this information may be recorded manually in the notes or automatically through the generation of digital files by the instrument.

Analysis of the observations, measurements, and results data is performed, typically using some additional software support, and often using specialist digital tools. Raw data is processed and evaluated, and often compared with data that has been generated previously by the researcher, their collaborators, or from the data repositories on the Internet. For fieldwork, the analysis stage may involve the preparation and analysis of materials gathered in the field, the construction of maps, and the analysis of raw data. At the end of the analysis stage there are usually some results generated that determine the next step in the research. More research may be required, for example, further collection or observation in the field, or a new experiment to perform. If the research is complete, the researcher may move onto the refining stage. Refining involves the preparation of materials for sharing or publication. For researchers, this may involve the preparation of presentations for meetings and reports for members of the research team and collaborators, and later by the production of scientific papers for the wider scientific community. The refining stage is where much of the creative thought and development of ideas occurs.

Publication and preservation are the later stages of the process where the record and the final products of the research are deposited in repositories for access to the wider scientific community, or alternatively locked away in a filing cabinet and forgotten. We will talk more about the preservation stage of the record-keeping life-cycle in Chapter 5.

The cartoons in Figures 3.2–3.5 provide some examples of how different researchers construct their experiment records and organise their work, including this element of reusing data from other scientists in their activities. The researchers in the examples are using differing combinations of

Figure 3.2 An example of recording the experiment process using paper (Synthetic Chemist).

Figure 3.3 Recording a flow chemistry process using the iLabber ELN (Synthetic Chemist).

Figure 3.4 Recording the experiment process using a customised Word template (Synthetic Chemist).

Figure 3.5 Recording the experiment process using the LabTrove ELN (Environmental Chemist).

paper-based and computer-based methods for capturing their records. These 'day in the life' cartoons are based on interviews with real researchers and show some of the similarities and differences between the processes and individual styles of research; the first three cartoons are different styles observed within a single department. The cartoons also highlight some of the problems that the researchers face in terms of when they are able to take notes and how they manage their data.

3.3 Learning to Keep Records

It is important that records of scientific activities are accurate, complete, and accessible. If the record is missing information, contains incorrect data, or is difficult to understand, then it cannot serve its function as a tool of learning, support, and memory for the scientist to whom it belongs, let alone to colleagues or the wider community who may need to make sense of it, or even worse if the ownership or reputation of the research is at stake. Unfortunately, researchers often receive little or no formal training in techniques for recording, analysing, storing, managing, or sharing their data.[8–11] This is true of both field and laboratory scientists, although as Canfield observes, at least the field scientists have the ample and accessible notebooks of notable 19th and 20th century field scientists to read as guidance.[12]

It is common for undergraduates in laboratory sciences to be provided with guidance on what information should be included within their laboratory notebooks and the structure of their formal reports as part of classroom learning or as formal guidelines for laboratory work. In teaching environments, the experiment record provides an insight into the effectiveness of learning, whether the student has understood the theory of the experiment, accurately followed the procedures, and understood why some aspect of the experiment may not have behaved as expected. The experiment record must therefore be an accurate and complete reflection of what actually happened, and not simply a regurgitation of the intended plan. The notes taken during experiments for learning may also form the basis for future research or instructions for techniques in future work, and therefore has a longer lifespan than may be anticipated at the time of creation. Outside of the educational environment in laboratory science, formal guidance is less common, and in fact many research laboratories do not provide in-house standards or recommended best practices for recording experimental data in laboratory notebooks.[13]

The education for taking field notes is rather mixed. Standard formats for what to record in notebooks are more common for disciplines such as zoology and field biology where the records are focused on recording the details of particular specimens that are collected. A classic example of this is the system developed by Grinnell for the recording of field notes and specimens for the Museum of Vertebrate Zoology.[14] The system includes a narrative journal for recording daily activities and observations; a specimen catalogue with specific information to enable the generation of a detailed label for each specimen in the museum; and a species account to record information and observations over different locations and times. Recording information and observations about a species at different locations and times ensures that variations by season and microhabitats can be identified. For other field scientists, such as geologists and archaeologists, some standard information may be collected such as the date, details of the locality, and weather conditions, but the information captured is generally less prescribed. Often, the education about note-taking is a few words of guidance at the beginning of the fieldwork, rather than the formal written guidance provided by some disciplines.

In some cases, the learning of record-keeping processes may be developed as part of membership of a scientific community of practice. For example, some research groups share best practices or standards amongst themselves, often with the development of local practices. However, most scientists will develop their own personal style of recording by trial and error and sometimes with help from a mentor.[11,15,16] The development of additional data management within the notes, such as a coherent numbering system, table of contents, or other indexing systems, may only come through learning from earlier mistakes.[17] These additional data management techniques are often imposed retrospectively because of problems encountered in relocating information at a later time. The external memory is only as good as the mechanisms that allow it to be searched without re-reading everything that was written. Even with good training and good role models, it can still take decades of trial and error to evolve a good record-keeping system.[18] Discovering that you have failed to record pertinent information when you go back to your notes can be a disaster, and is particularly a problem for fieldwork where a number of weeks may have been spent in the field before the problem is realised, by which time an accurate memory of events has faded.

If your own notes can be difficult to use, then it is even more difficult to make sense of the notes of others. From my own discussions with researchers, the majority indicated that they had not received formal training in record-keeping or data management. Many of them also mentioned the difficulty of taking over another researcher's work and how a lot of training was required before they were able to repeat the work. Part of the reason for this is that researchers often use a standard operating procedure, in other words, they perform the same research activity over and over again until it becomes almost subconscious, like the procedural memory of driving a car or the scripts we mentioned in Chapter 2. This is particularly true for laboratory work, where the same procedure or experiment may be performed repeatedly with very small changes. The consequence of this repetition is that not everything is written down because it is seen as 'obvious' and unnecessary duplication. The researchers themselves already 'know' how to do the procedure after their training and practice, and therefore do not feel the need to record all of the context information. The researcher may not even be consciously choosing not to record this 'tacit' knowledge. The consequences are that when one researcher takes over the work of another, the process is hard because the record they are trying to work from is missing significant information. New PhD students trying to repeat experiments from the literature have also raised the same issues of missing information in the protocols described, but the same problem is also encountered by experienced researchers attempting to replicate the results of others. A few months after completing research even the original researcher may struggle to recreate their own work as they move on in their research and their knowledge changes. The methods and protocols provided in the literature are often missing critical information because standard operating procedures vary between different groups and laboratories, but the differences are not documented. Usually there is no additional method information or specifics given as to materials used outside of the published account as a reference for others trying to repeat the research. As we discussed in Chapter 2, these problems confirm that important aspects of personal and organisational memory are not externalised in the scientific record when these details are 'routine' and 'obvious' to the author.

Differences in personal style can also affect 'understandability' for other readers of the notebooks. For example, if authors of the notebook use abbreviations or symbols to have particular meanings within the notes, it is unlikely that another reader will know what these mean without a key or glossary. One particular personal style may also choose to capture different information to another, leading to missing or ambiguous information. For example, one author may use a locality name or identifier, whereas another may use precise GPS coordinates. One author may not record the units for measurements when they capture data entries. That author may always use the same units and therefore not have any doubt about the precise value, but to another reader the numbers will be ambiguous and unusable. To manage these effects of personal style on record-keeping, tools that encourage consistency and prompt the recording of information that may be forgotten could improve the quality and reliability of the record. A digital tool could either prompt the capture of information in a specific format or even better ask the user what their preferred format is, for example, but allow another user to automatically convert between different units or values. As we will discuss in Chapter 8, the design of user interfaces for digital tools can remedy some of these problems by providing clarity and consistency for users, whilst poorly designed tools can actually exacerbate the problem.

3.4 Structure and Organisation of the Records

Although researchers have great individual freedom to choose how they create their records to best represent the objects and activities of their research, there are also disciplinary, institutional, and professional norms in how they undertake to do so.[17] Notebooks usually include not only data, but also details of reflective thinking, refinement of scientific theory, ideas, conjectures, plans, equipment and the corresponding apparatus set-ups, methods, observations, analyses, and results.[13,19] The information contained within is often, but not always, in temporal order. The structure of a notebook at a high level can provide an insight into the evolution of the research trail through the movement of research from activity to activity, location to location, specimen to specimen, or material to material depending on the discipline of the researcher.[20] For example, in fieldwork, the record will reflect the path taken between different locations and how understanding develops as a wider area is explored. Within the laboratory, the progression of experiments documents the changing ideas of the researcher and the directions taken to develop a new process or to explore new materials.

Institutions, organisations, and individual research groups may provide guidance for the structure and organisation of the notebook as a whole. For example, there may be requirements for front matter that identifies the author and owning institution of the research, the location of the laboratory or field locality, and information relevant to Intellectual Property (IP) concerns and financial sponsorship information such as the title of the grant proposal, the agency ID number, dates of support, and the name of the project lead.[16] There may also be a requirement to have a table of contents, for example, a list of experiments with short descriptive titles and chronological numbering to facilitate cross referencing. Some notebooks may also contain a glossary to capture the meaning of abbreviations and symbols. The majority of notebooks are organised chronologically, but in some cases the content may be organised by individual activity. For example, for experiments that are long-running and where multiple experiments may be running in parallel, it can be easier to organise the notebook by experiment, leaving space between them to be completed over the duration of the experiment. However, the practice of leaving space and filling it in later can be frowned upon, especially if comments are added after the event. For the integrity of the scientific record, particularly for IP protection, it is considered essential to ensure that any changes to the record are clearly indicated and dated. For example, mistakes are shown by drawing a line through the content but leaving it so it is still visible. If additional notes are needed where no space is available, then a cross reference between the original record and the new comments can be made. The rationale for this is that the original information may still be useful, whilst obliterated content gives the appearance of having been deliberately manipulated.[16] Digital tools for record-keeping have some advantages in solving some of these problems. For example, information captured at different times can be organised in a variety of ways. For example, entries in a digital notebook can be arrange chronologically like the paper notebook, but they can also be arranged by topic, so that all content relating to a particular project or experiment can be viewed together. Digital tools can also provide provenance information for the record, so that date and times of entries can be viewed, and content can be amended if required, but the original content can still be viewed, for example, to demonstrate that a change was genuine and the results have not been manipulated.

Table 3.2 shows the typical structure of individual records for a number of different disciplines. The entries in the table include examples of different structures that have been directly observed in use by scientists or that have been derived from various student guidelines mentioned. It can be seen that there are many similarities between the different disciplines, but field sciences and laboratory sciences show a number of key differences. For example, in the field sciences, 'location' is important, but the date is more significant to the lab-based notebooks. The timestamp has a long history of importance in scientific and engineering work, and is extremely important in providing proof for the protection of intellectual property.[21]

Not all disciplines or individuals make use of a formal structure in their notebooks, and instead choose to include only a minimum of standard information. For example, the notebooks I have seen authored by synthetic chemists have a very consistent and formal structure, however, the vast majority of notebooks authored by physical chemists show very little formal structure at all, although all included dates and often included titles. Occasionally, they included a formal record of an experiment with more structure, but the majority of the content was a narrative of the thought processes of the researcher and their experiences working on a particular problem together with some data. Thoughts about what they had discovered and how they were going to build upon it to refine their processes or designs was also included in the narrative. This more narrative style demonstrates the value of writing notes for reflection and aiding thinking as well as memory, a function that is also observed by researchers taking notes in the field.[27] It is unclear whether those scientists taking very formal notes with rigid structures are missing out on the opportunity for reflective thinking, but the different styles are likely to reflect differences in the nature of their research. We will discuss more about potential benefits and consequences of formal structures and the use of templates for record-keeping when we talk about designs of interfaces for digital tools in Chapter 9.

3.5 Searching and Retrieving Content

Although we may consider the primary purpose of notebooks to be for taking notes, for the records to be of value, they need to be easily retrieved. Many notable researchers from history spent time creating retrieval aids and reminders to help locate the contents of their many notebooks. For example, Faraday numbered his records and created retrieval sheets with a short description and reference to the numbered records, contents lists, and he even produced indexes for his indexes![28] In modern times, however, the use of a table of contents at the beginning of a notebook is quite rare, even though some guidelines recommend them.[17] Many researchers start recording new experiments on a blank page, making it reasonably easy to create a table of contents retrospectively.[13] In my discussions with

Table 3.2 Example notebook structures from a variety of scientific disciplines.

Discipline and source	Structure	
Life sciences[7]	• Title • Date • Goal • Abstract	• Protocol overview • Protocol steps (workflow) • Results • People involved
Field biology[22]	• Locality • Route of travel and map • Hours of observation • Weather conditions • Participants • Habitats and topography • Species seen and numbers • General impressions	**Catalogue** section of notebook includes locality and samples collected. **Species** account section of notebook includes species identification, where found, what it was doing, how it looks and sounds.
Field geophysics[23]	**Field:** • Name/Date • Diagram • Location Map • Apparatus ID • Tabular data	**Lab Reports:** • Objective of experiment • Diagram • Location Map • Apparatus ID • Tabular data • Calculations • Graphs • Discussion
Geology[24]	• Excursion title • Excursion objectives • Weather/conditions • Locality number • Location/grid reference	• Locality description • Observations • Measurements • Sketches with scale
Biology (lab)[25]	**Title page:** Name, address, course, phone number, project title or description **Table of Contents pages:** Date of experiment, descriptive title, page, signatures **Table of Protocols pages:** Protocol, date, page, signatures **Glossary:** Term, definition **Experiment records:** • Title • Date • Hypothesis	• Notes, reflections and literature references • Objectives (why, testing hypothesis, repeating old experiment, who proposed experiment?) • Protocol details (materials, chemical manufacturer, equipment, photos, drawing, experiment set-up) • Results, raw data, observations • Discussion, interpretation, and conclusions • Notes and reflections on outcome.
Physics[26]	• Title • Date • Name of lab partner	• Sketch of equipment • Tabular data • Graph of data

Table 3.2 (*Continued*)

Discipline and source	Structure	
	• Aims/background • Method and explanation of experimental decisions	• Comments on the Physics • Conclusions • Review of your work
Chemistry	• Title • Date • Short description • Risk assessment • Procedure including key observations	• Data collected • Results, analysis, and conclusion (included in a formal report)
Synthetic chemistry (paper notebook)	• Reaction scheme • Table of reagents • Hazards • COSHH entries for the significant or unusual hazards	• Planned procedure • Actual procedure with weights, times and observations • Analysis of the results
Synthetic chemistry (iLabber template)	• Author • Experiment ID • Title • Batch ID • Start Date	• Reaction scheme • Reagents • Safety info (manual) • Procedure • Processed results
Synthetic chemistry (Word-based notebook)	**Table of contents:** • Experiment ID • Experiment Title • Date • Reaction scheme • Links to experiment files • Links to other folders • Comments	**Experiment files:** • Experiment reference • Date • Title • Reaction scheme • Reagents and risks • Links to literature • Experiment procedure • TLC plate • Links to data and spectra
Environmental chemistry	• Title • Date • Composition • Form • Synthesis location • Conditions	• Tools • Processing • Description/observations • Measurements • Data
Physical chemistry	• Date • Title • Materials table (name, quantity, hazard)	• Precautions • Method • Disposal • Spillage Control

numerous researchers I have only come across one who mentioned creating a table of contents within a paper notebook. A table of contents is easier to create using digital tools, and multiple different kinds of tables of contents can be created. For example, a separate index can be created for activities or experiments, projects, field locations, and samples, providing an easier way to access related files. For example, Andrew, the synthetic chemist from Figure 3.4, who used Word to capture his experiment records, also created a table of contents document to keep a track of his individual experiment records and to provide a simple way to search and retrieve them. Many Electronic Laboratory Notebooks (ELNs), which we will talk about more in Chapter 4, provide a table of contents facility. However, these tools often only display a small amount of information about the experiment, making it difficult to find content based on the criteria that is most relevant. For example, for synthetic chemists, being able to search for and retrieve experiments based on the visual structure of the materials involved is desirable, and including such information in the table of contents of ELNs would improve their usefulness and usability.[29] For a paper notebook, such a chemical structure could be included in the table of contents, or alternatively in an index of materials, together with the page numbers of the experiments in which they are included. However, the manual creation of such indexes requires substantial effort. A simpler strategy that is often employed is to highlight key elements or topics within the entries, in such a way they can be easily searched if flicking through the notebook.[27] Highlighting with coloured markers, circling or framing regions, 'dog ears', and paper clips are also strategies used by researchers to make content easier to find.[5] Despite the potential usefulness of highlighting and annotation, guidance on the use and benefits of annotating paper notes is relatively rare.

3.6 Linking to Data and Other Records

Various forms of structuring and highlighting are commonly used in paper-based notes to link together different content throughout. The highlighting marks content as important and makes it easier to find. The use of the highlighting, together with other markings and symbols annotating the content, can be considered to be a paper-based form of metadata. Demarcations, such as horizontal lines and different pen colours, text started on a new page, categorisation with topic and date, and the use of symbols can also indicate relationships between content across multiple pages. Symbols often relate to the topic of the content and link together information about the same topic at different locations in the notebook.[5] To assist with finding information later, various 'tags' and symbols are often used to distinguish different snippets of the content.[30] Although the use of highlighting, symbols, and so on, tends to be individual preference and relatively informal, some formal note-taking guides do include standard use of such techniques. As an example, guidance on 'Keeping a field journal', from The University of Western Ontario, provides examples of ways to highlight important information.[31] For example, wavy lines are used to indicate common names of species described in the notebook, whilst straight lines indicate scientific names, and new locations and dates are also highlighted by underlining. These markings enable the key information to be picked out quickly and easily, without the need to read the entire text on the page. The guidelines also show an example of 'marginal symbols' for annotating text and linking between items, for example, a symbol with a dot within a square is used to indicate a photograph was taken at this point. This also provides an example of linking to external information and data, for example, the researcher could include the name of the file for the photograph at the same point.

Where information for the record is collected on structured forms, metadata can be stored on both the forms and within the notebook. For example, field archaeologists typically capture most of their excavation data on paper forms, often called context sheets. Context sheets typically contain numerical identifiers for locations and archaeological contexts. The context sheets act as a central repository for the team, whilst individuals capture references to these identifying 'metadata' within their own notebooks. Every element of the archaeological excavation has an ID, for example, the locality, the trench, each layer in the soil, and each archaeological find, plus other metadata such as a geographic reference. These identifiers provide a link between the localities, archaeological contexts on the context sheet, and to the notes made in the notebook. The notes themselves provide valuable additional information and other elements of interest such as stratigraphic context, colour descriptions, and interpretations, but also provide links to other external information such as photo IDs, drawings, plans, and maps. Within analytical laboratories it is also common to record important information on forms. For example, in an analytical laboratory, a form may be completed as a sample moves through the laboratory and undergoes different tests. Identifiers and metadata from the forms may also be included in the content of individual notebooks to enable a link to be made between different elements of the record, from the official record on the form to the less formal, but still useful, notes in the notebook.

Our investigations of the note-taking behaviour of chemistry researchers seems to show that more experienced researchers have more highlighting and formatting in their notebooks than those who are less experienced. This observation suggests that the inclusion of more formatting and 'marking up' of content may develop with expertise or experience in note-taking and would fit well with the idea that much of note-taking experience comes from developing a personal style and learning from earlier mistakes. We found the following types of highlighting and information were commonly observed in the notebooks of the more experienced researchers:

- Underlining of titles and individual elements such as important information or results
- Horizontal lines
- Names of files
- Tabular information/data
- Bullet points

- Different coloured inks
- Double underlines and double horizontal lines
- Circling and framing
- Use of all capitals for emphasis
- Annotations
- Timestamps
- Sketches, graphs, and diagrams

In contrast, less experienced researchers used much less emphasis in their notebooks, although titles and 'metadata' style information, such as date and name, were often underlined. Numbered and bulleted lists were also used by some, but tabular data and sketches were rarely included. Some symbols were used, but these were typically 'arrow' types symbols indicating flow or conclusion from a previous statement in the text. Interestingly, the more experienced researchers also appeared to use a less formal structure in their note-taking, using a much more diary-like style, with more personal information with detailed descriptions of set-ups, ideas, observations, and other reminders, and often details of 'internal discussion' or thinking, although this did vary by discipline. Even the more experienced researchers in synthetic chemistry still used a very formal style.

3.7 Images in Record-keeping

Notebooks are not just about the creation and keeping of textual and numerical notes. One of the benefits of paper for recording is that it is easy to create sketches and diagrams that can be used to add valuable context to the text and numbers. Drawing is used extensively in both laboratory and field notebooks. In the laboratory, drawings are often made of equipment set-ups so that they can be repeated, but also sketches may be made of specimens and other samples, and visual elements of test results. In the field, sketches are frequently used to record features of a locality for orientation and to highlight particular features within the environment, as well as specimens and artefacts. Although digital photography makes the capture of images much simpler, in many cases, a hand-drawn image is used as a companion to images captured through other mechanisms to enable annotations and provide clarity. Drawings also have the advantage of being able to convey information more selectively, which often enables information to be more effectively communicated than with a photograph.[32] The most important information can be highlighted in a drawing in a way that cannot easily be done with photographs. Although photography is ideal for catching fast moving events and details, drawing can capture the whole picture within a single image and also represent elements that can't be captured in a photograph, such as maps, paths, and movements over space and time.[33] Figure 3.6 shows an example of how additional information can be conveyed in a drawing compared to a photograph. The sketch shows a cross section of a geological locality, with representations of different rock

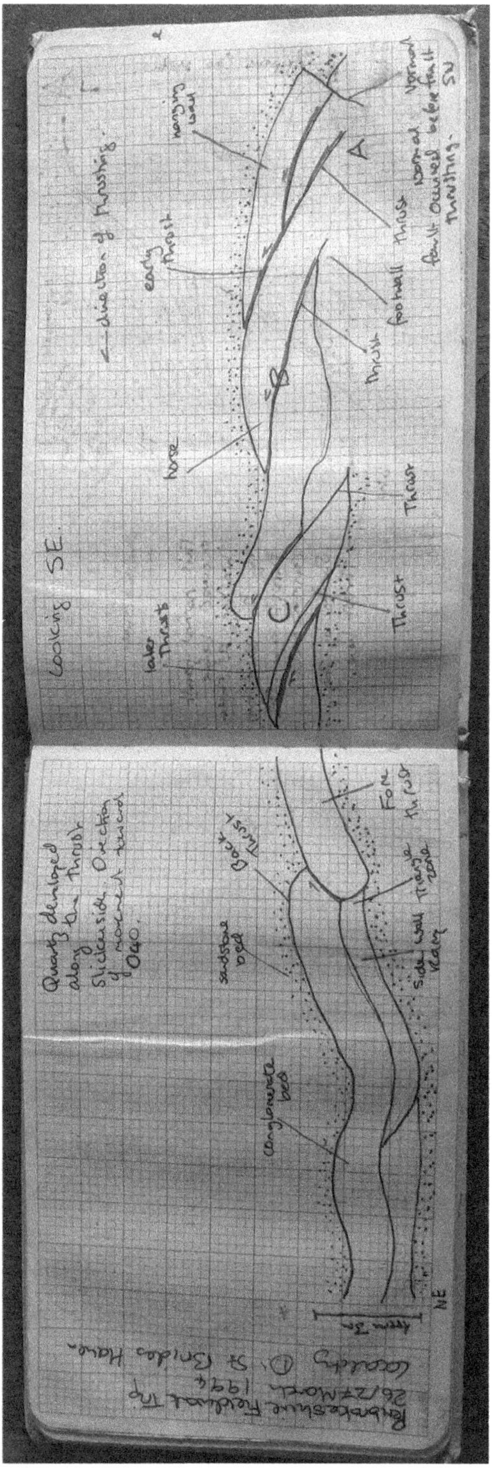

Figure 3.6 A drawing of a locality in Pembrokeshire showing rock types and details of geological structures such as direction and timing of movements that would be difficult to represent using a photograph.

types and how the rocks have moved in relation to each other over time. Although a photograph could be annotated with this information, the sketch enables the key information to be clearly picked out, and the whole of the large geological feature to be captured in a single image.

3.8 Record-keeping Behaviour

The details about the structure of notebook records, processes involved, and different notations described in the previous sections provide some insights into record-keeping behaviour and how they differ amongst researchers. However, there is more to record-keeping than just the nature of the 'inscriptions' themselves. In fact, as we will explore in this section, record-keeping can be a profoundly personal activity.

Shankar regards the act of capturing notes and the daily routine of 'writing up' as a ritual norm of scientific society.[11] The following of a prescribed structure within a notebook could be considered a part of that ritual, together with the details of when and where the record is constructed. The specifics will depend on the nature of the work and the habits of any particular discipline. There is a wide variation in personal and local practice with individual micro-cultures within science in terms of when and where notes are taken.[17] For example, laboratory scientists who run one experiment per day may record every step in the notebook as they go along, but those whose experiments are performed over longer periods of time may update their notebooks only on a weekly basis instead.[34] Field workers may take notes as they work, or only at particular points, such as when a specimen or artefact is found, or only after the whole day's work is complete.

As someone who has studied a number of different scientific disciplines at different times, I would agree that at least some of my own experiences of note-taking could be described as ritual. In laboratory work this comes from the formal structure and the routine process of planning, conducting, and analysing the results of an experiment. As a geologist, record-keeping definitely has a ritual element to it, with notes, sketches, and measurements captured in the field during the day and each night a process of ensuring that the notes are correct and legible. A key aspect is also the 'inking in' of details captured in pencil in the notebook (if it had been raining) and on the geological maps constructed during the day to permanently fix what was observed. No sleep is allowed until the details have been properly preserved in ink. Although the process is individual, this ritual of inking in was often completed as part of the group. In archaeology, note-taking seems to be less of a formal task with the notebook there to capture the details that you individually feel are important and to act as reminders of tasks. The capture of the notes was typically associated with particularly significant events during an excavation, such as beginning excavation of a new soil layer, the taking of measurements for a survey, or the discovery of a 'small find' or special artefact. The formal recording takes place on centrally stored forms such as the context sheets and small finds records. Details captured in the notebook are more informal and are there to help track down the correct information if something has gone wrong, for example, the same context ID has been used twice, or to provide additional contextual information.

Other archaeologists I have spoken to assert that the notebooks are not such an important record of an excavation unless it belonged to the project leader, but that they are more of a trophy to be stored on the shelf. The notebook represents proof that you were there and a part of the experience. This feeling that the notebook is a trophy or treasured item is common, particularly when associated with a special experience such as participation in an expedition or unique activity, or as an embodiment of important work and discovery.[35] This attachment to notebooks is not unique to archaeologists, and old notebooks can routinely be seen gracing the shelves of both laboratory and field scientists. For many scientists, notebooks can have a second life as a source of reminiscence or for reference to past activities. For example, geology notebooks frequently have a second life as a reference for future excursions with staff using their own notebooks as guides for fieldwork expeditions when educating their own students. This use of the notebook as 'proof' of having been involved in a project or completing work also came out in the discussions of why we take notes in Chapter 2.

Notebooks also provide others with an insight into not only the research of the author, but also into their life and personality, what Yeo calls an 'ego-document'.[36] Reading through the notes allows the reader to understand how the author developed their ideas and how simple observations mature into deeper insights.[37] The notes can also give insights into the nature and reality of scientific endeavour at that point in time, giving historical context as well as educational value. It is debateable whether digital tools for record-keeping can match paper notebooks as personal trophies and treasures of experience. Part of this may be to do with the physical nature of the notebook and partly due to the nature of paper, where hand-written content and hand-drawn images add personality and life to the narrative on the page. Not only can you hold notebooks in your hand, and flick through the pages, but the years of neatly lined up notebooks on a shelf represent the physical evidence of a researchers experience and scientific prestige.[38] Although it may be difficult to see how a digital memory stick or files on a computer screen can quite match up, we will discuss more about how the evidence of digital contributions and relative prestige can be represented in the virtual world through trophies, badges, and rankings in Chapter 11.

3.9 The Private Nature of Notebooks

Whether you consider your notebooks to be private or not may be influenced by your role and the nature of your research. For example, Wilson found that only principle investigators on a project viewed laboratory notebooks and associated records as a shared resource, whereas others relied on information to be converted into public resources such as talks, articles, and reports in order to access it.[13] In some environments, it is more common for supervisors and colleagues to review the notebooks.[34] In fact, for laboratories concerned with quality control or IP protection, the need to review notebooks is part of standard operating procedures, often requiring that every page of the notebook must be individually reviewed and signed off.

We tend to think of the notebook as being the place where all notes and associated activities are recorded in the laboratory or in the field. In reality though, many researchers may actually be recording a lot of their work on scraps of paper or in 'rough' notebooks and then later discarding them or transcribing them when a better or more complete story has formed in the researcher's mind. For example, Shankar describes the note-taking behaviour of one researcher who uses a rough notebook for writing equations and taking notes during an experiment, and then later writes-up the 'main record' into his 'main notebook' in the correct order once everything is understood.[15] Neatness of handwriting may be an indicator that researchers are creating a 'main notebook' because the record was created from notes or from memory after the event compared with rougher notes made during an activity.[39] I have observed that scientists commonly make use of scraps of paper to make notes whilst carrying out experiments, particularly for capturing measurements or formulating calculations as they go along. This behaviour is also quite consistent with the described habits of archaeologists in the lab, where researchers tend to take notes as they perform the analysis on scraps of paper. These notes are retained only temporarily and are discarded at the end of the process when the structured data has been captured digitally. There are risks with this approach of using scraps for 'temporary memory' in terms of losing a part of the research record. Captured information is discarded at the end of the process, so there is no record of whether the information or calculations were correct. It is also easy to make transcription errors when the measurements are entered into the 'final' notebook or converted into digital form.

Sharing of data and research materials before publication is a hot topic in science, as mentioned in Chapter 2, and as will be discussed in more detail in Chapter 11. The benefits of sharing include more transparency in research, and the opportunity for better collaboration and new discoveries, but the downside is the potential 'theft' of data and the loss of IP protection. Sharing of data is one thing, but I have observed how the mere idea of sharing of 'raw' notes can provoke a hostile reaction from researchers. Although researchers may expect their notes to be viewed by a manager, supervisor, or funder, the idea that other people outside of a select few might be able to access them typically provoked a negative reaction. This was particularly true in the context of using digital tools, such as ELNs, through which it is very easy to enable sharing with supervisors, colleagues, and others. For the researchers that got upset, the idea of a tool where sharing was even possible was enough to mean they felt would not be allowed to use it. Although this reaction was most severe from researchers whose work was commercially valuable, the idea of sharing notes prompts anxiety from those whose research is not constrained by IP concerns. Although intellectual property is a concern to many, and protection of one's commercial aspirations has historically been the main concern in terms of 'secrecy' within science, there is a second concern that is less well documented. This second concern centres around the issue of perceived 'readiness' of the information. The notes within a notebook are typically raw and unpolished. Researchers have commented that they would not want to let anyone else see their notebook because they view it as a personal record and a 'work in progress'. Publications or formal reports are the finished items that are suitably 'polished' and are therefore ready to be exposed to others. In contrast, notebooks contain personal thoughts, opinions, mistakes, and wrong paths that researchers are not willing to share. In particular, researchers who work independently were less likely to feel that sharing their notebooks in order to receive feedback on their notes would be valuable. In contrast, those that appreciate the value of easily being able to share their notes were already working collaboratively. These researchers were actively working with other groups on their research, often across different locations, and therefore already shared their work and needed to receive input from an external group to help them. Researchers who supervised others also saw the benefits of sharing for keeping tabs on the research of others, even though they were reluctant to share their own work!

For those working collaboratively and with paper, the logistics of sharing and audit are actually rather difficult. Sharing of paper notebooks is limited by the physical nature of the medium. In order to share a notebook, we must be at the same location, so I can give you the notebook, and while you are reading it, I cannot be using it. The same is not true of digital tools, which enable the records to be shared with many people, in different locations and time zones, all at the same time without preventing the author from continuing to add to it. However, perceptions of the personal nature of the notebook can still hinder the sharing of the content and, as indicated above, different user roles may have contradictory requirements.

Some groups expect their colleagues, students, or other collaborators to view and make use of their notes, whilst many researchers only share their notebooks with their supervisors. It may be that the researchers who expect to share their notes with others produce better quality notes than those who are keeping the notes only for themselves. Adherence to defined guidelines or standards may also improve note-taking quality. If funding or employment is dependent on adherence to precisely defined guidelines for keeping notebooks, then it is more likely a quality notebook is being kept and the researcher is likely to be less resistant to sharing. The following of guidelines and a particular structure may also mean the researcher feels the notes are less personal and private. As mentioned before, there is a possibility that a rigid structure may also change the nature of the notes to make them less reflective and potentially include less narrative, leading to the inclusion of less personal information.

3.10 Conclusion

In this chapter, we have examined how the way in which researchers structure their notebooks is to some extent based upon their experience and the nature of the science that they are working on. Although at a high level, the needs for different disciplines appears the same, there are some differences that need to be considered in the design of digital tools. In particular, the differences between the information that may be recorded between a formal structure and a more diary style notebook.

In Chapter 4, we will take a look at the history of the transformation from analogue to digital in science, including the digitisation of instruments and the development of technologies for storing and managing data. We will take an in-depth look into the development and role of digital tools for record-keeping, in particular the use of electronic notebooks in both the field and laboratory. We will also take a brief look at the impact of computational science and working *in silico*.

References

1. A. Tabard, J. D. Hincapié Ramos and J. Bardram, in *Proceedings of the 2012 ACM Annual Conference on Human Factors in Computing Systems – CHI '12*, ACM Press, Austin, 2012, p. 3051.
2. C. N. Klokmose and P. Zander, in *Proceedings of COOP 2010, Computer Supported Cooperative Work*, ed. M. Lewkowicz, P. Hassanaly, V. Wulf and M. Rhode, Springer, London, 2010, p. 119.
3. L. Arnstein, R. Franza, G. Borriello and S. Consolvo, *IEEE Pervasive Computing*, 2002, **1**(3), 13.
4. M. G. Van Kleek, M. Bernstein, K. Panovich, G. G. Vargas, D. R. Karger and M. C. Schraefel, in *Proceedings of the 27th International Conference on Human Factors in Computing Systems – CHI 09*, ACM Press, New York, 2009, p. 687.
5. P. Brandl, C. Richter and M. Haller, in *Proceedings of the 28th International Conference on Human Factors in Computing Systems – CHI '10*, ACM Press, New York, 2010, p. 599.
6. R. Yeh, C. Liao, S. Klemmer, F. Guimbretière, B. Lee, B. Kakaradov, B. Stamberger, A. Paepcke, in *Proceedings of the SIGCHI Conference on Human Factors in Computing Systems*, ACM Press, New York, 2006, p. 571.

7. P. M. Scholl, K. Van Laerhoven, in *ACM International Joint Conference on Pervasive and Ubiquitous Computing Adjunct Publication – UbiComp '14 Adjunct*, ACM Press, New York, 2014, p. 1381.
8. Committee on Science, Engineering, and Public Policy (COSEPUP), *On Being a Scientist: A Guide to Responsible Conduct in Research*, The National Academies Press, Washington, 3rd edn, 2009.
9. E. T. Mitchell, PhD Thesis, University of North Carolina, 2010.
10. A. Swan and S. Brown, *The skills, role and career structure of data scientists and curators: an assessment of current practice and future needs. Report to the JISC*, 2008. Available at http://ie-repository.jisc.ac.uk/245/1/DataSkillsReport.doc (accessed May 2018).
11. K. Shankar, *J. Doc.*, 2009, **65**(1), 151.
12. M. R. Canfield, in *Field Notes on Science and Nature*, ed. M. R. Canfield, Harvard University Press, Cambridge, 2011, pp. 1–18.
13. D. F. Wilson, PhD Thesis, University of Glasgow, 2011.
14. J. Ilerbaig, *Am. Arch.*, 2010, 73, 463.
15. K. Shankar, *J. Am. Soc. Inf. Sci. Technol.*, 2007, **58**(10), 1457.
16. F. L. Macrina, *Scientific Integrity: Text and Cases in Responsible Conduct of Research*, ASM Press, Washington DC, 4th edn, 2014.
17. K. Shankar, in *Proceedings of the 2nd Conference of the History and Heritage of Scientific Information Systems, ASIS&T Monograph Series*, ed. B. Rayward and M. Bowden, Information Today Inc, Medford, 2004, p. 118.
18. A. Behrensmeyer, in *Field Notes on Science and Nature*, ed. M. R. Canfield, Harvard University Press, Cambridge, 2011, pp. 89–108.
19. C. Bird, C. Willoughby and J. G. Frey, *Chem. Soc. Rev.*, 2013, **42**, 8157.
20. F. L. Holmes, *Isis*, 1987, **78**(2), 220.
21. M. Cardenas, in *Proceedings of the 2014 American Society for Engineering Education Zone IV Conference*, ed. L. Gossage, P. Marayong and M. Dyrud, American Society for Engineering Education, Washington, 2014, p. 96.
22. Available at http://gk12calbio.berkeley.edu/lessons/fieldnoteguide.pdf (accessed Sep 2018).
23. Available at http://cires.colorado.edu/people/sheehan.anne/lab.html (accessed Mar 2015).
24. Available at http://www.geos.ed.ac.uk/undergraduate/field/fnb/ (accessed Mar 2015).
25. Available at http://blog.lib.umn.edu/jveldof/calculator/BiologyLaboratoryNotebooks.pdf (accessed Mar 2015).
26. Available at http://labs.physics.dur.ac.uk/skills/skills/labbook.php (accessed Sep 2018).
27. B. Heinrich, in *Field Notes on Science and Nature*, ed. M. R. Canfield, Harvard University Press, Cambridge, 2011, pp. 33–48.
28. R. Tweney, *Bull. Hist. Chem.*, 1991, **11**, 51.
29. R. Cooke and M. C. Schraefel, 2004, Available at https://eprints.soton.ac.uk/id/eprint/259251.

30. M. Van Kleek, G. Bernstein, K. Panovich, G. Vargas, D. Karger, M. C. Schraefel in *Proceesings of the 27th International Conference on Human Factors in Computing Systems – CHI09*, ACM Press, New York, p. 687.
31. Available at http://instruct.uwo.ca/biology/320y/fj.html (accessed Sep 2018).
32. J. Kingdon, in *Field Notes on Science and Nature*, ed. M. R. Canfield, Harvard University Press, Cambridge, 2011, p. 137.
33. J. Keller, in *Field Notes on Science and Nature*, ed. M. R. Canfield, Harvard University Press, Cambridge, 2011, p. 161.
34. P. Jones and C. Nemeth, in *Ambient Intelligence for Scientific Discovery: Lecture Notes in Artificial Intelligence*, ed. Y. Cai, Springer, Berlin, 2005, vol. 3345, p. 152.
35. R. Kitching, in *Field Notes on Science and Nature*, ed. M. R. Canfield, Harvard University Press, Cambridge, 2011, p. 76.
36. R. Yeo, *Notebooks, English Virtuosi, and Early Modern Science*, University of Chicago Press, Chicago, 2014, p. 22.
37. J. Perrine and J. Patton, in *Field Notes on Science and Nature*, ed. M. R. Canfield, Harvard University Press, Cambridge, 2011, p. 234.
38. P. Naskrecki, in *Field Notes on Science and Nature*, ed. M. R. Canfield, Harvard University Press, Cambridge, 2011, p. 210.
39. C. Hoffman, *Sci. Context*, 2013, **26**(2), 279.

CHAPTER 4

Digital Tools and Electronic Laboratory Notebooks

4.1 Introduction

There can be no doubt that major technological advancements over the last 150 years, from the incandescent light bulb to the smartphone, have made a significant difference to the way that we live, work, and interact with others. Technology has always been an important part of the development of science, with new instruments and devices providing new ways to measure and observe, and facilitating the computation of ever more complex calculations. Computers are now pervasive in our homes and work environments, and very few scientists, in or out of the laboratory, will have no interaction with digital tools as part of their work. Computers are used through all stages of the research cycle, from searching for information to help plan research, capturing measurements and observations through a myriad of digital recording devices, to using software to analyse data. Digital tools are used to produce materials for publication or sharing and all this digital information needs to be stored and organised on digital devices so that it can be found again later. There is a long history of invention in science, from the development of the earliest tools for measuring time and performing astronomical calculations, scientists have often needed to create their own instruments and devices to further their research rather than relying on existing technology. Tool making today often involves the creation of devices making use of readily available electronic components or the development of scientific software for processing data and performing calculations and simulations. In fact, much science now happens entirely *in silico*, capturing data electronically and using the data to develop models and simulations to generate or make predictions about the properties or behaviours of materials in the real world.

Recording Science in the Digital Era: From Paper to Electronic Notebooks and Other Digital Tools
By Cerys Willoughby

Published by the Royal Society of Chemistry, www.rsc.org

In this chapter, we will take a brief look at the history of the transformation from analogue to digital in science, including the digitisation of instruments, and the development of technologies for storing and managing data. We will also take an in-depth look into the development and role of digital tools for record-keeping, in particular the use of electronic notebooks in both the field and laboratory. We will also take a brief look at the impact of computational science and working *in silico*.

4.2 Digitisation and Science

Although some may argue that the 'information age' began with the invention of the printing press and the ready availability of the printed word, the development of the computer has played a central role in the transformation of society into the 'information society' since the 1940s.[1] Computers have not only transformed the way we work, replacing manual methods of accounting and record-keeping, but also contributed to significant developments in science by enabling the high speed and accurate use of methods that had previously been manual, difficult, and error prone, such as performing complex calculations and processing large quantities of data.

The pace of technological development of the computer has been rapid, facilitated by the development of smaller, faster, and cheaper circuits, together with the exponential growth of processing speeds. The massive pace of technological advancements in technology is neatly highlighted in an article by Gerola and Gormory from the early 1980s.[2] The article describes the set-up and use of technology at an IBM research and development laboratory over the previous 20 years, and not only highlights just how rapid the development of technology was during that time, but also how much the technology and the way we use it and has progressed beyond what the authors would have imagined could have been developed in the next 20 years. Some of the hardware technology that the authors discuss in the article seems to us to be almost antiquated, for example, describing the magnetic disk technology for data storage as being like a phonographic record player with a magnetic record. However, the technology described was advanced for its time, for example, their network infrastructure was cutting edge with connectivity between over a thousand CPUs to other labs and offices at 140 sites around the world, at a time when Ethernet had only recently been commercially introduced and standardised, and Wi-Fi would not be developed for another decade. The major concern for researchers in the laboratory was raw computing power; high processing speeds are needed to carry out complex and accurate numerical calculations for the creation and validation of scientific models. Once validated, models can quickly and easily be applied to a multitude of problems, in the same way as real-world laboratory experiments. The more processing power, the more calculations can be performed, and the more models or applications can be tested. The end results of these large-scale computer calculations are large quantities of numeric data that are difficult to make sense of, so the next most important technological development for the researchers in the laboratory were graphical displays that enabled the data to be visualised as a graph or a map. The graphical displays not only enable large quantities of data to be understood, but also provided a way to assess the data by a direct appreciation of the form on the screen. For example, a model examining the formation of galaxies generates a stream of data that is difficult to assess in numerical form, but representing the data on a graphical display enables the model outcomes to be compared with the actual shapes of galaxies seen through a telescope. Although the generation of graphical representation on the screen may seem trivial now, the ability to be able to view graphics and to be able to manipulate objects in 3-D, on screens that previously only displayed text, was a hugely important development. Despite the excellent communication capabilities between different labs at the time, sharing and collaboration between sites was not seen as an important technological development for the researchers. The vast majority of collaborations appear to be restricted to the lab at which the authors were located, with limited collaboration between sites, a pattern that appeared common amongst similar organisations, perhaps reflecting the perception of scientific research as being an individualistic pursuit until relatively recently.

Technology has continued to advance rapidly over time, and the capabilities of hardware, software, and communications are significantly enhanced beyond what Gerola and Gormory could have imagined at the time. Computers have advanced to the point where the majority of us carry one in our pockets with vastly superior capabilities than early supercomputers, in terms of both speed and processing power. These computers are of course smartphones, but even many smartwatches can outperform the early supercomputers. Today, fully functioning computers have been developed that are smaller than a grain of rice and can be embedded in any device; collecting data, performing calculations, acting as a controller, and communicating with other devices through a variety of protocols. Although Gerola and Gormory had some basic e-mail and file sharing capabilities, the early Internet was still taking shape, and even the term 'Internet' would not be coined for another decade, along with the development of the World Wide Web. The development of the Internet, and the Web in particular, have had an enormous impact on the way that we work, particularly when it comes to finding, sharing, and, of increasing importance, storing information.

Calculations, analysis, modelling, and data representation are still some of the most important elements of computer use in science, but computers are now ubiquitous in science, even for disciplines that are not intensely quantitative. Computers are used in every stage of the research process, from planning a research activity, finding appropriate literature and data sources, capturing, preparing, and analysing data, and for writing reports, publishing papers, and communicating with collaborators. At the very minimum, most scientists are using email, word processors, spreadsheets, and the Web, with many more using more specialist systems such as databases, analysis and modelling tools, content management systems, repositories, and reference management software. We will talk more about the future of technological development in science in Chapter 12, but in this chapter, we will take a more in-depth look at some of the key technologies used in science, and in particular those that relate to how we capture, manipulate, manage, and store the scientific record.

4.3 Capturing the Data: Digitisation and Automation of Instruments

Data is an essential raw material of science. A fundamental activity of science is the collection of data, and once collected data is combined, transformed, and analysed to test our hypotheses and form the basis of our scientific understanding. Historically, data would be collected by hand through sample collection, measurement, and observation in the laboratory, observatory, and field. Although much data is still collected manually in this way, an increasing amount of data is collected automatically in digital form. The scale of data collection using digital instruments varies enormously from a simple digital wireless thermometer, to the complex and expensive telescopes and satellites used for Earth and space observation. For large-scale projects in particular, access to digital instruments such as telescopes with digital cameras, satellites, and sensor networks has facilitated new ways of conducting research and significantly increased the amount of data that can be collected and that needs to be processed and stored. For example, satellites for Earth observation were first launched in the 1970s and have provided previously impossible to access information about meteorology, vegetation, atmospheric gases, ice cover, and other environmental changes over time; whilst more recently embedded sensor networks have been able to provide automated environmental monitoring at a more local scale.[3]

For smaller projects, the majority of instruments that are used in the laboratory or in the field are now digital. Many instruments still require users to transcribe the data into their notebooks or directly into a computer. However, an increasing number, especially those that produce multiple data readings over space or time, collect or transmit the data in digital form so that it can be viewed and manipulated in a computer later. In fact, scientists using such instruments may never see the raw data because pre-processing of the data can take place within the equipment itself.[4] Data may be generated in the form of small files of tabulated numbers, digital images, videos, or enormous files with complex data and readings taken at millisecond intervals. Examples of digital instruments used in the field include equipment involved in mapping and surveying such as GPS units and total stations, digital calipers, geophysics equipment, microphone and digital audio recorders, meteorological equipment, sensors, hand-held devices, and digital cameras. In the laboratory, a majority of measuring instruments such as balances, thermometers, spectrophotometers, refractometers, viscometers, and more are digital. In addition to these analytical instruments, other devices used in the preparation of samples for experiment or analysis, such as stirrers, ovens, and centrifuges, may also be digital. These digital devices can not only produce data in a digital format, but also have the potential to be remotely controlled by computer. Sensors and hand-held devices such as barcode or radio-frequency identification (RFID) readers may also be used and the data captured automatically. We will discuss more about sensors and similar technologies in Chapter 12.

Many instruments in the laboratory are now controlled by computer, and many tasks that would have been performed manually by scientists in the past are now entirely performed by computer-controlled instruments and related technology. Automation within the laboratory has a long history frequently driven by innovations in industrial production, although before the widespread availability of electronic components in the 1940s, most automation was limited to devices constructed by individuals for specific tasks, for example filtration, percolation, and washing.[5] The development of electrical components enabled the creation of many devices for analysis and measurement, for example, electrical equipment for conductivity measurement, photoelectric cells for automatic titration, and chart recorders, photocells, and timers for stills and fraction collectors. Even before the development of the computer, instruments were able to capture data in printed format and plot results, for example, the recording potentiometer described by Bovie in 1915 that enabled the rates of a reaction to be plotted as a graph when different acids or bases were added to a solution under test.[6]

The addition of computers to laboratory instrumentation came quickly, with the development of a combination of mass spectrometer and computer in the early 1950s by the petrochemical industry, which used them to automatically generate mass and intensity information in a tabular form for hydrocarbon compounds, replacing a formerly labour intensive manual activity that was needed to generate a table by hand.[7] Later comparisons could be made with previously measured pure compounds with the mass spectra data recorded on punched-cards from early reference databases, such as the API RP44 reference database. By the mid-1950s, automated commercial instruments, such as the Technicon blood analyser, were being offered to medical laboratories that could for the first time enable a sample to be added at one end and a few minutes later the results of the analysis would be emitted at the other end.[5]

Today, almost all automated systems in the laboratory are based around computers. For industrial laboratories, automation aided by software is commonplace; the computer either directly controlling the instrument or managing the control of a set of instruments, stepping up from 'bench-automation' to 'lab-automation'. Such laboratory level automation often involves the use of robots, pneumatic systems, and conveyer belts to move samples between instruments and even between different rooms. Although many instruments have interfaces for control built into them, separate computers can include software that can perform more sophisticated control of multiple instruments to control the flow of an experiment or test protocol. For example, in continuous-flow isotope ratio mass spectrometry (CF-IRMS), an instrument in which chemical reactions take place is connected to a continuous-flow interface which in turn is connected to a mass spectrometer.[8] The computer-controlled connection of the three instruments enables the reaction to use a smaller quantity of sample materials and the results to be monitored in real time. Computer automation of instruments and other lab equipment can speed up processing and laboratory efficiency, but there are challenges that make automation difficult. A lack of standards between different suppliers of instruments means that it is difficult to set-up connections between different instruments or a controlling computer using the correct protocols. The controlling software needs to be able to recognise the instrument and communicate with it, and this often involves the need to create new software to integrate it into the analysis flow, requiring specialist skills including electronics and low-level programming.[9] Large-scale automation also has a major impact on the way that work is done within the laboratory, changing not only the tools and processes that are used, but also the way that people work and their roles within the organisation.[10] These changes in many ways account for the differences between industrial and academic or research and development laboratories, because the focus of activity has changed. Industrial labs, such as clinical testing and pharmaceutical production, seeking to reduce costs and improve quality have often looked towards manufacturing as a guide, leading to increases in the amount of automation, changing the roles and skills that are required in the lab. For example, there may be less emphasis on sample preparation and other manual tasks, and more focus on data management and quality control.

4.4 Storing Data: Databases

Both raw and derived data once collected need to be stored. At the heart of much scientific software is a database of some kind. Databases are important for both the medium term and long-term storage of the record and facilitate sharing of data collated within an organisation or collected by an external organisation, such as the many repositories of data and literature available to search on the Web. Databases actually originated before the invention of the computer, but are also intimately linked with the development of machines to process large quantities of data that took too long to process by manual means. The first development came from Herman Hollerith, one of the founders of the Computer-Tabulating-Recording Company that later became IBM. Hollerith devised a tabulating machine to calculate the results from the 1890 US Census, significantly decreasing the amount of time it took to process the census and saving millions of dollars in the process.[11] The census data was stored on punched-cards that were processed automatically in their millions by the tabulating machine. As the benefits of using tabulating machines and punched-cards became apparent, their use in both business and science grew. Punched-cards remained a dominant storage material, along with paper tape, until magnetic tape began to take over in the 1950s.

In the original tabulating machines, calculations were performed as the cards were added to the machine to create the totals needed for the census analysis. For other kinds of uses, it became necessary to be able to access individual records from a punched-card archive and to be able to sort the records in a meaningful way. This led to the development of needle-sorting punched-cards, where punches on the edges of the cards could be used to find records that matched particular criteria, essentially an early form of search. One example was the API RP44 mass spectral database, mentioned previously, that was made available on needle-sorting punched-cards in 1953.[7] Another example, from 1973, describes the development of a system to manage the storage and retrieval of biological samples.[12] The system enables not only the location of the sample in a freezer rack to be identified, but also to be able to answer common questions that the laboratory workers had, such as what tests had been completed, and whether all the planned analyses had been carried out. The punched-card in this system contains 'metadata' about the type of sample, when and where the sample was taken, when it was sent to another laboratory, an identifier for the subject, a location, and a section for remarks that can be used to record additional information such as the volume of the sample. Pairs of holes on the punched-card represent the tests that need to be done. No holes punched means a particular test is not needed, one hole punched in the pair means the test has not been completed, and the second hole in the pair punched means that the test has been done. There is also a hole for indicating all tests are complete. Identities for the subjects are a numerical number with a set of numbers each with a hole for hundreds, tens, and units. So for example if the ID number is 273 a hole is punched in '2' in the '100s' box, '7' in the '10s' box, and a '3' in the 'UNITS' box. Dates and other numerical information are handled in a similar manner.

Once the punched-card has been set up with the information to identify the tests that need to be done, they are included in the stack with the other punched-cards in no particular order. When the analyst is ready to perform a particular test, they can push a needle, similar to a knitting needle, through the upper hole of the selected test. When the needle is lifted, all of the punched-cards that do not need to be tested are removed, leaving behind the cards with the specimens that do need to be tested. As shown in Figure 4.1, the analyst can then push the needle through the lower test hole in the remaining cards and raise the needle. The cards on the needle are

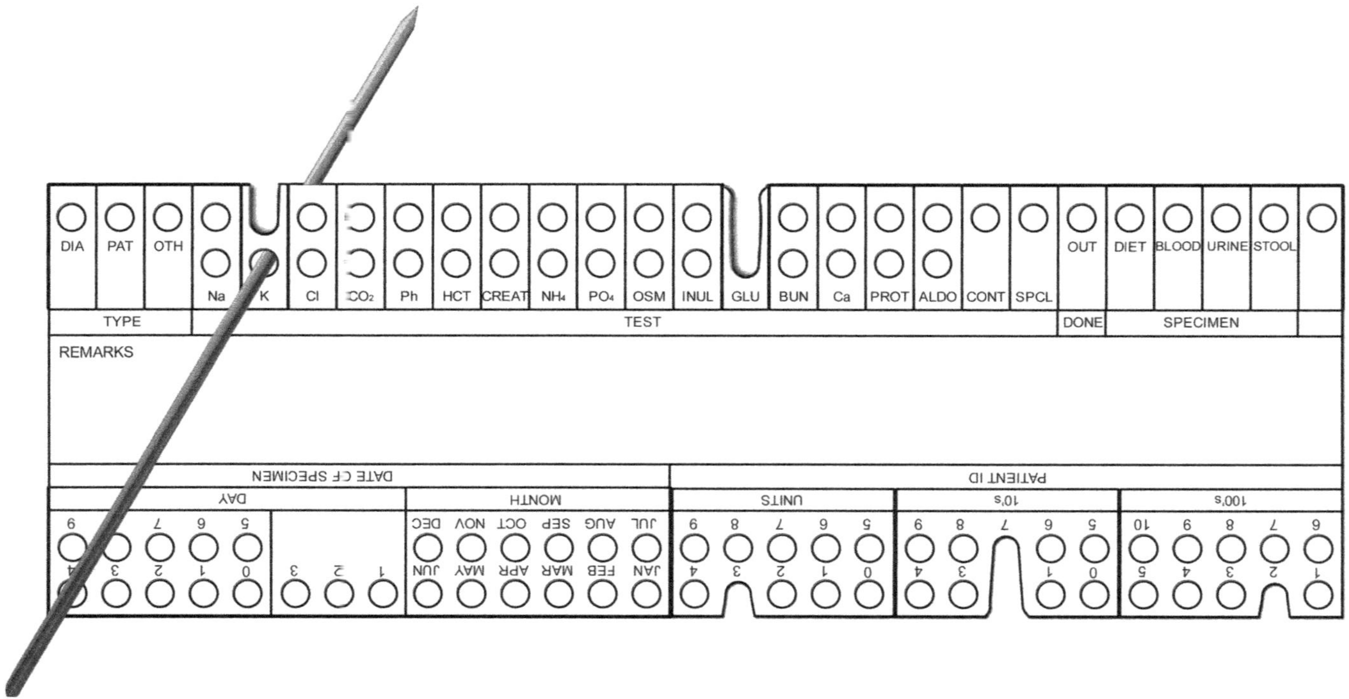

Figure 4.1 Simplified version of a punched-card for capturing test results in the lab to show the needle sorting technique. The test has a patient ID, and two tests selected, with one already completed. The needle is being used to find samples that need to undergo the test for potassium levels.

the specimens that need to be tested, and those specimens that have already been tested are left behind. As the test is completed, the results are recorded on the card, and the second hole for the test is punched. If all tests are now complete, the 'DONE' hole is punched. The same needle-sorting strategy can be used to find samples based on the subject's ID or the date the sample was taken, potentially using multiple needles at the same time to select the correct cards. The location information for the sample is coded onto the punched-card and manually inspected when samples are being located. Although relatively unsophisticated, the needle-sorting system has a lot in common with a more modern Laboratory Information Management System: the registering of a sample at the beginning of the process, the flow of the record between the different tests, the capture of results at the end of each test, and the built-in sample storage management. It also provides an insight into familiar issues in data management of the design and flexibility of the data structure and the constraints between managing data and the workflows of the user. Although the analysts had to deal with new procedures to manage the punched-cards through the test process, there were also benefits from the system in rapidly being able to filter the records and find samples.

Punched-card systems were reasonably effective with small quantities of records, but took up large quantities of storage space, and were difficult to search with large quantities of data. Magnetic tape enabled much more data to be stored, but records could only be accessed sequentially and not individually. This made magnetic tape useful for businesses such as banks that used batch processing to process transactions overnight, but was less useful for attempting to access individual records with up-to-date information in real time. It was not until the 1950s, with the invention of the spinning magnetic disk, that individual records could be rapidly and easily accessed. To locate an individual record on the correct part of the disk, an index was required, leading to the development of the first file systems. At this point, each software application needed to handle access to the data records individually, leading to a duplication of effort and problems of data corruption caused by errors in the data handling code.[13] This problem was overcome by the development of Database Management Systems (DBMS) in the 1960s. These DBMS handled the complexities of enabling access to the data and maintaining its integrity on behalf of programmers, facilitating the effective organisation of records on storage media and handling much larger quantities of data. The DBMS required that a schema was created to define the structure of data before it could be added to the database and provided a meaningful way to navigate between items in the database, so that all information could be retrieved. These early DBMS systems were relatively inflexible, making it difficult to add new structures or queries for the data once the database had been built and populated.

The development of Relational Database Management Systems (RDMS) in the 1970s, improved the usability of the database by adding functionality to perform queries on the data itself, providing advanced search capabilities and enabling reporting on the data. These new capabilities were achieved by the development of query languages, including Structured Query Language (SQL) which became the standard language used with the majority of relational databases. Query languages provide a way to program the design of specific structures for data to be held in, for the retrieval of data based on user-defined criteria, and also commands to add and update data within the database. Although relational databases have been the mainstay of both businesses and scientific applications for decades, the focus on ever increasing quantities of 'big data' and the movement of technology and storage online has begun to favour non-relational databases, or 'NoSQL'. NoSQL databases have been around for many years, but their development and popularity has been propagated by the growth of large Web companies, such as Google, Amazon, and Facebook. The shear amount of data that these companies handle presents huge challenges of scale. Datasets are massive, requiring vast quantities of storage, but the data are also very diverse and difficult to model. Solutions need to be able to rapidly retrieve and manipulate the data and the needs of these have driven revolutions in not only database and data management systems, but entire hardware infrastructures and architectures, particularly focused on distributing data across multiple servers.[14] Such architectures require appropriate software that can manage the transfer and manipulation of data in parallel across hundreds, or even thousands of machines. In contrast, the only way to scale RDMS is to partition database tables across multiple database servers. This is a very difficult process to manage and also makes querying and data retrieval much less efficient, reduces data integrity, and adds new problems such as load balancing that can be very complex to solve.[13]

There are a number of advantages of NoSQL databases. Greater amounts of data can be stored within their file systems and the design of the data structure does not need to be carefully planned in advance. The flexibility of the data structure enables the database to be able to handle changes in the way that data is captured and manipulated over time. It is even possible to delay the creation of a schema for the data until the point the data is read. The structure of the data can be different for different data objects, so that only information that is relevant is stored. Although these databases are called NoSQL, many of them do in fact include query languages that can be used to retrieve the data in the same way as RDMS. The wide variety of ways that data can be held in different NoSQL databases means that different databases are useful for specialised purposes, for example, databases that store graph information, such as Resource Description Framework (RDF) triple stores, are useful for working with data that describes relationships between objects. As a result, there has been a proliferation of different kinds of databases that are optimised for different tasks that are relevant to scientific activities, including handling and storing 'Big Data', network data, and data from connected devices including the Internet of Things.

4.5 Managing the Laboratory: Laboratory Information Systems

Laboratory Information Systems (LIMS) are software systems that are designed to manage laboratory data, workflows, and processes, and are intimately linked to automation. Early LIMS were bespoke systems, typically developed as part of the automation activities of any particular individual laboratory. Commercial LIMS began to become available in the early 1980s from a number of different suppliers, together with the first use of the acronym LIMS.[15] LIMS are primarily focused on the capture, storage, and reporting of information from the laboratory. They are of particular value in laboratories involved in routine testing and production because they support an end-to-end process where requests and samples are registered in the computer and downloaded onto the appropriate instruments and then tracked across the various activities of the lab. For example, in an automated clinical chemistry laboratory, a request for a particular analysis is made and entered into the computer. When the specimen from the patient is taken, it is prepared, analysed, and validated. The results of the analysis remain associated with the request and patient data, and an interpretation of the data is reported.[10] Test results together with specimen information are stored in a repository to enable easy and fast retrieval of the important clinical data. If an analytical process is not automated, then the LIMS system is typically involved at the beginning and end of the analytical process when the samples are registered, tests schedule, and completed, but not in the intermediate stages at the bench where the analysis is completed.

LIMS with extended features are popular with laboratories with a focus on Quality Control (QC), Quality Assurance (QA), and highly regulated production such as clinical laboratories, pharmaceutical production, and the manufacturers of medical devices. Such features include the ability to enforce Standard Operating Procedures (SOP) and integration with other business and production systems, including Management Execution Systems (MES), Corrective and Preventative Action (CAPA) systems, and Enterprise Resource Planning (ERP) systems that include facilities for managing accounting, client information, procurement, project management, and manufacturing.[10] LIMS also often provide functionality or links to other useful sources of information, such as repositories of SOPs, sample storage inventories, and chemical inventories. LIMS enables not only tracking of individual samples and requests, but also the capability to analyse the throughput of work in the lab, and identify areas of concern for compliance, safety, or efficiency.

A related system to LIMS are Laboratory Execution Systems (LES). LES have some overlap with the functionality of LIMS, but are typically used to support analysts in analytical testing laboratories to adhere to SOPs and to reduce errors made by directing the steps that the analyst must take in performing a test. LES may connect to instruments to ensure they are calibrated correctly, and to capture data from the analytical process. LES may store the procedural methods within its own database or retrieve them from an external system, such as a LIMS, when needed.

4.6 Digitising the Record

The rise of technology has made the keeping of notes both easier and more complex.[16] Scientists are using a variety of digital tools such as instruments, sensors, microphones, and cameras to capture data for their research. These tools may capture huge quantities of information in seconds. This data can then simply be uploaded to an appropriate storage device. However, there is a risk that the quantity and easy availability of such data may give researchers a sense of completeness when recording their research and discourage them from making separate notes. Although large in quantity, digital data often retain little context and are often missing information about the where, how, and why of its collection. For example, although data collected by an instrument may contain some additional descriptive information such as the type and configuration of the instrument and date and time of readings, the context of the data, such as what materials are being analysed, sampling strategies, processes used to create the materials, and why the data is being captured cannot be recorded by the instrument. Data files alone cannot provide the context of the research, and if they are not adequately linked to detailed notes about the research, they become useless. A data file without any context cannot be understood or reused, and if the notes themselves are lacking in important context, then the same problem occurs. Initially, a researcher may remember the relationship between the data collected and the research, but over time the relationships and connections between different digital elements tend to break down when thorough notes are not available to bring them together.[17] Even when adequate notes are captured in paper notebooks, it can be challenging to link together the hand-written context with the digital files. One of the most common ways of solving this problem has been to print out the data from the digital device and to literally stick it into the paper notebook in the correct location. Other solutions are to diligently record file names and file locations in the notebook, so that the data can be tallied with the correct notes at a later time. This strategy is reasonably effective, so long as the file locations never change (unlikely over time) and there are no problems of duplicated file names. To manage links to large quantities of digital data in paper notebooks in this way requires great patience and skills in data organisation.

An example of just such a problem comes from archaeology, where creating links between paper notebooks and data captured from both the field and laboratory analysis is a difficult problem. A single excavation may generate 30 000 photographs, geophysical reports, character data, reports, text descriptions in a database, and structured data such as X-ray, multi-spectral, 3D database, point cloud, and geometry data, that may be stored in spreadsheets, databases or other repositories. Whilst some of the data may be stored online, much of it is distributed in inaccessible locations including personal computers, in private repositories, and even on memory sticks. The manual creation of documents that contain the links to the data would be valuable and ideally the link would be bidirectional and generated at the point the data file is created, however there are a lack of tools that enable this to be done effectively.[18] A potentially better solution to the problems of managing notes and digital data is to take notes electronically and provide direct links to data within these digital notes.

Creating digital notes means different things to different people. Although it might be assumed that creating digital notes means using some kind of electronic notebook on a computer or mobile device, many people actually think first about capturing notes in a traditional paper notebook with a digital pen or even just transcribing notes from a notebook into a word processor![19] Before the advent of either the digital pen or electronic notebooks, it was possible to create a digital image of a notebook page using a scanner. Historically, such scans generated very large files that were of limited use because each page would still have to be read manually to extract the content. Optical Character Recognition (OCR) technology enables scanned writing to be translated into text on a computer, making it possible to translate the notes into an editable form that could usefully be incorporated into other documents or searched to find specific sections. However, OCR of sufficient quality and resolution to read hand-writing is only a relatively recent development, and it is still somewhat error prone in its transcription of hand-written material. In this era of the digital camera, it is now possible to capture a digital image of a notebook page in a relatively small file, but the problems of making effective use of the image once captured remain.

A more popular tool that bridges the gap between paper and the digital domain is the digital pen. The digital pen automatically captures the notes as they are written and can transfer the notes in an editable format to digital storage as the notes are being written or at a later time. The digital pen has many advantages for capturing notes because it captures both hard copy and electronic notes, and is easy and familiar to use. The pens can also capture sketches, as well as text-based notes, and these are uploaded as images that can be incorporated in other documents at a later time. Digital pens are just as portable as pens and pencils, which makes them ideal for use in environments where larger digital tools such as computers, Personal Digital Assistants (PDAs), and tablets are less practical or more fragile, such as in the field or in certain laboratories. There are, however, some downsides to digital pens. Digital pens record in a linear fashion, so that information is captured only in the order it is written, which means they are not appropriate for capturing content from forms where the forms will be completed in a different order, such as the context sheets used for archaeological digs or forms for recording laboratory tests.[20] The problem of recording in a linear fashion also makes it difficult to go back, add, or correct previously written notes. The most obvious disadvantages compared to an ordinary pen and paper is that the digital pen is generally more expensive and they require batteries.

4.7 Electronic Notebooks

For many people, taking digital notes means to capture notes directly onto a computer or a mobile device using generic software, rather than a specialist tool. There are many different applications that can be used to capture digital notes, including a large number that have been designed for the purpose, but at the most basic level, a word processor on a computer can be used. As mentioned previously, the vast majority of scientists are already using word processors as part of their everyday activities, making the tool familiar and easy to use for creating notes in the format of choice. When a word processor is used to capture notes, the user needs to not only chose the format in which the notes are recorded, but also to derive a way to organise the notes. Should a single document be used for all notes like a journal, or should notes from different days or different activities be stored as separate documents and organised in some meaningful way within the file system? There are many different ways to organise documents, for example, in chronological order, alphabetical order of title, and ordering by subject. It is easiest to create information in chronological order, as we typically do in notebooks and journals, but it may make more sense to organise information using a different methodology depending on the purpose of the record or the individual preference of the researcher. For example, information organised by subject, such as in the common-place books we discussed in Chapter 2, can make information easier to retrieve, whereas information ordered chronologically can be more difficult to search. As we also discussed in Chapter 2, over time it may be easier to remember 'what' an activity was about rather than 'when' the activity was carried out, which can be supported by organising materials by subject or if a meaningful title is provided for a file. On the other hand, data files typically have generic names relating to the date the file was created or a timestamp, so organising materials by date can be a useful way to keep data and notes in documents connected together. The advantage of word processors is that they provide the flexibility for the user to make those decisions for themselves, and to make changes as their process for note-taking evolves over time. Additional content such as images, calculations, tables, and links to other files or content online can be added using the majority of word processors, increasing their value for storing related information together in an easy to use format. Other files that cannot be added directly to a document in the word processor can be stored on the local file system, servers, or using an online service such as Microsoft OneDrive, Google Drive, Apple iCloud Drive, or Dropbox. The majority of these services provide some free storage space and ways of sharing the content with other users.

An alternative to the word processor for capturing digital notes is to use journaling software or note-taking applications, with numerous options available, including Microsoft OneNote, Evernote, GoodNotes, and WritePad. These applications usually have limited capabilities, but are similar to those of word processors and may have built-in functions to organise the files, share documents, encrypt the data, and to store online for access from multiple devices. Note-taking software may also have useful features such as including timestamps and be searchable to help with retrieval at a later stage. Electronic notebooks are available for use on a computer, but are also frequently available for use on mobile devices such as phones and tablets. In some cases, content such as text and sketches can be captured by using a stylus rather than keyboard to enter content on the screen on these devices. Electronic notebooks on mobile devices overcome some of the portability problems of computers and laptops by being relatively small and light. An added benefit of the mobile devices is that is it often possible to directly capture other media such as photographs, audio, and video within the notebook itself, saving time and effort in switching between applications, uploading, and embedding such content. Journaling software can be very useful for capturing notes electronically for field researchers if they do not need to capture information in a special format. PDAs and tablets with GPS are often used for data collection in the field, and are particularly useful for providing a way to access digital information such as maps, photographs, and geophysical data that can be used in mapping or to pinpoint a research location. An example of such use is the GeoPad rugged electronic notebook that was developed for use in teaching geological fieldwork, and is used by several universities in the USA.[21,22] The tablet PC is not only equipped with OneNote for capturing data input in the field, but also with GPS capabilities, Geographic Information System (GIS) software for interactive mapping and 3-D visualisation, spreadsheets, photo-handling software, and other tools likely to be of value during the fieldwork.

Mobile devices, however, do have similar problems to computers, for example, screens can be difficult to read in the sun, mobile devices can be fragile and expensive to replace, and they also have a need for a power supply. It can be difficult to extract data stored on mobile devices for use with other applications at a later stage, especially if it is stored in a proprietary format. The electronic notebooks also tend not to be suited to the capture of large amounts of data. There are, however, some specialist devices and laptops that are extremely rugged for use in diverse field environments, and screen readability and battery life on mobile devices is also improving all the time.

Basic electronic notebooks like these are ideal for situations where great flexibility is required and ad hoc notes are collected. The generic electronic notebooks mentioned above are useful for capturing unstructured information. They are, however, less suitable in situations where more formal notes need to be captured and in situations where security and privacy are of concern. They also have limited specialised functions, so are less likely to be suitable for making calculations and integrating content from external sources, including the kinds of data that might be produced by instruments in the field or in the laboratory. For these situations, more specialised kinds of digital notebooks are required.

Blogs and wikis are in some ways quite similar to the general electronic notebooks described previously, except that the content of the notes is stored online. Blogs are designed to be personal diary-style systems, with posts entered and displayed in reverse chronological order. Wikis, on the other hand, are more suited to creating pages that may be updated over time, particularly by multiple authors and can act as a central repository of information. The more collaborative nature of wikis also means they are more likely to provide access control, timestamping, and version control features that are useful for collaborative working. Wikis may therefore be more useful as repositories of team information such as standard operating procedures, and other shared documents. There are examples of more specialised note-taking software applications that are based on both blogs and wikis, and we will discuss two of these in Chapter 6, the blog-based LabTrove and the wiki-based openwetware.[23,24]

4.8 Capturing the Record in the Field: Electronic Field Notebooks

Electronic notebooks designed specifically for fieldwork are not very common and tend to be designed with a specific task in mind. Frequently, the solutions are a hybrid integration of paper notes and digital assets making use of a digital pen. One example is the ButterflyNet system facilitating the capture and access of data for biologists in the field.[25] In common with many scientists, field biologists organise their work using paper notebooks, but then analyse the results using computers, leading to a need to convert their paper-based notes into digital form. In the ButterflyNet system, the biologists collect notes on paper using a digital pen to capture them in digital form. Digital photographs are captured with a 'smart' camera that allows the images to be annotated on the fly with a stylus, and samples are collected with barcode tagged bags that enable each sample to be uniquely identified. Digital data is automatically associated through the use of timestamps or manually by the creation of a link using the smart camera. The user can draw a standard symbol in the notebook to indicate a 'hot spot' after the capture or selection of a digital photograph and can be prompted to confirm the link between the notes taken and the photograph. The tagged specimens can also be associated with notes or photographic data by including notes directly in the sample bag or by including the barcode of the sample in a photograph. Once the data has been uploaded to a computer, a browser application can be used by a researcher to search, access, and view their own data and also data from shared notebooks. Although these biologists felt that paper was still a necessary part of the solution to capture notes digitally in the field, the use of ruggedised computers makes it more practical to implement fully electronic notebooks in the field, such as the GeoPad described previously.

Another example of an electronic notebook for the field is the ENVIT Field Notebook system, designed for researchers in environmental science. In the ENVIT Field Notebook system, an electronic journal on a ruggedised mobile computer has been integrated with sensors to enable the collection of water quality and other environmental data with geolocation information more simply in the field.[26] A water quality probe captures information automatically, whilst data can be captured manually using the notebook software. Data captured manually by users and the automatically captured data are combined and stored on a database. The database can be integrated with GIS data to enable visualisations, and then later all the data can be transmitted to computer storage for further processing and sharing using wireless technology.

4.9 Capturing the Record in the Laboratory: Electronic Laboratory Notebooks

The specialised tool for capturing notes digitally in the laboratory is the Electronic Laboratory Notebook (ELN). There are many different types of ELN that serve different purposes and different audiences. ELNs aimed at a research and development audience are more likely to have functionality that supports protection of intellectual property and collaboration, whereas an ELN for an analysis or production environment are more likely to have functionality for compliance and traceability. There are ELNs that provide specialist extensions that support particular disciplines or requirements of particular industries such as drawing tools for chemical structures for users in chemistry or sequence aligners for analysing DNA, RNA, and proteins for bioinformatics and clinical use. There are also ELNs that are non-specialised and designed to be more flexible and enable use by scientists in any discipline or even working across disciplines. These non-specialised ELNs can be useful in academic environments where practices are more diverse and less prescribed than within industry, and where the varying needs of individual researchers may be more significant.[23]

The development of ELNs came later than the development of LIMS, and many early ELNs were seen as extensions to the functionality of LIMS to enable better management of the laboratory.[27] An example of one of the very earliest applications labelled as an ELN in the 1980s was RS/1.[28] Labelled as an ELN by some, RS/1 was a sophisticated statistical package with data management, data analysis, statistical analysis, modelling, and 2D and 3D graph and plot production, together with reporting functions to support decision making.[29] RS/1 also included a built-in programming language, RPL, that enabled users to create procedures in a language that resembled simplified English. These procedures enabled repetitive operations to easily be carried out, reducing manual effort, ideal for repeating analyses on captured data, and enabling automation through the use of a batch mode. RS/1 also integrated with a variety of tools to support use in the development laboratory and pharmaceutical production environment. RS/Discover, for example, provided an interface for guiding users through creating and optimising their experiments and processes, including graphics to display relationships between factors and results. A variety of tools were provided for monitoring and quality control, and even a tool aimed at pharmaceutical and biotechnology companies for managing large-scale clinical trials. RS/1 was also able to collect data automatically from instruments, providing some similar functionality to LIMS. Although claimed by some to work just like a paper notebook, it is less clear how much of the all-important context information traditionally captured in the paper notebook actually got entered into the system.

The driving force behind the development of early ELNs were the pharmaceutical and chemical industries that needed to be able to track activities in the laboratory for quality assurance, compliance, and in particular to protect their Intellectual Property (IP).[30] Earlier electronic notebook systems focused on the concept of the 'paperless office' had been developed and used by research and development organisations, but these were found to be lacking in the capabilities desired by scientific industries. A number of hybrid ELNs were generated that captured notes electronically, but then required the users to print them out, in order to be signed and witnessed for IP protection and archival purposes.[31] In 1997, the Collaborative Electronic Notebook Systems Consortium (CENSA) was established to address these insufficiencies in the ELN market.[32] CENSA was a collaboration between a number of key players in the pharmaceutical and chemical industries, academic institutions, software development organisations, and national laboratories. CENSA sought to design an architecture for an ELN specifically for the storage of experimental data. One of the ideas of CENSA was to build a standard architecture to which modules could be added for a variety of disciplines including chemistry, biology, and materials science. CENSA also envisioned a solution that would use specialised and standardised 'hooks' to enable integration with other applications. CENSA also considered there to be two types of ELN: a personal electronic notebook (PEN) that represented an electronic version of the paper notebook, and a collaborative electronic notebook (CEN) that would enable communication and information retrieval within teams.[33]

CENSA had some ambitious plans for ELNs, but a key development for the wider adoption of ELNs in industry was the acceptance of electronic records for regulatory and patenting purposes. In order to achieve this acceptance, technology is required that provides timestamping, digital signatures, and mechanisms to prevent the unauthorised manipulation of the record. Once this technology was accepted as being reliable, then it became more feasible to remove the paper notebooks from the laboratory and to replace them with technology. Another driving factor within industry during this time was the need for greater collaborative facilities, in part as a reaction to the increasing prevalence of outsourcing across continents.[31] An example of an ELN developed with collaboration in mind was the Pacific Northwest National Laboratory (PNNL) ELN.[34] The PNNL ELN was a Web-based general purpose ELN that enabled discussion between distributed users by allowing users to add replies to notes recorded in the notebook. These 'sub-notes' appeared indented below the original notes to support a threaded discussion between team members. The ELN would also allow users to set up notifications, so that they could be emailed about updates to individual pages or whole notebooks on the system. The Web-based architecture of the ELN software enabled users to access their notes from any machine with a browser, both inside and outside the lab.

The way that ELNs capture data and information has changed over time as ELNs have evolved into broader solutions, beginning with very rigid structures with a focus on reaction registration during the 1990s and becoming more flexible over time.[31] The driver for these early rigid systems was the implementation of ELNs to facilitate compliance to standard operating procedures; the rigidity mandated by the strict electronic script that needed to be followed by the user and verified by the system.[35] Over time, the design of ELNs has become more flexible and input can be provided through the use of generic word processor documents or Web-based user interfaces. With the current proliferation of ELNs, there are a mixture of flexible generic ELNs that enable the capture of unstructured information, whilst the majority of commercial ELNs still make use of domain specific input forms and templates to support laboratory workflows. The use of structured and unstructured records for recording experimental activity reflect the nature of the science under investigation. Structured records are more likely to be produced in contexts where relatively repetitive or routine actions are performed, for example, in analytical and clinical testing or where a new chemical entity is being developed as a planned iteration on an existing entity. Unstructured records are more likely where the research is more experimental and less routine, for example, the creation of completely new chemical or biological entities, or where new processes are being developed. The ELNs that are focused on more structured data have more of an overlap with LIMS, although the LIMS focus is more on the samples moving around the laboratory than the experiments themselves. ELNs are better at handling multiple diverse experiments on samples that generate a variety of different output data. More recently, ELNs designed specifically for the Quality Control market, such as qELN, have brought back more of the rigidity, as flexibility for these systems is undesirable.[30,31] In these environments, standard forms are used to capture data and to direct the procedure during the analysis or experiment and therefore templates within the ELN system can be configured to look exactly like the paper forms already in use.

Some ELNs are based on Semantic Web technologies, such as Resource Description Framework (RDF) and metadata ontologies, which we will discuss in more detail in Chapter 10. The advantage with using these technologies within ELNs is that the formalised metadata provides additional meaning and context to the data added to the ELN. The metadata also enables the content to be machine-readable, facilitating interoperability between different systems such as repositories and publishing tools.[23,34,36–39] In Chapter 6, we will take a more in-depth look at ELNs, look at some specific examples, and consider both their benefits and potential drawbacks when compared with traditional paper notebooks.

4.10 Working *In Silico*

As mentioned previously, the majority of scientists are already using some form of digital tools to process and analyse their results. These tools range from the very simple, such as using spreadsheets to create graphs and generate summaries, to highly complex tools used for statistical analysis and modelling. Data generated through scientific activities may need to undergo some form of pre-processing before analysis, and there are a variety of tools for specific disciplines and data types to assist with these processes, for example, OpenRefine[40] and Trifacta.[41] Data cleansing tools can be used to validate data and to correct and remove data with a variety of problems, including data that are inaccurate, corrupted, invalid, inconsistent, or duplicated. Data reduction can be used to simplify data and reduce the quantity of data that needs to be analysed, and can include activities such as averaging, producing summaries, sorting, encoding, scaling, and filtering. Data can also be transformed in various ways to standardise the data, structure it for analysis, or to augment the data by combining it with external information or aggregating together data from different sources. Off-the-shelf software exists for both general and domain specific analysis, but many scientists produce their own digital tools for processing and analysing their own data, often using scripting languages such as R or Python.

For disciplines such as particle physics, astronomy, and life science analysis, where the data is 'born digital' *via* telescopes, sensor networks, and other high-throughput instruments such as gene sequencers, much of the human scientific endeavour is carried out through interaction with digital tools for processing and analysing the data. For other scientists, their research is carried out entirely *in silico*, in other words, using computers, as their data is derived from analysis of existing data, or experiments using modelling and simulation. For example, biological and chemical elements, such as molecules, proteins, cells, and potential chemical compounds can be modelled using computers to investigate potential drug interactions to search for safer or more effective drugs, or to investigate the relationships between genetics and disease.

The term 'workflow' is typically used to describe repeatable experiments or analyses carried out *in silico*. These workflows represent the computational tasks required to capture or take digital data and manipulate it to produce the final results at the end of the process and may combine a number of different tools or services to do this. In theory, a workflow could be the documented instructions of how to do this manually, but usually represents some kind of automated process that enables the procedure to be repeated again with the same or new data. A workflow could be a collection of scripts, each performing a particular task, which are called in a particular order to carry out part of the workflow. For example, there may be separate scripts to retrieve data from storage, perform pre-processing tasks, transform the data, and analyse the results. There are also a number of 'workflow management systems' (WFMS), such as KNIME,[42] Kepler,[43] Galaxy,[44] Pipeline Pilot,[45] Vistrails,[46] and Taverna,[47] that enable users to build up their own workflows using hundreds or even thousands of prewritten modules. WFMS may be targeted to a particular discipline, such as Galaxy and Kepler, or be more generic, providing functionality appropriate across scientific disciplines. Many of the WFMS also provide access to additional discipline capabilities *via* third-party extensions and Web services. WFMS provide relatively simple tools for creating workflows with complex analytical capabilities that are repeatable. Using WFMS workflows can be automated, re-run, re-used, and even shared, without the need for strong programming skills and the complexities of setting up a specific software development environment.

4.11 Conclusion

Technology has had a significant impact on the way that science is conducted and the way we capture our records. Electronic notebooks provide ways to capture notes digitally and to more effectively manage the links between notes and digital data. Although adoption of electronic notebooks, especially ELNs, is good in industrial settings, the majority of academic researchers are still using paper notebooks and having to manage the disconnect between paper-based notes and digital data. We will talk more about the benefits of Electronic Notebooks and how they facilitate scientific research in Chapter 6, in particular focusing on Electronic Laboratory Notebooks. We will also look at some examples of the ELNs currently available and look at how some groups make use of them. We will also explore some of the potential disadvantages of ELNs compared with paper notebooks for capturing the research record and take an in-depth look into barriers to the adoption of ELNs.

In whichever way data is captured, processed, transformed, or analysed, the capabilities provided by digital tools enables more data to be generated and analysed in a shorter space of time than traditional methods of data capture and analysis. As we will discuss in Chapter 5, the proliferation of digital data leads to some big challenges in terms of not just storing the data, but also managing it, and ensuring that the details about how the data was created do not get lost. Data without the context provided by the rest of the research record, such as the methods and background recorded in a notebook, can be difficult for anyone other than the creator of the data to understand, and as a result, make it difficult to use, and most importantly, to replicate. We will discuss the issues and challenges around research reproducibility and how data management practices can help. We also take a look at the role of metadata, standards, identifiers, and provenance in data curation and their importance for discoverability and reuse. We will also examine the vulnerability of digital data and the steps that need to be taken to ensure that data is preserved for future access and reuse.

References

1. M. S. Mahoney, *Ann. Hist. Computing*, 1988, **10**, 113.
2. H. Gerola and R. E. Gomory, *Science*, 1984, **225**, 11.
3. C. L. Borgman, J. C. Wallis and M. S. Mayernik, *Comput. Support. Coop. Work*, 2012, **21**, 485.
4. C. L. Bird, C. Willoughby and J. G. Frey, *Chem. Sci. Rev.*, 2013, **42**(20), 8157.
5. K. Olson, *J. Lab. Autom.*, 2012, **17**(6), 469.
6. W. T. Bovie, *J. Med. Res.*, 1915, **32**(2), 295.
7. M. A. Grayson, in *The History and Heritage of Scientific and Technological Information Systems*, ed. W. B. Rayward and M. E. Bowden, Information Today, Inc., Medford, 2004, p. 190.
8. A. Laugier, T. Steele and F. Falco, *Lab. Autom. News*, 1997, **2**(1), 22.
9. M. C. Carvalho, *J. Lab. Autom.*, 2013, **18**(4), 328.
10. G. S. Streitberg, P. T. Bwititi, L. Angel and K. A. Sikaris, *J. Assoc. Lab. Autom.*, 2009, **14**(2), 94.
11. W. R. Aul, *Think*, 1972. Available at https://www-03.ibm.com/ibm/history/exhibits/builders/builders_hollerith.html (accessed Oct 2018).
12. D. L. Sternlight, *Lab. Med.*, 1971, **2**(6), 34.
13. G. Harrison, *Next Generation Databases: NoSQL, NewSQL, and Big Data*, Apress, Berkeley, 2015.
14. P. Fox and J. Hendler, *Science*, 2011, **331**, 705.
15. G. A. Gibbon, *Lab. Autom. Inf. Manage.*, 1996, **32**(1), 1.
16. M. R. Canfield, in *Field Notes on Science and Nature*, ed. M. R. Canfield, Harvard University Press, Cambridge, 2011, p. 1.
17. A. Behrensmeyer, in *Field Notes on Science and Nature*, ed. M. R. Canfield, Harvard University Press, Cambridge, 2011, p. 89.
18. G. Earl, 2012, personal communication.
19. J. Perrine and J. Patton, in *Field Notes on Science and Nature*, ed. M. R. Canfield, Harvard University Press, Cambridge, 2011, p. 211.
20. C. Warwick, C. Fisher, M. Terras, M. Baker, A. Clarke, M. Fulford, M. Grove, E. O'Riordan and M. Rains, *Lit. Linguist. Computing*, 2009, **24**(2), 211.

21. P. A. Knoop and B. van der Pluijm, in *The Impact of Pen-based Technology on Education: Vignettes, Evaluations, and Future Directions*, ed. D. Berque, J. Prey and R. Reed, Purdue University Press, West Lafayette, 2006, p. 103.
22. Available at https://serc.carleton.edu/research_education/geopad/whoisusing.html (accessed Sep 2018).
23. K. A. Badiola *et al.*, *Chem. Sci.*, 2015, **6**, 1614.
24. Available at http://openwetware.org (accessed Sep 2018).
25. R. Yeh, C. Liao, S. Klemmer, F. Guimbretière, B. Lee, B. Kakaradov, B. Stamberger and A. Paepcke, in *Proceedings of the SIGCHI Conference on Human Factors in Computing Systems*, ACM Press, New York, 2006, p. 571.
26. E. R. Vivoni, R. Camilli, M. A. Rodriguez, D. D. Sheehan and D. Entekhabi, in *Hydraulic Information Management*, WIT Press, Southampton, 2002, p. 275.
27. R. D. McDowall and D. C. Mattes, *Anal. Chem.*, 1990, **62**(20), 1069.
28. M. Delaney, *J. Chem. Educ.*, 1987, **64**(1), 29.
29. L. L. Heun and M. S. Lajiness, *SAS Conference Proceedings: SAS Users Group International 16*, New Orleans, 1991, p. 1048.
30. R. R. Pavlis, *ELN and the Paperless Lab*, Labtronics Inc., 2011.
31. K. Taylor, in *Collaborative Computational Technologies for Biomedical Research*, ed. S. Ekins, M. A. Hupcey and A. J. Williams, John Wiley & Sons, Inc., Hoboken, 2011, p. 303.
32. R. Lysakowsk, in *Proceedings of the 19th Annual International Conference of the IEEE Engineering in Medicine and Biology Society*, IEEE, 1997, p. 2659.
33. M. Zall, *Chem. Innovation*, 2001, **31**(2), 14.
34. J. D. Myers, E. S. Mendoza and B. L. Hoopes, in *Proceedings of the IASTED International Conference, Internet and Multimedia Systems and Applications*, ed. H. Hamza, Acta Press, Anaheim, 2001, p. 334.
35. R. C. Hice, *Scientific Computing*, 2009, Available at https://www.rdmag.com/blog/2009/05/roadmap-clear-definition-eln (accessed Sep 2018).
36. S. J. Coles, J. G. Frey, C. L. Bird, R. J. Whitby and A. E. Day, *J. Chem. Inf.*, 2013, **5**, 52.
37. T. Talbott, M. Peterson, M. Schwidder and J. D. Myers, in *Proceedings of the 2005 International Symposium on Collaborative Technologies and Systems*, IEEE, 2005, p. 136.
38. M. Borkum, C. Lagoze, J. G. Frey and S. J. Coles, in *Proceedings of the 2010 IEEE Sixth International Conference on e-Science*, IEEE, 2010, p. 316.
39. G. Hughes, H. Mills, D. De Roure, J. G. Frey, L. Moreau, G. Smith and E. Zaluska, *Org. Biomol. Chem.*, 2004, **2**(22), 3284.
40. Available at http://openrefine.org (accessed Oct 2018).
41. Available at http://www.trifacta.com (accessed Oct 2018).
42. M. R. Berthold, N. Cebron, F. Dill, T. R. Gabriel, T. Kötter, T. Meinl, P. Ohl, K. Thiel and B. Wiswedel, in *Data Analysis, Machine Learning and*

Applications, ed. C. Preisach, H. Burkhardt, L. Schmidt-Thieme and R. Decker, Springer, Berlin, 2008, p. 319.

43. I. Altintas, C. Berkley, E. Jaeger, M. Jones, B. Ludascher and S. Mock, in *Proceedings of the 16th International Conference on Scientific and Statistical Database Management (SSDBM'04)*, IEEE Computer Society, Los Alamitos, 2004, p. 423.
44. J. Goecks, A. Nekrutenko and J. Taylor, *Genome Biol.*, 2010, **11**(8), R86.
45. Available at http://accelrys.com/products/collaborative-science/biovia-pipeline-pilot/ (accessed Oct 2018).
46. Available at https://www.vistrails.org (accessed Oct 2018).
47. K. Wolstencroft, R. Haines, D. Fellows, A. Williams, D. Withers, S. Owen, S. Soiland-Reyes, I. Dunlop, A. Nenadic, P. Fisher, J. Bhagat, K. Belhajjame, F. Bacall, A. Hardisty, A. N. de la Hidalga, M. P. B. Vargas, S. Sufi and C. Goble, *Nucleic Acids Res.*, 2013, **41**, W557.

CHAPTER 5

Managing and Preserving Digital Research Data

5.1 Introduction

Data collection is a fundamental aspect of all scientific activity. Whether we are in the field or the laboratory, we are making observations, collecting or creating materials, taking measurements, making calculations, and computing results in some form. Data ranges greatly in its size and scale, with relatively small amounts of data produced by most individual researchers to vast quantities of data from huge international collaborations such as the Human Genome project and the European Organisation for Nuclear Research (CERN). Data can be both digital and physical, but the proliferation of digital devices, instruments, and sensors used in both everyday life and science are substantially increasing the production of digital data. Data is not only essential to support and back-up the scientific claims made within a single research project or analysis, but can also be reused and repurposed using new methods or in new disciplines to generate new results that were never envisaged during their original collection.

In the last couple of decades, computing power has continued to grow, whilst storage has become cheaper, and the cloud has facilitated the ability to perform both data processing and storage online where multiple groups can access and interact with the data. Computer simulations have become a standard tool for investigating science where practical experiments of theory are impossible to carry out, for example, in mathematics, physics, and astronomy, but also in biomolecular, environmental, and meteorological science. These simulations not only generate large quantities of data that need to be processed and stored, but also require huge quantities of data from various instruments and sensors in order to perform their

Recording Science in the Digital Era: From Paper to Electronic Notebooks and Other Digital Tools
By Cerys Willoughby

Published by the Royal Society of Chemistry, www.rsc.org

calculations. Coupled with the increasing speeds of computer power over time and the ability to network computers to further increase computing power, simulations are producing rapidly increasing quantities of data, leading to the term 'data-intensive science'.

Developments in digital instrumentation have also led to sensors that can capture huge quantities of digital information that need to be stored and subsequently analysed. For example, astronomical surveys, such as the Sloan Digital Sky Survey produce enormous quantities of data that have to be managed. The Sloan Digital Sky Survey has been producing imaging and spectroscopic data to create a map of the universe since 1998, using a dedicated 2.5 m telescope in New Mexico to capture more than 150 TB of data used by researchers across the world.[1] Even experimental and analytical sciences can produce large quantities of data, for example, high-throughput gene sequencing and particle physics experiments can generate terabytes of data every day.[2,3] Such 'big data' enables science to be conducted in new ways by analysing the vast quantities of data for patterns, but also leads to challenges in terms of how such large quantities of data can be stored and managed, particularly to make it available for use by others.

Although research data management tends to be discussed in terms of 'big data', even relatively small amounts of data must be effectively managed to ensure that research data can be found, can be used, is not lost over time, and does not become obsolete. In some disciplines, it is possible to replicate experiments in order to reproduce the data if it is lost, but in many sciences, such as climatology and astronomy, it is simply not possible to reacquire the data because the data is dynamic or because of the enormous costs involved in repeating the capture. There are a number of challenges involved in data management. For big data, data volume is a major challenge to overcome, and largely relies on the capability of hardware and software to handle these large volumes. Large quantities of data take up large amounts of space for storage, but also require large quantities of memory and special handling in order to be opened and processed by software for viewing, analysis, and visualisation. For all scales of data, major challenges in their management occur because data are created using many different techniques and in many different formats, and once created, data is analysed in many different ways. In some disciplines data may always be produced using the same format and analysed in the same way, with relatively few variations that can make it easier to design systems for managing the data, but in other disciplines, such as chemistry, both experiments and data can be highly variable, and the nature of both can also change significantly over time.[3]

We tend to think of 'data' as structured numerical information, usually in digital form, but data in scientific contexts is in fact much more diverse. The definition of 'research data' varies across different organisations and disciplines, but in general can be considered to be all the information collected as part of scientific activities that contribute to the production and validation of research findings. This information includes both digital and non-digital information, and contains not only observations, measurements, and findings, but also materials required to reproduce the original research results such as methodologies, protocols, workflows, survey materials, models, algorithms, and code. Research data may be in the form of tabulated numerical data, but may also be in the form of images, graphs, photographs, videos, audio recordings, datasets, geospatial data, maps, and other documents, including discipline specific materials such as spectra for chemistry and protein and genetic sequences in biology. Physical items such as samples, specimens, and artefacts are also considered part of research data, together with materials that describe them. Research data may be in the raw form in which it was collected, in a transformed form, for example, after it has been cleaned, combined, or summarised, and also data that has been analysed and in its final form for use or publication. The notebooks, both laboratory and field, used to capture information about the methods and environment of data capture are also considered research data, together with any other materials used to collect the primary research data, including forms. Research data is the information that we collect and record as part of the scientific record, but has not historically been included as part of the published record. However, this is now beginning to change.

Before the advent of digital technologies and digital instruments in particular, primary research data would be collected manually by transcribing data into paper notebooks or onto forms. The main methods of organising and storing these materials would be to place them into filing cabinets or stack them on shelves. As instruments began to produce digital output, a common way of combining the paper-based record of the research and the digital output has been to produce paper-based printouts of the data that have then been attached directly to the notebook, ensuring that the data and the context about how and why the data was captured are stored together, and therefore easy to find. In the short term this solution is reasonably effective, but eventually, and especially when the notebooks are used in challenging conditions, these additional materials are likely to become detached and lost. For many scientists, there has been a growing separation between collected data, analysed data, and the context captured in the notebook over time, as more information is collected and analysed in digital form. It is less practical to print out the data in an appropriate format and attach it to a notebook, and new ways have to be found to manage the links between the data and the physical elements of the record. This proliferation of data in digital format creates new challenges in terms of managing, storing, and preserving the information to ensure that it can easily be found and utilised in the future, by both the creator of the data and others, as we shall discuss in this chapter.

5.2 Reproducing Science

Historically, data has been either kept by the individual creator, kept within locked cupboards in a department, or in some cases kept in an institutional library making the data accessible to relatively few people.

Although it is easier to share digital data and store it in locations that are accessible to others, in the majority of cases, the digital data is just as inaccessible as paper-based data. Digital data is typically stored on personal computers, USB keys, private servers, and often at best institutional storage behind firewalls. As discussed previously, the inaccessibility of research locked up in paper notebooks or hidden away in filing cabinets means that time is being wasted on repeating experiments that have already been tried and failed. An additional consequence of inaccessible data, and other related resources, is that the research cannot easily be reproduced by others.

Reproducibility is a cornerstone of science; if research findings cannot be replicated by others, then the credibility of the research is in doubt. Historically, research has been published as journal articles that typically contain a high-level account of the research methods and summaries of the data generated with an interpretation of the results, but very rarely are the full data and detailed methods included. For example, in a 2016 study, only one of 441 biomedical articles provided a full experiment protocol, and none had provided access to raw data.[4] A lack of access to both the raw data generated by the research, and the details about how the research was conducted, makes it difficult, if not impossible, to reproduce. This lack of transparency calls into question the credibility of the research and makes it harder to identify research where mistakes have been made or even where there may have been deliberate fraud. Research that uses computational techniques also suffer from this lack of transparency because it is rare for the computer code and raw data that generated the published results to be made accessible.[5] We will talk more about the reasons why scientists might be reluctant to share and how they can be encouraged to do so in Chapter 11.

More recently, there has been a drive towards both improving transparency and encouraging engagement in science by making data associated with journals openly public. For some areas of science there have been a number of data repositories available for many years that enable the sharing and safe storage of data. There are a variety of discipline specific and region specific examples, including the Archaeology Data Service that has been storing heritage and archaeological data for more than 20 years;[6] the National Center for Atmospheric Research (NCAR) Research Data Archive, which primarily holds atmospheric and oceanographic data together with model outputs from atmospheric and geosciences research;[7] the DNA Database of Japan (DDBJ) that provides sharing and analysis services for life sciences;[8] the European Bioinformatics Institute (EMBL-EBI) that has a number of different repositories for various biological and chemical data;[9] and another European repository called NOMAD that holds data related to materials science.[10] The majority of repositories are domain specific, but some generalist repositories also exist, although many of these are commercial. Examples of generalist repositories include figshare, which contains content ranging from science and technology to creative writing;[11]

the Open Science Framework that is open source and not for profit;[12] Zenodo that was launched by CERN to facilitate open science;[13] and the Dryad Digital Repository that accepts any science and medical-related content.[14] There are also various national data services that typically provide access to datasets, tools, and best practice guides together with regional dataset registries that provide access to important datasets and registry services. Examples include the US National Data Service,[15] the Australian National Data Service,[16] the UK Data Service,[17] and the EU Open Data Portal.[18]

Journals have long accepted data related to publications in the form of supplementary material as optional, but increasingly inclusion of raw data as supplementary material is becoming mandated. In the last few years, there have been an increasing number of journals that either accept or have been created for the purpose of publishing articles that describe datasets for reuse. There are, in fact, now over 100 data journals that accept data papers, spanning across topics including chemistry, ecology, health sciences and medicine, genomics and molecular biology, geoscience and earth system science, and archaeology, although only a very small number are exclusively dedicated to data papers.[19] Advantages of formal publication of datasets over simple deposition in a repository is that they are not only accessible and reusable, but also discoverable, citable, and peer-reviewed. The majority of the journals that publish data papers rely on third-party repository providers, and delegate the responsibility for the storage and management to other organisations, with the majority relying on data citation to provide the link between the publication about the dataset and the data itself. Publishers of ordinary journal articles and data papers will deposit their supporting data in one of the types of repositories mentioned previously, and a permanent link is made between the journal and the dataset using a data registry such as DataCite to assign a unique digital identifier to the data.[20] We will talk more about DataCite and unique identifiers later in the chapter.

Even when the data is published, there are a number of factors that can hinder the discoverability and usefulness of the data. Data alone is very rarely meaningful enough to be effectively interpreted and thoroughly assessed, even by an expert, and therefore it is essential that supporting documentation is included with the data. Factors that can prevent the reproducibility of both traditional and computational research include:[21–23]

- A lack of documentation and annotation for the data, including missing information about inputs, intermediate steps, and outputs.
- Missing important computational steps including the details of parameter settings and configuration files.
- Third-party dependencies, such as Web services, are no longer accessible.
- Example data, input data, and intermediate data are missing.

- Use of proprietary software and use of proprietary scripts at intermediate parts of the process or specific software infrastructures that need to be installed locally.

To counter these problems, it is not only necessary to include documentation about the datasets, but also to provide primary data and sources, scripts, software source code, and step-by-step protocols.[24] Whether research data is to be made available outside of an organisation or not, it is also important to ensure that data is properly annotated, managed, and preserved, so that is it discoverable, accessible, understandable, and kept safe.

5.3 Research Data Management and Digital Curation

As we increasingly use computers to not only generate and analyse data, but also to discover data and integrate datasets, it is necessary to ensure that the research data is understandable by both humans and computers. Humans can cope with some incompleteness or ambiguity in information, but a computer cannot, and therefore the way that data is described and managed needs to take account of these limitations. Effective data management processes must be followed to ensure that the data is discoverable, and its provenance accurately described so that its origin, history, contents, and context can be understood by both humans and computers.

Much scientific data is neglected until the point it needs to be shared or added to a repository for publication or preservation. Common problems are that the datasets are not structured in a way that makes them suitable for sharing, they have no documentation associated with them, and the contents of the files, such as column and row names, may contain abbreviations or codes that are now mysterious and completely meaningless to anyone other than the creator of the data. Although it is possible to retrospectively structure, fix the contents of files to make them intelligible, and to add information to the data to describe it later on, it is much easier to do these things at the point the data is created. Fixing these problems later also means that potentially difficult work needs to be done at a point in the project when researchers are moving on or work on a new project is already underway, making these efforts less of a priority and more of an annoyance. A better alternative is to manage the research data through a series of processes that are designed to ensure that data is properly described, documented, and preserved from the very beginning of a project.

A set of principles known as the FAIR Guiding Principles published in 2015, has been developed to provide straightforward guidance to teams to ensure their data and related resources are reusable. FAIR is gaining traction in encouraging change in the way that research data are managed to ensure their preservation for the benefit of future science.[25] FAIR stands for Findable, Accessible, Interoperable, and Reusable, and considers both human and machine perspectives to ensure that data can be used by both humans and by software that has been developed to manipulate it. These principles address the best practices in data management and stewardship across all communities to ensure that digital assets are properly collected, annotated, and preserved so that the scientific record is accessible, findable, and reusable now and in the future. As more and more data is produced in digital form, so more and more software is developed to manage it, and the data becomes more and more diverse. Without steps to properly describe the data and make it interoperable, then the challenges of the data deluge become ever more significant. The use of rich metadata to describe the data, provide additional context, and define limitations on use, is an essential part of the guidance of the FAIR principles. The principles advocate the use of rich metadata, persistent unique identifiers, provenance, and standards in the form of communication protocols, metadata formats, and vocabularies, to ensure that data is findable, accessible, interoperable, and reusable.

The FAIR Principles have been widely adopted across the world as an important foundation for the needs of data sharing in open science, but of themselves only represent guidelines and do not mention specific technologies or provide a specific standard to follow.[26] The authors of the FAIR Guiding Principles do, however, provide a number of examples of technology implementations that make use of the principles. These examples, such as Dataverse, FAIRDOM, ISA, Open PHACTS, Worldwide PDB (wwPDB), and UniProt, are varied, but primarily manage the storage and discovery of different scientific datasets and their metadata.[27–32] Of the examples provided, the majority make use of persistent identifiers or URLs that can be used for citations, standard formats, and vocabularies for both the data and metadata, and provide both human and machine accessible interfaces for both. It is still early days for the FAIR Principles and they are the subject of much research, but no doubt more developments will be made in technology to support them in the coming years, both in open and closed research. The FAIR principles stem from the concept of 'digital curation' which emerged some years before. We will talk more about digital curation, metadata, identifiers, provenance, and standards in the following sections.

Digital curation represents a range of processes that relate to the management of data over the entire lifecycle of the data, including initial concepts, maintenance, and ultimately data preservation. Digital curation processes also add value to the data by adding additional context that aids in both discoverability and usability of the data. Curation as a term has long been used in libraries, galleries, and museums to reflect the role of the 'curator' in selecting, organising, presenting, and taking care of items in a particular collection. Digital curation is similar in that the role of the digital curator is to choose the information that will be shared and preserved, to organise it so that it can be found and understood, and to ensure that it is appropriately preserved and maintained for accessibility. Organisations with a focus on data preservation and management, such as the UK's Digital Curation Centre (DCC) and the Australian National Data Service (ANDS) have created their own guidelines and lifecycle models to act as tools for educating curators, researchers, and organisations to understand the processes involved and to develop their own roles, responsibilities, and methods for undertaking digital curation.[33,34] The lifecycle models for data curation tend to be focused on the actions related to the preservation of the data, often from the perspective of the preservation and curation actions within a repository, and with less focus on the data before it is ready for preservation. From the perspective of the researchers and organisations producing the data, curation processes are just as important before the data is deposited in the repository to ensure that the data is appropriate and ready for both preservation and sharing.

Figure 5.1 shows the processes involved in the data research management lifecycle, based on the DCC's Curation Lifecycle model. The processes begin with planning before the creation of data. Once the data has been created, various processes can be followed to ensure that the data is ready for use during all points in the lifecycle. When the time comes to preserve or publish the data, it can be appraised, and the appropriate data can be selected for use or disposal if it is no longer needed. The processes for preservation tend to be thought of as processes owned and carried out by owners of the data repository, but this is not necessarily the case, and work has to be done before data can be uploaded to a repository. We will talk more about preservation and the processes required later in the chapter.

Figure 5.1 Data research management lifecycle.

Ideally, digital curation, data management, and digital preservation should all be considered from the beginning of a project. Strategies should be developed for how the information will collected, how it is to be stored and managed, for how long it will be retained, and any restrictions that need to be applied to the data, such as managing sensitive or confidential data. Funders of scientific research are increasingly requiring that a formal Data Management Plan (DMP) is completed by research teams to document how data will be managed both during and after a project. DMPs typically cover how the major digital curation processes will be managed and is often considered as part of the appraisal of project proposals, particularly for publicly funded work. The DMP not only specifies how the data will be created, managed, and preserved, but also expectations for sharing the data so that it can be reused in the future. The DMP may also ask the researchers to investigate existing sources of data for their potential reuse in the project, in order to encourage reuse and avoid duplication of effort in data collection. In addition to funders, organisations such as universities and regional or disciplinary data centres may also provide DMP templates, examples, and tools. For example, the ESRC DMP is an example of a DMP guide and requirements from a funder;[35] DataOne provides examples of DMPs from different disciplines to meet the requirements of the National Science Foundation (NSF),[36] the Digital Curation Centre provides the DMP Online tool,[37] and the DMPtool was developed by an international collaboration of universities and other organisations.[38]

When new data is generated for a project, it is important to consider how it will be captured and handled. Questions that might be asked in a DMP include:

- What is the nature of the data?
- How will the data be collected?
- Are there any issues of copyright or intellectual property rights with the data?
- Are there any ethical issues associated with the data; is it sensitive or private?
- How will licensing and consent for the data be handled?
- What volumes will be generated?
- How will the data be processed?
- What format will it be captured in?
- What formats are most suitable for sharing in your community?
- What metadata will be captured?
- Will standard or interchangeable formats be used?
- How will quality be ensured?
- How will the data be secured and backed-up?
- What are the risks to the data and what disaster recovery procedures are in place?
- How will the data be organised and managed?
- How will the data be shared?

- Where will it be archived and how will archive data be selected?
- How will the data be citable?
- How long will it be retained?
- How will unwanted data be disposed of?
- Who has responsibility for the different stages and tasks in the process?

As data is collected and analysed, it needs to be managed to enable it to be safely stored and effectively utilised during the project, but also to simplify the process of preparing it for sharing and preservation. If the data is captured in paper-based or analogue forms, it needs to be digitised or transcribed to capture it in appropriate formats. Data also needs to be checked and validated to ensure the data quality. The data that is created, and new data generated as a result of processing and analysis, needs to be associated with a variety of metadata that describe the data and provide additional context. Some metadata may be captured as a part of the data creation process, whilst other metadata may need to be added at later stages in the process. More information about the metadata needed will be described in the following sections. Before the data can be published and shared, it needs to be documented to ensure that it is usable by others. Data will need to be prepared before it can be archived and shared. For example, the data may need to be cleaned, consolidated, aggregated, and anonymised if it contains sensitive or personal information. Decisions need to be made about exactly which data will be preserved and shared, and where it will be stored. There will also be decisions about who it will be shared with, who owns the copyright, and what licensing will be used to determine how that data can be used. It may also be necessary to change the format of the data or data objects before they can be deposited in a repository, if it has not already been captured or packaged in a standard format. We will talk more about why it may be important to restructure the information before deposition in a repository when we talk about preservation later in this chapter. Once data has been deposited in a repository, it undergoes further cycles of evaluation, selection, migration, checking, preservation, and storage to ensure the long-term preservation of the data until such time as the data is no longer required and can be disposed of. For some materials, policies or legal requirements may mean they require a secure destruction to ensure sensitive or confidential data cannot be accessed or misused.

5.4 Metadata for Curation

Metadata is an essential part of the curation process and is often described simply as 'data about data'. Metadata provides a simplified representation of what may be large and complex information. This representation is presented in a structured format that can not only be utilised by machines, but can also be understood and usable by humans. There are several kinds of metadata that can be used for different purposes, as well as to describe different properties of the research data. Metadata is most commonly thought of as information that can be used for discovery and description, but it can also be used to identify the information, organise it, to enable interoperability, and to support data management, control, and preservation. Metadata is particularly valuable for use in record-keeping and it is an essential part of systems that are designed for the purpose of capturing and managing the scientific record.

Metadata has its origins in the catalogue card system of libraries, designed to help users to find books and other materials within their collections. Library card catalogues, and their digital equivalents, hold standard information about the materials within the library. This information typically includes authors, title, publisher, publisher's location, edition, copyright date, subject, number of pages, a unique identifier such as an International Standard Book Number (ISBN), and the shelf location. Where space for the description is no issue, the catalogue or database may also include information such as multiple subject headings, topics, genre, a short description or 'blurb', ratings, media time, and even a link to an electronic version. It is clear to see how access to such information about the library materials can facilitate discovery. The metadata enables the user to find materials related to a particular subject, by a particular author, from a particular date, or even more usefully in digital catalogues, to be able to locate materials based on keywords. Metadata is nothing new in science either, with observations on natural phenomena, such as weather, ocean conditions, and astronomical information, historically captured in tabulated form, with column or row labels representing the metadata elements of the information being captured.

Metadata is a surprisingly big part of the digital world and vast quantities are generated with many of our everyday activities. Metadata is generated with every file that is created, every photo that is taken, every phone call made, and every social media interaction. For example, for a Microsoft Word file, information is captured about the author of the file, the date it was created, the last time it was accessed, how many times the file has been revised, total editing time, number of pages, how many words are in the document, and more. For every photo taken with an SLR camera, around 100 metadata properties in the form of Exchangeable Image File (EXIF) data can be generated describing details of the camera, lens, and software used, the specific settings used on the camera when taking the picture, the internal software settings, dates, time, GPS locations, and the format of the image itself.[39] For every mobile phone call made, metadata is captured describing the caller and recipient, potentially the location of both, and the time and duration of the call. The metadata captured for a social media interaction varies depending on the platform, but will typically include the identity of the author, the application used, dates and times, location, and often others involved in the interaction such as a person being replied to or a person present in the same location at the same time as the author.

Much of the metadata described above is automatically generated, but it is also possible for humans to add additional metadata to help describe and organise the content. For example, the EXIF data for a photograph is automatically captured when the shutter is pressed on the camera and embedded in the digital image file, but if a photograph is uploaded to a photo sharing site such as Flickr, then more context metadata can be added. Figure 5.2 shows an example of metadata in the Flickr interface, where you can add additional information about the photograph, such as a title, description, tags, privacy, and names of people who appear in the photograph. There are also options to share the picture with a relevant group or to add it to an album. All these elements represent metadata that can be used to describe the photo, but they also enable discovery. These elements also provide ways to group and organise the information; the people, tags, albums, groups, and location metadata provide ways of collecting and viewing related photographs without any additional effort. Because metadata is machine readable, software can perform sensible functions

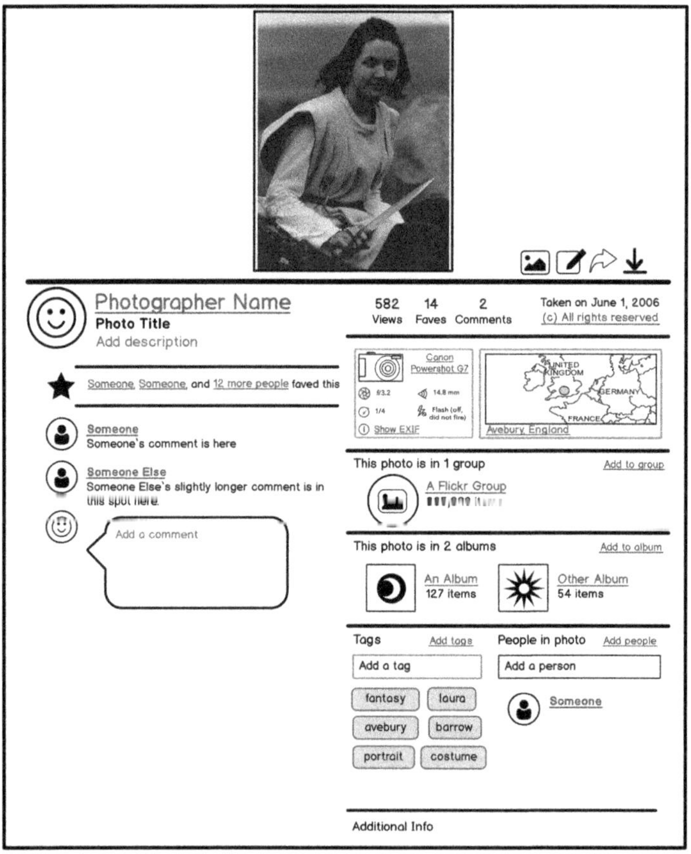

Figure 5.2 Different kinds of metadata in the Flickr interface.

using them, but they are also human-readable enabling the user to make direct use of the information as well.

Metadata can be stored with the data (for example, the embedded EXIF data in photographs) or separately in databases or in metadata registries, where the metadata can be used to facilitate discovery of the datasets within a repository.

5.5 Types of Metadata

Descriptive metadata describes an information object, such as research data or an associated resource. The examples of metadata that we described previously, such as the metadata describing a Word document or the elements describing a photograph, are examples of descriptive metadata. Descriptive metadata is primarily used for the discovery and identification of information objects. Descriptive metadata often provides context information that helps with the discovery of data and also enables information objects to be assessed for relevance for a particular use. The context may be descriptive information about what the information object contains, how it was produced, and its relationship to other data, but it can also contain information about what is needed to use it, for example, software dependencies or permissions. Without adequate description metadata, context about the data will not be available, potentially rendering it impossible to interpret, understand, and use.

There are other types of metadata that are useful in a data management and digital curation context. In addition to descriptive metadata, the most common types of metadata are structural metadata and administrative metadata, although sometimes different types of metadata are discussed separately, for example, technical metadata, preservation metadata, and rights management metadata. Structural metadata describes the relationships between files in a dataset and provides information about how they are organised and their relative locations. For example, structural metadata that describes a scanned copy of a printed book may include details about how the files are grouped into sections, chapters, and the order of pages within the book, along with the physical and logical relationship between the individual files. Technical metadata describes the format of the individual files, for example, the type of file, encoding, character sets, languages, and compression; as well as properties such as the date the file was created, the file dimensions, and the software, version, and settings used to create the file. Administrative metadata provides information about the management of data, including the provenance and source of the data, intellectual property restrictions, and access rights. The technical metadata is often viewed as being a part of the administrative metadata.

Finally, preservation metadata contains additional information of importance in describing the storage and preservation actions that have been carried out on the data. Technical metadata and administrative metadata are often considered a part of the preservation metadata, as they describe details about the file format and the technical environment that is required to store and use the files, the provenance trail of the data, and access rights. Other preservation metadata is primarily added at the point of preservation or during the preservation cycle whilst in a repository and provides information about what has happened to the files during preservation. Such metadata includes information about changes and actions that have happened to the files, for example, changes made as a result of migrating files, including a checksum for each file and information about how it was generated so that the authenticity of the information object can be checked and verified in the future. Different types of metadata may be added to an information object at different stages of the curation lifecycle. For example, descriptive metadata can be added when an object is created, but as described, preservation metadata is more likely to be added later when the information is ready for archive, and in response to actions that occur during the preservation part of the lifecycle such as transformation to a standard format or migration to a new software version. For certain applications, metadata may be collected about the use of an information object. For example, on Flickr, a photograph has information captured about the number of times it has been viewed, favourited, and commented on. In a repository, metadata may be captured about the number of times the item has been viewed, downloaded, by whom, and under what license.

There is some overlap between certain types of metadata, and some standards and schemas for capturing metadata may focus on just one type or provide the option to add all of the different kinds needed to adequately describe the information object and its lifecycle. A large number of schemas have been developed to standardise the metadata that is captured for information objects, with many originating from the information sciences, but also from computer science and scientific domains. Schemas specify how the metadata is encoded, the values that may be used, and the format those values can take. In some, a controlled vocabulary is used to limit the specific contents of the metadata values, enhancing the interoperability of the metadata. For example, a controlled vocabulary could be used to restrict the list of possible metadata values for file type, such as 'text', 'image', 'video', or 'audio', that can be selected for a particular data file. Software for managing metadata can be used to verify that one of the allowed formats has been chosen. Such constraints in the values that can be assigned to metadata can help to prevent ambiguity, for example, preventing the use of multiple terms with a similar meaning such as '.doc', 'txt', or 'rtf'. Controlled vocabularies can also come with standard definitions so that the exact meaning of a particular value is specified. Having standard definitions for terms can be particularly useful in scientific domains where the same term may have quite a different meaning in different disciplines, again preventing ambiguity in the meaning of the metadata value. Metadata schemas are commonly encoded using eXtensible Markup Language (XML), but they may also be encoded using attribute-value pairs, text, comma-separated values (CSV), Resource Description Framework (RDF), and other mark-up languages.

Schemas for descriptive metadata supporting the discoverability of information objects are often focused on specific subject domains or formats of objects, but one of the most widely adopted standards is the generic Dublin Core Metadata Element Set, often known as the 'Dublin core'.[40] This is a schema of 15 standard metadata terms, also formalised as the International Standard ISO Standard 15836:2009.[41] Since 2008, more elements have been added to this list to generate the currently preferred Dublin Core Metadata Initiative (DCMI) Metadata Terms list.[42] Like other metadata standards that focus on documents, the Dublin Core metadata is similar to library catalogue card metadata, containing elements describing creator, title, subject, date, publisher, format, identifier, and language. Other elements describe useful additional information such as access rights, licenses, audience, version, related and reference resources, and provenance information.

The Dublin Core standards have been used as the basis of numerous other metadata schemas both inside and outside the scientific domain. An example from a scientific domain is itself a widely used standard in biological science, the Darwin Core.[43] This is a standard for describing and sharing information about both living and preserved biological specimens based on their taxa. The standard provides a standard list of term definitions and enables the capture of extended metadata about the specimen beyond identification such as sex, measurements, environment, location, condition, relationships to other organisms, and behaviour. Useful additional metadata for understanding the provenance of the samples, such as details of the collector, owning institute, sampling protocols, field notes, and administrative data such as access rights, are also included. The schema is also extendible so that organisations can include additional metadata that would be relevant to share. Figure 5.3 shows an example of how the Darwin Core can be used to provide metadata for a data file about palaeontological data. The CSV data file contains no metadata, but the content of the XML Darwin Core file defines the structure of the data so that a mapping can be made from the values in the data file to the fields defined in the XML. For this example, you can determine that each row in the data file contains the ID number of the fossil in the catalogue, the scientific name of the fossil, the location the fossil was found, and the geological context. Using the URLs in the XML file, you can also find a definition for the metadata terms, if they were unclear.

Another schema for biological science is the Genome Metadata standard, which consists of over 60 different metadata elements to describe bacterial genomes.[44] The schema includes mostly descriptive metadata, describing information about the study organism and its host, genetic sequences and phenotypes, and additional provenance and administrative information about the study project.

(a)

```
123,Ichnogenera,Constitution Hill,Llandovery Series
124,Monograptus turriculatus,Rheidol Gorge,Llandovery Series
125,Pteronitella retroflexa,Wren's Nest,Wenlock Series
126,Pyramidalia simplex,Torquay,Devonian
127,Spirifer undiferous,Newton Abbot,Devonian
128,Trinucleus fimbriatus,Builth Wells,Caradoc Series
129,Rhynchonella cuneata,Wren's Nest,Wenlock Series
130,Calymene replicata,Girvan,Llandovery Series
131,Sphaerirhynchia,Much Wenlock,Wenlock Series
132,Rhynchonella cuboides,Torquay,Devonian
```

(b)

```
<?xml version="1.0" encoding="UTF-8"?>
<archive xmlns="http://rs.tdwg.org/dwc/text/"
  xmlns:xsi="http://www.w3.org/2001/XMLSchema-instance"
  xmlns:xs="http://www.w3.org/2001/XMLSchema"
  xsi:schemaLocation="http://rs.tdwg.org/dwc/text/ http://rs.tdwg.org/dwc/text/tdwg_dwc_text.xsd">
  <core rowType="http://rs.tdwg.org/dwc/xsd/simpledarwincore/SimpleDarwinRecord" ignoreHeaderLines="1">
 <files>
   <location>http://data.abrgf.org/download/specimens.csv</location>
  </files>
  <field index="0" term="http://rs.tdwg.org/dwc/terms/catalogNumber" />
  <field index="1" term="http://rs.tdwg.org/dwc/terms/scientificName" />
  <field index="2" term="http://purl.org/dc/terms/Location" />
  <field index="3" term="http://rs.tdwg.org/dwc/terms/GeologicalContext"/>
  </core>
</archive>
```

Figure 5.3 Example of using a metadata standard to describe a data file containing data about a fossil collection. (a) CSV data file with no metadata. (b) A Darwin Core metafile providing metadata to describe the meaning of the fields in the CSV data file.

There are a number of standards for describing geographic metadata, including two international standards, ISO 19115 and ISO 19165, which are currently under development.[45,46] ISO 19115 contains several parts with guidance on using metadata to identify and describe digital geographic data, images, and gridded data, and includes an XML schema. ISO 19165 extends ISO 19915 to include requirements for the long-term preservation of digital geospatial data including the capture of preservation metadata, provenance information, and technical metadata.

An early scientific metadata format for scientific datasets was the Directory Interchange Format (DIF), originating in earth sciences, and now maintained by NASA.[47] The schema includes descriptive elements about the creation of a dataset including locations, geological units, instruments, and temporal information, but also provenance metadata about the collecting organisation, personnel, and the data centre. The standard also includes other elements including a Digital Object Identifier (DOI), citation details, references, and access rights. A handy template is available with the schema specification to make the standard easier to follow and both are provided in XML.

Another example of a discipline specific descriptive metadata standard is the MIDAS Heritage UK Historic Environment Data Standard for recording information on archaeological sites and artefacts, and other cultural sites such as architecture, shipwrecks, parks, and battlefields.[48] The standard includes the minimum levels of information required for recording heritage assets and also provides guidance on sharing and preserving this knowledge. An interoperability toolkit is provided to support the sharing of heritage data, and this includes an XML schema conforming to the standard, a validation tool, and a Web services protocol to enable resource discovery in MIDAS Heritage compliant databases.

There are numerous other scientific metadata standards and schemas available, with specifications to support sharing of particular data objects within a specific domain. There is no 'one-size fits all' schema, even within disciplines, and many projects design their own specific schemas completely from scratch or as extensions to existing standards. A problem with this approach is that the metadata generated is not likely to be compatible and interoperable between different standards, and tools are not readily available that support them. However, it is possible to create mappings between different schemas, for example, using Extensible Stylesheet Language Transformations (XSLT), which can be used to convert from one XML format to another. The Digital Curation Centre keeps a list of metadata standards and related tools, ordered by discipline.[49]

Many of the metadata standards include a mix of metadata types but some, specifically designed for improving interoperability and data discoverability in repositories, have a focus on administrative and preservation metadata. An example of a standard designed for interoperability between repositories is the Metadata Encoding and Transmission Standard (METS) schema.[50] This XML-based schema includes structural information about how a data object is organised, including files included in the object, their type, identifier, and relative location. This standard also supports administrative metadata including intellectual property rights statements; technical metadata about the file such as file type, size, timestamps, orientation, compression, and information about the software used to process the file; together with preservation metadata such as information about the object's lifecycle and history. The standard also allows the inclusion of descriptive metadata or linking to an external file holding descriptive metadata. A similar standard targeted at preservation metadata is the PREMIS Data Dictionary for Preservation Metadata.[51] This standard includes objects, actions, agents, and rights statements to describe the materials and activities that occur during preservation in a repository. The World Wide Web Consortium maintains a standard known as PROV that uses a controlled vocabulary to enable the interoperability and exchange of provenance information on the Web.[52]

There are also a smaller number of standards that are designed to describe metadata useful for records and record-keeping. The ISO has published the ISO 23081 series of international standards providing requirements for metadata useful for archives and records.[53] The first standard of the series specifies requirements about what metadata should be captured, what it should describe about the record, and the business processes for managing records. The standard also provides guidelines on selecting metadata based on business need, roles and responsibilities, the functions of metadata in records management, and the creation, management, and maintenance of metadata.

Although scientific records are increasingly being captured in digital form, for example, within Electronic Laboratory Notebooks (ELNs), the use of proprietary formats is a hindrance to the sharing and reuse of these records. Existing standards for metadata are focused on sharing and describing datasets, and tend to be either very generic or very domain specific. Although these standards and vocabularies may be useful for inclusion in tools for capturing scientific records, particularly for the capture of data from instruments or machines, they are not very suitable for describing the important contextual information captured in our notebook records. Although currently unusual, notebooks are sometimes included as part of the supplementary data for publications and it makes sense that with the increasing publication of data that the notebooks behind the data will be published or linked to as well. Open Science activities are also leading to the sharing of 'live' notebooks, such as the publicly available Open Source Malaria notebooks.[54] With a large number of ELNs facilitating open science based online, and with increasing pressure to publish notebook contents, metadata standards will be needed to ensure that the notebooks and their contents can be discovered, managed, shared, reused, and preserved for the future. One proposed metadata schema for the content of ELNs is the elnItemManifest schema that provides metadata elements for descriptive, administrative, and structural metadata.[55] The schema can be applied to both paper and digital notebooks, but ELNs should have an advantage as much of the metadata can be captured automatically, and the user can be prompted to provide the more descriptive elements for their notebooks. We will discuss more about interfaces for capturing metadata and the differences between machine-generated and human-generated metadata in Chapter 10, and we will also talk more about Open Notebooks in Chapter 11.

5.6 Provenance and Identification

Provenance describes both the origins of a piece of work or object and also the record of who has had responsibility for it and what has happened to it since its creation. A provenance trail captures information about the people or organisations involved in generating and transforming the data, details of the sources of the data, and processes that have been used to generate the data or transform it throughout its lifecycle. For scientific records, provenance can therefore be used to provide additional context and background to a piece or work or data, and act as a guide to its quality and authenticity. Provenance is also important in the context of reusing data because it enables the origins of the original data to be tracked, together with any processing or transformation that the original data has undergone during its reuse. It also enables proper credit to be given to the originators of the data. For scientific data, it can also be important to understand how the data was collected and any pre-processing the data has undergone before the data can be integrated with other datasets for reuse, or to understand how the data was analysed in order to assess the validity of the findings. Within repositories, information about preservation steps such as transformations to new formats and migrations can be also captured as part of the provenance metadata to indicate how the data has been maintained over time. Without adequate provenance data, it becomes impossible to answer questions such as 'Who created the data?', 'What processes were used to create it?', 'How has it been transformed, when, and by whom?', 'Are these datasets derived from the same raw data?', and 'Is this data in the same format it was when it was created?'. Without being able to answer these questions, it also becomes difficult to reproduce and validate results.[56] Tools that generate, transform, and preserve digital materials need to have the capability to capture these kinds of provenance information and ensure that the information is maintained and accessible along with the original file.

Much of the information about the who, how, and what was involved in the creation and transformation of digital materials can be captured automatically by tools that have been designed to do so. For example, an ELN, or another tool designed to capture scientific activities, is likely to be able to capture details of the author of the research and the details of changes made to the research record. Changes to the research record could be timestamps for updates made to the record or comments added. If the tool enables version control, then version information may be available to indicate the number of edits made and the changes between a new version and a previous version. Instruments may capture important provenance information in their data files such as timestamps and configuration settings, although these may not be made visible within other digital tools. If data from an instrument is processed using software, then the software may also include information about the software version and configuration settings used for the processing.

Such provenance information is helpful for reproducing results, reusing, and preserving data, but other information may be needed for completeness that is less easily automatically collected. For example, to reproduce the results of an experiment, details about the specific materials or organisms used may be needed, but may not be recorded by the user or automatically collected by the software tool. Specific details that might not be captured automatically include details such as batch numbers, expiry dates, suppliers, the equipment used, environmental conditions, and software versions used to perform calculations. Another challenge can be that different information is likely to be captured within different tools, and the metadata values to describe the provenance may be inconsistent and lacking in interoperability.

Provenance is an essential component of trust in research data. Without adequate and reliable provenance information, the authenticity of materials cannot be scrutinised leading to a lack of trust and endangering the reproducibly of the research. The provenance record in itself is important and requires details of its own provenance in order to ensure that the records are genuine and have not been tampered with. An important element of trust and provenance is the use of unique identifiers to provide a unique reference to a particular version of some data or research object. Digital files can easily be copied, shared, and manipulated, and it can become difficult to work out which is the original or definitive version, as anyone who has worked on a document with multiple authors over email can confirm. Assigning unique identifiers to records and datasets (and including versioning information) can significantly help, particularly when the records or data are stored in a known location that can be referenced using the unique identifier. When data is published on its own or as part of a formal publication, it is usually necessary to assign a unique identifier and register it with a data registry of some kind, whilst publications themselves are also typically assigned their own unique identifier. Using a unique identifier enables the data to be citable alongside any published literature and enables attribution to be provided to the contributors of the data.

In order to be reused, data needs to be discoverable, and assigning a unique identifier to the data is an effective way to ensure that the data can be found, both when it is first published, and in the future when it may be less visible. Assigning a unique identifier also ensures that it can be found even if the location in which it is stored changes or the original links to the data get broken. A formal DOI is a persistent unique digital identifier for an object. A DOI name identifies an object on digital networks and can be assigned to physical, digital, and abstract objects. The DOI name does not identify the location of the object, but through the use of a 'Handle System' is able to be resolved to the location where the object may be accessed. The DOI system and principles for the creation, registration, and administration of DOI names have been standardised by International Standard ISO 26324:2012.[57] A number of different Registration Agencies (RAs) exist that are responsible for allocating DOI name prefixes, registering DOI names, and providing infrastructure to allow users to declare and maintain metadata about their registered objects. The different RAs provide services to register DOI names for different kinds of objects. For example, Crossref provides DOIs for scholarly information, such as journal articles, conference proceedings, and books;[58] DataCite provides DOIs for research data including scientific datasets;[59] and there is even an RA for TV and movie assets, the Entertainment Identifier Registry (EIDR).[60]

These formal DOIs have a standard format with a DOI prefix that provides information about the registering organisation and a suffix that identifies the individual object. The suffix may just be a sequential number, but can also have elements in it that are generated by the system where the DOI is registered. DOIs for different types of object all use the same structure. All DOIs begin with '10.' to identify them as DOIs. The following is the DOI for a literary source, the doctoral thesis of Stephen Hawking from 1966, but made openly available by the University of Cambridge: 10.17863/CAM.11283. The '17863' is the registrant prefix, in this case, the University of Cambridge Repository system. The 'CAM.11283' suffix is also generated by the repository system as the unique identifier. An example dataset, in the form of microarray data, comes from the University of Edinburgh's Datavault system: 10.7488/6813c777-c45c-4841-8a61-6e5eb367afa6. This shows quite a different structure for the suffix, whilst '7488' is the registrant number for the university. Any piece of literature, dataset, or even physical object that is to be shared with a wider audience should have one of these DOIs to ensure they can be found. The person who registers the data is responsible for ensuring that the metadata information about the resource is kept up to date. If in the future the location of the resource is changed, for example, the data is moved to a new repository or a web resource changes URL, then the metadata can be updated to reflect the new location. Because the DOI is used to reference the object, the DOI can always be used to find the original object, no matter how or where it has subsequently been used.

It can also be useful to have a way to uniquely identify individuals who have contributed to a piece of scientific research for the purposes of provenance. The ORCID service can be used to create identifiers for individuals that can then be linked to data objects as well as traditional published outputs such as journal articles.[61] The structure for the ORCID ID is also based on a formal structure defined in an International Standard, this time, ISO 27729:2012, which specifies the International Standard name identifier (ISNI) for the disambiguation of public identities.[62] ORCID IDs are randomly generated by the ORCID registry and are displayed as a URI, for example, my ORCID ID is: https://orcid.org/0000-0003-1721-9212.

5.7 Preserving the Record

For the benefit of individual researchers, for the creation of an organisational memory, and for the wider scientific community, it is essential to preserve the scientific record and to ensure it is accessible to the appropriate audience. Research data hidden away in filing cabinets or stored on a personal computer are inaccessible and therefore not available for the benefit of the community, making it difficult or even impossible to reuse them or to gain insights from the knowledge they contain. However, inaccessibility of research data is not the only problem. Research data are vulnerable to loss and degradation in whatever format they are created in. Data on paper are vulnerable to the ravages of time, but digital information is also vulnerable. Accidental deletion, overwriting information, and hard drive failure are problems that many of us have experienced. Deliberate tampering, virus attack, and malicious destruction are also potential problems. Even where back-ups are created and security measures taken, digital data is still vulnerable to degradation and obsolescence as the physical media that stores the data gradually degrades over time. Magnetic tapes lose their magnetism and oxidise; optical media such as CDs and DVDs are damaged by high temperatures, sun-light, and scratches. USB flash drives have a finite lifespan based on how carefully they are stored and the number of times they are accessed. And the average hard drive starts to wear out after only four years, as well as also being vulnerable to high temperature, dust, and disk crashes.[63]

A process known as 'bit rot' or 'data decay' also causes a deterioration in integrity of digital data over time as the individual binary 'bits' within a data file get flipped due to degradation of the storage medium or by the passage of cosmic rays through memory. Even a single flipped bit can cause a file to become lost, corrupted or unreadable, and there is no way to know it has happened without being able to compare to the original file. A common way to detect bit rot is to create a checksum for a file when the file is first added to the system. The checksum represents the contents of the data file and is used to check that the complete and correct file was received by the system. If a new checksum is created for the file at any point, it should be the same as the original file; if the file has been changed or corrupted, then the checksum will be different. To protect files from bit rot and loss through hardware failure, multiple copies of a file can be taken to provide a back-up and additional care can be taken to store the back-up copies in a suitable environment. Periodically, files can be copied to fresh media to prolong their life and reduce the chance of bit rot. The status of files can be monitored by regularly generating a new checksum for each file and comparing it to the original. Producing a checksum, along with auditing tools and user authentication, can help to ensure that data inside of digital tools have not been manipulated, forged, or substituted. Digital preservation techniques such as migration inevitably alter the data and using a combination of the checksum together with the provenance and context of the data can help to confirm the authenticity of the data.

A harder problem to solve is digital obsolescence. Digital obsolescence can take a number of forms. Media obsolescence occurs when the data is stored in a format that is no longer compatible with the computers available. For example, it is becoming increasingly rare to find computers that come with an optical drive (CD or DVD drive), especially on the laptop computers that most of us use for our research. If you want to access data on a floppy disk, tape, or laser disk, the chances are you will need to locate an original computer to go with the reader device to be able to read the data. Not only do the devices that read the data become obsolete, but the data format used by operating systems and software change over time, meaning that the files may still be impossible to read without an original computer with the correct hardware, software, and operating system, even if a reader device could be used with a modern computer.

Accessing 'heritage data' residing in old formats and on obsolete hardware can be very difficult, although often the results can be rewarding.[64] Data rescue has been used to bring legacy scientific data from both paper-based and near-obsolete digital formats into valuable use. For example, many disciplines can benefit from historical data to understand the relationship between various components of the earth system and changes over time, particularly for studies of changing environments and climate. There are a variety of historical sources that are of value to these studies and provide important information informing us about changes in biodiversity, ocean levels, levels of chemicals in the atmosphere, and land use. However, these sources are often paper-based records, photographs, or obsolete digital formats that need to be converted to a contemporary digital format in order to include them in current research and simulations. One such example is the Global Oceanographic Data Archaeology and Rescue (GODAR) project, begun in 1993 to rescue historical oceanic data that existed only in paper, analogue, or obsolete digital forms.[65] It was estimated that around half of the oceanic temperature profiles taken were not available in data repositories and that it would take decades to acquire new digital data that would enable the study and modelling of long-term ocean variability. As a result of the programme, the number of available oceanic temperature profiles from before the project has tripled, contributing valuable information to the scientific study of the relationship between the ocean and climate.

Small volume data from individuals and small teams, sometimes known as long-tail data, is also often at risk because it is often not considered to be reusable for other purposes, is not stored in shared repositories, and does not adhere to standard formats used by larger scale data collection.[66] The original collectors of the data may also move on or retire, taking with them both the know-how about how to access the data and the background context to the data that is needed to make sense of it. The importance of rescuing long-tail research data before it is lost has even led to the formation of awards to encourage the preservation of earth observational data.[67] Rescuing data that is at risk can be difficult and may require significant expenditure if original equipment needs to be sourced and repaired. However, for paper-based materials that have been well preserved, crowd sourcing can be used to extract important data with much success. An example of this is an international collaboration, RECLAIM, that seeks to capture weather, climate, and marine data from historical ship logbooks.[68] The work included a project that made use of volunteers from the Zooniverse citizen science portal to transcribe historic weather data from hundreds of thousands of pages of ships logs from Arctic survey and whaling ships.[69]

Using digitisation and transcription techniques can help save paper-based and analogue data at imminent risk of obsolescence, but a great deal of effort is required to keep digital research data accessible and usable over the long term. Data in digital form needs to be migrated to new formats in order to protect them from becoming obsolete. Where updated versions of the software exist, a file from a previous version can be opened and saved in the latest version. Often when a new software version is released, functionality is included for compatibility with previous versions to enable migration to new formats. Such processes can be manual and time-consuming, but some programs may allow batch processing to convert multiple files to the new version at the same time. There may be potential for some data loss during migration of data to a new version. For example, it is quite common for display and formatting information to be lost from word processed or spreadsheet documents when they are converted to a new version or to a different software package. Such display and formatting information can have a significant impact on the usability of a particular data document, and so trying to maintain this information can be important. One way to do this is to convert data files into open standard formats that make them less vulnerable to becoming obsolete and that also standardise the contents. For word-processed files, conversion to Rich Text Format (RTF) or Portable Document Format (PDF) formats can retain formatting and layout information. PDF, in particular, is commonly used as a preservation format because it maintains the styles, formatting, and layout of documents by including all this information within the file, in addition to the text-based and graphical contents. The PDF standard has been an open format managed by the International Standards Organisation since 2008 and is well supported by the software community. For tabular data, such as information from tables and spreadsheets, Comma Separated Values (CSV) are commonly used as a standard export format. This simple format is also well supported and used, with many programs providing functions that enable the import and export of data in that format.

The benefit of converting to a standard format is that any migration processes should need to be completed less often, because standard formats should not undergo fundamental changes that render the files obsolete. Updating the file for each new version of the software, on the other hand, is an ongoing process. It also does not solve the problem of obsolescence if the originating software goes out of existence. It is also common for major software versions to make significant changes to the underlying data files, in some cases, making it impossible to migrate from the previous version to the new version. This is especially common in specialist software with proprietary formats, where the underlying file formats may be frequently changed to handle new functionality in the software. There are a number of standard formats for scientific information that can help tackle the problem of obsolescence for research data, including general formats for data files, data from instruments, geographic data, and also discipline specific data formats such as astronomy, biological sciences, chemistry, and environmental sciences. A list of scientific data formats can be found on the File Formats wiki.[70]

An alternative method of continuing access to digital data formats over time is to use software emulation, in other words, to write software that can interpret or use the original data and display it as the original hardware and software would have done. The technique was pioneered by computer games enthusiasts who wanted to continue playing obsolete games on new devices or to port them to new hardware platforms. Although emulation is possible for running legacy software and displaying data, and in theory is closer to the original experience of accessing the data, in practice it can be difficult to achieve. In order to create an emulator of the original software, the original code must be available to port to a new platform, or alternatively, the specifications of the original data structure are needed in order to write new software that can display the data.

Obsolescence is an issue that is often only considered once a project has been completed and materials are being considered for deposition in an archive of some sort, whether that is an institutional repository or an external repository for a specific discipline or in association with a publication. Before deposition, it may be necessary to convert files to a standard format or to the latest version of the software that supports them. Once deposited in the repository, it may become the responsibility of the repository managers to prevent degradation of the data and obsolescence, but where this is not the case, then responsibility for ensuring the ongoing accessibility and relevance of the data needs to be established.

5.8 Conclusion

Research data management is an important activity, not only so that the creators of the research data are able to find the data and use it, but also so that during or after the end of the project others are able to access, understand, and reuse the data. It takes more than access to the datasets themselves to be able to make sense of them, and following good practices, such as the FAIR principles, can help to ensure that data is discoverable and reusable. Metadata is extremely valuable in terms of discoverability of research data, but also provides additional context that can facilitate the reuse and trust in the data once found. It is also essential to ensure that research data and scientific records are carefully preserved to avoid their loss in the long term.

Electronic Laboratory Notebooks and other digital tools can facilitate the management of research data, including providing ways to organise, structure, store, and annotate the data using metadata. In Chapter 6, we will go into more details about the functionality of ELNs and provide some examples of different types and their use. In Chapter 10, we will take an in-depth look at metadata and consider some of the challenges of ensuring that adequate and appropriate metadata is generated by scientists to describe their research data. We will discuss the 'burden of curation' and 'metadata friction', and look at ways that interfaces for digital tools can be designed to make these tasks easier.

References

1. Available at https://www.sdss.org/dr14/ (accessed Oct 2018).
2. C. Goble and D. De Roure, in *The Fourth Paradigm: Data-Intensive Scientific Discovery*, ed. T. Hey, S. Tansley and K. Tolle, Microsoft Research, Redmond, 2009, p. 137.
3. J. G. Frey, D. De Roure, M. C. Schraefel, H. Mills, H. Fu, S. Peppe, G. Hughes, G. Smith and T. R. Payne in *Proceedings of First International Workshop on Hypermedia and the Semantic Web*, ed. D. Millard, 2003, Available at https://eprints.soton.ac.uk/258790/1/htsw03-frey.pdf (accessed Oct 2018).
4. S. A. Iqbal, J. D. Wallach, M. J. Khoury, S. D. Schully and J. P. Ioannidis, *PLoS Biol.*, 2016, **14**(1), e1002333.
5. C. Hutton, T. Wagener, J. Freer, D. Han, C. Duffy and B. Arheimer, *Water Resour. Res.*, 2016, **52**(10), 7548.
6. Available at http://archaeologydataservice.ac.uk (accessed Oct 2018).
7. Available at https://rda.ucar.edu (accessed Oct 2018).
8. Available at https://www.ddbj.nig.ac.jp/index-e.html (accessed Oct 2018).
9. Available at https://www.ebi.ac.uk/services (accessed Oct 2018).
10. Available at https://www.nomad-coe.eu (accessed Oct 2018).
11. Available at https://figshare.com (accessed Oct 2018).
12. Available at https://osf.io (accessed Oct 2018).
13. Available at https://zenodo.org (accessed Oct 2018).
14. Available at https://datadryad.org (accessed Oct 2018).
15. Available at http://www.nationaldataservice.org (accessed Oct 2018).
16. Available at https://www.ands.org.au (accessed Oct 2018).
17. Available at https://www.ukdataservice.ac.uk (accessed Oct 2018).
18. Available at https://data.europa.eu/euodp/data/ (accessed Oct 2018).
19. L. Candela, D. Castelli, P. Manghi and A. Tani, *J. Assoc. Inf. Sci. Technol.*, 2015, **66**(9), 1747.
20. Available at https://www.datacite.org (accessed Oct 2018).
21. K. Belhajjame, J. Zhao, D. Garijo, M. Gamble, K. Hettne, R. Palma, E. Mina, O. Corcho, J. M. Gomez-Perez, S. Bechhofer, G. Klyne and C. Goble, *Web Semant.*, 2015, **3**, 16.
22. M. R. Berthold, *Reproducibility and KNIME*, 2015, Available at https://www.knime.com/blog/reproducibility-and-knime (accessed Oct 2018).
23. D. Garijo, S. Kinnings, L. Xie, Y. Zhang, P. E. Bourne and Y. Gil, *PLoS ONE*, 2013, **8**(11), e80278.
24. K. Baggerly, *Nature*, 2010, **467**(7314), 401.
25. M. D. Wilkinson *et al.*, *Sci. Data*, 2016, **3**, 160018.
26. B. Mons, C. Neylon, J. Velterop, M. Dumontier, L. O. B. Da Silva Santos and M. D. Wilkinson, *Inf. Serv. Use*, 2017, **37**(1), 49.
27. Available at https://dataverse.org (accessed Oct 2018).
28. Available at https://fair-dom.org (accessed Oct 2018).
29. Available at http://isa-tools.org (accessed Oct 2018).
30. Available at https://www.openphacts.org (accessed Oct 2018).

31. Available at https://www.wwpdb.org (accessed Oct 2018).
32. Available at https://www.uniprot.org (accessed Oct 2018).
33. Available at http://www.dcc.ac.uk/resources/curation-lifecycle-model (accessed Oct 2018).
34. Available at https://www.ands.org.au/working-with-data/data-management/overview (accessed Oct 2018).
35. Available at https://www.ukdataservice.ac.uk/manage-data/plan/dmp-esrc (accessed Oct 2018).
36. Available at https://www.dataone.org/data-management-planning (accessed Oct 2018).
37. Available at https://dmponline.dcc.ac.uk (accessed Oct 2018).
38. Available at https://dmptool.org/ (accessed Oct 2018).
39. Available at http://www.cipa.jp/std/documents/e/DC-008-Translation-2016-E.pdf (accessed Oct 2018).
40. Available at http://dublincore.org/documents/dces/ (accessed Oct 2018).
41. Available at https://www.iso.org/standard/52142.html (accessed Oct 2018).
42. Available at http://dublincore.org/documents/dcmi-terms/ (accessed Oct 2018).
43. Available at http://rs.tdwg.org/dwc/ (accessed Oct 2018).
44. Available at http://enews.patricbrc.org/faqs/2-genome-data-and-tools/genome-metadata-faqs/ (accessed Oct 2018).
45. Available at https://www.iso.org/standard/53798.html (accessed Oct 2018).
46. Available at https://www.iso.org/standard/67325.html (accessed Oct 2018).
47. Available at https://gcmd.nasa.gov/add/difguide/whatisadif.html (accessed Oct 2018).
48. Available at https://www.historicengland.org.uk/images-books/publications/midas-heritage/ (accessed Oct 2018).
49. Available at http://www.dcc.ac.uk/drupal/resources/metadata-standards (accessed Oct 2018).
50. Available at http://www.loc.gov/standards/mets/mets-home.html (accessed Oct 2018).
51. Available at http://www.loc.gov/standards/premis/ (accessed Oct 2018).
52. Available at https://www.w3.org/2001/sw/wiki/PROV (accessed Oct 2018).
53. Available at https://committee.iso.org/sites/tc46sc11/home/projects/published/iso-23081-metadata-for-records.html (accessed Oct 2018).
54. Available at http://malaria.ourexperiment.org (accessed Oct 2018).
55. S. J. Coles, J. G. Frey, C. L. Bird, R. J. Whitby and A. E. Day, *J. Cheminf.*, 2013, **5**(1), 52.
56. C. Hansen, C. R. Johnson, V. Pascucci and C. T. Silva in *The Fourth Paradigm: Data-Intensive Scientific Discovery*, ed. T. Hey, S. Tansley and K. Tolle, Microsoft Corporation, Redmond, 2009, p. 153.
57. Available at https://www.iso.org/standard/43506.html (accessed Oct 2018).
58. Available at https://www.crossref.org (accessed Oct 2018).
59. Available at http://www.datacite.org (accessed Oct 2018).
60. Available at https://eidr.org (accessed Oct 2018).
61. Available at https://orcid.org (accessed Oct 2018).

62. Available at https://www.iso.org/standard/44292.html (accessed Oct 2018).
63. Available at https://www.backblaze.com/blog/how-long-do-disk-drives-last/ (accessed Oct 2018).
64. R. E. Griffin, *GeoResJ*, 2015, **6**, 92.
65. S. Levitus, *Data Sci. J.*, 2012, **11**, 46.
66. L. Hsu, K. A. Lehnert, A. Goodwillie, J. W. Delano, J. B. Gill, M. A. Tivey, V. L. Ferrini, S. M. Carbotte and R. A. Arko, *GeoResJ*, 2015, **6**, 108.
67. Available at https://www.iedadata.org/community/data-rescue-initiative/ (accessed Oct 2018).
68. C. Wilkinson, S. D. Woodruff, P. Brohan, S. Claesson, E. Freeman, F. Koek, S. J. Lubker, C. Marzin and D. Wheeler, *Int. J. Climatol.*, 2011, **31**, 968.
69. Available at https://www.zooniverse.org (accessed Oct 2018).
70. Available at http://fileformats.archiveteam.org/wiki/Scientific_Data_formats (accessed Oct 2018).

CHAPTER 6

A More In-depth Look at Electronic Laboratory Notebooks

6.1 Introduction

In Chapter 4, we looked at the impacts of technology in science, and in particular the development of informatics systems and the digitisation of note-taking. Although electronic notebooks can offer some significant advantages over paper notebooks for storing, organising, and managing data, there is often a resistance to their use, and their adoption can be disruptive.[1] In this chapter, we will focus on Electronic Laboratory Notebooks (ELNs), although many of the less specialised capabilities and benefits of ELNs also apply to the more generic electronic notebooks such as OneNote. Throughout their history, there has been an expectation that electronic notebooks will ultimately completely replace paper notebooks in the laboratory, with commentators pointing out the many advantages of electronic notebooks over paper notebooks.[2,3] Indeed, there are many benefits of ELNs, which we will discuss in detail below, but so far at least, there has not been the complete revolution some anticipated. Although in many industries and disciplines ELNs have been widely adopted, this is by no means ubiquitous, and in many environments ELNs appear to have made relatively little impact with the trusty paper notebook still tool of choice for capturing the research record. There are a number of reasons for this, which we will discuss in more detail later, but first we will discuss the many benefits of electronic notebooks over paper notebooks. We will also take a look at some examples of different types of ELNs and how their functionality may suit different environments and audiences. We will

Recording Science in the Digital Era: From Paper to Electronic Notebooks and Other Digital Tools
By Cerys Willoughby

Published by the Royal Society of Chemistry, www.rsc.org

also discuss some examples of ELN use in practice. Finally, we will go back to the issues of paper notebooks *versus* ELNs and discuss some of the reasons why users may be reluctant to use them and other barriers to their adoption.

6.2 Benefits of Electronic Notebooks

Although scientists have used paper notebooks successfully for centuries, there are many drawbacks with paper, even before we try to integrate notes captured on paper with our digital data. Paper itself is not an indestructible material and precious research data can easily be lost if a paper notebook is damaged by chemicals in the laboratory or if the laboratory catches fire. Paper can also be damaged by water and humidity, strong light, chemical fumes, extremes in temperatures, and dust.[4] In the field, loss or damage to a paper notebook is even more likely. It is all too easy to lose your grip on your notebook whilst clambering over rough terrain or wading through a river, only to watch distraught as your notebook disappears downstream. Or you might put the notebook down for a few minutes whilst you collect a sample, only to return to find a hungry cow munching on your work! Some field researchers keep their notes in a lose leaf form, or like archaeologists and ecologists complete paper forms in the field that are stored in folders. These individual pieces of paper are even more at risk as they can easily be misplaced or even disappear on a strong breeze.

For some, the fragility of paper is overcome by keeping duplicate notebooks, using carbon copy paper, or by only writing up their notebooks in a relatively safe environment, such as the office for laboratory work, or at the base for fieldwork. This is not a perfect solution to the problem because duplicating work or writing up later can not only lead to additional work, but may also introduce errors into the record. During manual transcription, it is easy to make mistakes whilst copying, for example, failing to copy over part of the notes, misreading the original notes, or substituting numbers when copying data. When notes are written up at a later time and at a different location to the original activity, there is also a risk that errors may be introduced as information is misremembered or not remembered at all. Although it is still possible to lose or damage a device such as a tablet or laptop computer, the electronic format has the advantage over paper in that it is much easier to create duplicate copies, different versions, and back-ups that can be retrieved in the event of loss or damage of the device itself.

Paper is also not suitable to take into all environments. For example, in some laboratory environments, paper dust is a contaminant that could negatively impact the outcome of experiments.[5] In a wet lab environment, paper can spread contamination between the laboratory and the outside, and in clinical laboratories, paper could also be responsible for transferring pathogens in and out of the laboratory.[6] Although portable electronic devices may not be suitable in all environments, for example, in locations with strong magnets, and may also spread contaminants, a device that always remains within the lab environment is unlikely to pose a contamination risk.

As we have mentioned before, a common solution to the challenge of linking digital data with the paper notes is to print the data out and stick it in the relevant place in the notebook. However, printing out documents to include within a paper notebook is not environmentally friendly, and unless the paper and ink are of archive quality, they are likely to degrade with time. Although photo prints in an album stored in the dark might survive for more than 100 years, outside of this environment, the lifespan of certain ink and paper combinations is actually much less than this.[7] We know that paper notebooks can survive for hundreds of years if properly cared for, but as previously mentioned, printouts stuck into notebooks are much more vulnerable and can easily become detached and lost, as well as degrade. Long-term storage of paper records and notebooks can also be expensive because they can take up large amounts of space.[8] Although electronic records also need to be stored, even back-up tapes have a much smaller footprint compared to paper, and storage in the cloud requires no local storage space to be utilised at all.

In order to maintain a paper notebook as a legal document, the notes must have every paged numbered, signed by the author, and witnessed by another member of the team.[9] Despite such precautions, it is still possible for notes to be manipulated after they have been signed or even be entirely fabricated.[10] It is much harder to fake electronic records. Electronic notebooks typically provide timestamps, provenance trails, and revision histories, together with robust electronic signatures that are significantly more reliable for legal defence and compliance purposes.[6] As mentioned previously, one of the key milestones in the history of ELNs was the introduction and acceptance of digital signatures for Intellectual Property (IP) and compliance purposes. The introduction of this technology into ELNs also makes the witnessing process much faster and effective because the witness does not actually need to be in the same physical location as the notebook. The witness can access the electronic notebook whenever and wherever is convenient to assess the contents and to add their signature, rather than needing to arrange a time and location to hand over the notebook that suits both the author and the witness. The witness can be prompted to sign when the notebook is ready, and this gives the opportunity for any potential problems with the author's activities to be picked up sooner. As mentioned previously, a paper notebook can only be used by one person at a time, whereas the ELN allows both the author and the witness to be using it at the same time. The ELN can also provide different ways of viewing the information and specify limitations on access. For example, the author is allowed to add new content and make modifications, whilst the witness is only allowed to review, comment, and sign the pages of the notebook.

Perhaps the biggest problem with paper is that it is not suitable for sharing, leading to a loss of knowledge for other members of the team and the organisation as a whole. Completed notebooks are usually locked away in filing cabinets or archive storage where they are inaccessible and potentially vulnerable to fire, water, or even insect damage.[11,12] If the organisation does not have a strict policy of handover, a researcher or employee may take the notebook with them when they move on, making it completely unavailable for further use. Even when notebooks are readily accessible, it can be difficult to make use of the records within. Paper notebooks are difficult to search, even if the owner has been diligent and created and updated a table of contents. Entries in the notebook may be difficult to read and may make use of personal abbreviations or symbols that have no meaning to the reader. The content may also be incomplete because the author of the notebook chose not to capture it or to duplicate their own tacit knowledge about their research. For example, in a set of repeated experiments, the full details of the experiment may only be captured once. As we have already discussed, often only the person who wrote the notebook can actually decipher it, and over long periods of time even the author may have trouble recalling any missing information and making sense of the notes. The addition of data collected and analysed can help with the results, but if the data was in digital format then it may be difficult to track it down and match it up with the notes in the notebook. One of the main consequences of this lack of ability to share and search within notebooks is the likelihood that research will unnecessarily be repeated. The same mistakes will be made by others and failed experiments will be repeated, leading to an unnecessary duplication of effort. This problem is exacerbated because it is relatively unusual for failed research to be published, and therefore the majority of failed research is locked up in inaccessible paper notebooks. Many researchers are reluctant to share their research, for reasons that we will discuss in more detail in Chapter 11, but they are particularly reluctant to share failed research, even though it may be of significant benefit to others to avoid repeating the same mistakes. Electronic notebooks can overcome the majority of these problems. The knowledge that is captured inside an electronic notebook can be shared with other users, so that the knowledge is available to the wider team, the organisation, for cross-organisational collaborations, for funders, and even for access to the public for open science research. The knowledge contained in the electronic notebooks can easily be searched, usually in a variety of ways. For example, information about chemistry experiments in an ELN can be searched based on information such as the materials used in the experiment, the title created by the user, but also by the date the experiment was carried out and by whom. The knowledge contained in the ELN is also not lost to either the organisation or individual when they later move on. Many ELNs provide functionality that either enables the conversion of the ELN notebooks into a format suitable for transportation and printing such as PDF, or even in a format that enables them to be imported into a new ELN. This means that the owning organisation can keep the notes in the

ELN, but the author can also keep a copy for themselves, if organisational policy allows it.

Collaboration is a major benefit of ELNs, although this can make those engaged in research that has significant intellectual property or commercial value rather nervous. Although ELNs are beneficial for collaboration by enabling sharing of the contents of the digital notes and data, and are used by many interested in open science, this does not mean that every ELN allows unchecked access. ELNs also implement security and authentication to ensure that only authorised users have access to the content. ELNs also usually have some mechanism of assigning particular rights to particular authenticated users, so that only those individuals who have the right to view or edit the content have the ability to do so. Collaborators may be restricted to a local team, but the user rights can be configured to reflect the hierarchy of the organisation itself and the groups within it. The ability to share research across sites is even more valuable for cross-organisation groups where members are not collocated, and are therefore unable to meet regularly in the same location to view the notebook and discuss the research. Collaboration in the ELN extends beyond just the ability to view and edit the notes and data captured within. The majority of ELNs with collaborative functions also enable discussion about the research by allowing users to add comments to the pages of the notebooks, or even to individual sections, files, or images within a notebook. For example, some ELNs provide the functionality to annotate images, so that comments can be made in the form of annotations to highlight areas of interest on images.

Another useful feature of ELNs for collaboration is access to user and project information. Users can often find information about other users within the ELN, what they are working on, recent updates, and project information. Access to such information can provide a way for other collaborators to keep informed about what is going on or for new members to learn about the team and project. Depending on the way that an organisation or team uses an ELN, there may be additional shared information stored in the ELN, such as meeting minutes, presentations, shared protocols, and literature references that may have value to the wider team. Some ELNs can be used for effective project management by enabling tasks to be assigned to members of the team and their progress monitored, or to enable updates on the project to be automatically communicated to the project stakeholders through notifications.

Taking notes electronically can save effort by reducing unnecessary duplication, not just of experiments across a team, but also effort in recording the research. For example, when experiments are repeated with the same or very similar methods, then the contents of a previous experiment, such as the methods and the format of the results, can be duplicated. In the duplicate record, the method can be easily modified for any small changes and only new information needs to be captured. This means that the complete information is always available for viewers of the experiment who may want to understand or duplicate the experiment themselves, but also reduces the effort for the experimenter in creating their records. This also reduces the chances of missing information or errors in transcribing methods between records.

A major advantage of electronic notebooks over paper notebooks is the ability to incorporate other types of information within them. Although in a paper notebook it is possible to attach printouts and to glue in photographs, these items may get detached over time. Manual referencing of digital files can be added to a paper notebook, but these are reliant on the names and locations of the digital files remaining constant over time. Within an electronic notebook it is usually easy to create links to digital information in the notes, which can then be opened separately, or even to embed content such as tabular files, photographs, audio, and video directly into the notes. If the linked files are stored remotely then there may still be challenges in keeping links up to date if names and locations change, but it is likely that the software can warn about broken links and make it easier to fix them. In many ELNs, the associated files can also be stored together with the notes that reference them, keeping all the related resources together.

Electronic notebooks can also integrate with other tools. This means that they can integrate with instruments enabling data to be captured automatically and therefore reduce the risk of transcription errors. Even when they do not integrate directly with an instrument, the data can be extracted in digital form and incorporated directly in notes in the form of tables, graphs, plots, or other images. Integration with instruments can also support more advanced functions, such as enforcing standards and Standard Operating Procedure (SOP) compliance by enabling the user to be prompted to perform certain tasks or to ensure that tasks cannot proceed before a prerequisite action has been completed.[8,13] We will discuss more about standards and compliance in Chapter 7. Instrument integration also enables users to capture parameter configurations of instruments, to perform calibrations, and even potentially to drive the instruments in order to control an experiment.

Electronic notebooks can also connect to other repositories of information. For example, internal repositories may include chemical inventory systems or inventories of sample storage locations that enable users to locate materials needed for an experiment. Connections to inventories can even enable ELNs to automatically update inventories based on the measured materials used during an experiment. External data sources may include databases of data that can be used for analysis and planning the experiment, repositories of scientific literature, dictionaries of technical terms, and reference databases such as properties and safety information about materials. Materials and samples tracking can also potentially be used to automatically reorder materials when they get low or to provide warnings if 'best before' dates have expired, or if other actions are needed to maintain samples effectively. ELNs can also themselves act as repositories of information, enabling notes and data to be stored together in the same location making them easier to link together, easier to search, and easier to manage in the long term.

ELNs are also able to facilitate the formal publishing of data. For industry, this may be through the production of regular reports for use by management, and within academia it can be through the production of journal papers or presentations. Some ELNs have tools that automatically generate reports or that aggregate content for publication, but even without these tools specifically built in, the data management and search facilities enable information and data to be pulled together relatively easily for such purposes. Content that is available in digital form can then be copied directly from the ELN for inclusion in new documents or exported in a suitable format for further manipulation.

Electronic notes in general can be a benefit for helping to organise notes and to remind users of important elements to record. For example, Naskrecki, in *Field Notes on Science and Nature*, self-confesses to being disorganised and constantly losing information written down on bits of paper, losing the entire notebook, writing down cryptic codes for samples whose meanings are later forgotten, and forgetting to record essential information entirely.[14] To overcome these problems, he captures everything as digital notes in a structured format, so that the content can be found again, and so that he is prompted to capture information he might otherwise forget. Another example, from our experiences at Southampton, was a chemistry postgraduate with severe dyslexia who struggled with using a paper notebook because the format is so reliant on text and with retrieving content by relying on memory. He used digital photographs extensively within his notes in the ELN to supplement the text that was written, and also made extensive use of metadata to help organise the notes and facilitate future retrieval of specific experiments.[6] In addition to text being easier to read than handwriting, electronic notebooks also usually include tools such as text formatting and spell checkers that can make text more legible, particularly if the developers have taken accessibility into account. For example, users with special needs may need to be able to adjust the size of the text or make the contrast between the text and the page higher if they have impaired eyesight, whilst dyslexic users often find their ability to read text is improved if they can change the colours of the background and text. For those with significant eyesight impairment, a screen-reader can even be used to read digital notes. ELNs often come with built-in dictionaries that can help correct mistakes and prompt the user for correct spellings of scientific terms and materials. These are capabilities that would be impossible for a paper notebook.

Another feature impossible for a paper notebook is the ability to perform calculations. A common task for many scientists is to perform calculations during their activities inside and outside the laboratory. Having the functionality to perform calculations within an ELN, or even have the ELN automatically complete calculations based on the context of the activity is very useful, saves the researcher time, and is less error prone than plugging numbers into a calculator and transcribing them back out. As previously mentioned, I have observed that many researchers actually write their calculations on scraps of paper and then later transcribe the results to the notebook. This means that the calculation and data used within it, is not recorded, making it impossible to later find out if a mistake was made! The ELN can also make use of common reference information to help with calculations, for example, the molar mass for molarity calculations and densities of materials for calculating viscosities. The greater reliability of calculations performed by software, and the actual capture of all the relevant data and formula used, can reduce the amount of time and effort required when the notes need to be reviewed by another member of the team, a supervisor, or an auditor.

6.3 Electronic Laboratory Notebook Examples

There are numerous different electronic notebooks, and ELNs in particular, available on the market. The number of ELNs available on the market increases all the time, although the market is quite volatile and ELN developers come and go over time, as some go out of business and others get bought by competitors. For example, Southampton University's ELN market study, based on previous surveys of ELNs, identified 103 different ELN systems, but less than 75% of them were active just a few years later, at the time of the survey, because many had been discontinued and others replaced after being purchased by larger companies.[15] ELN offerings also change over time, as new functionality is added, but also as the perceived needs of the markets change. With these factors in mind, this section will not go into extensive detail about the individual ELNs and their functionality because any individual ELN's functionality, offerings, and even existence are guaranteed to change over time. In this section, we look at a number of different ELNs that are currently available, but we will not perform an assessment of whether one ELN is better than other. Instead, we will discuss the functional differences and intended audiences to provide some insight into which kinds of systems may better suit particular organisations, disciplines, or researcher circumstances. There are many ELN options available, with a great deal of overlap in functionality. The inclusion of any particular ELN in this section is not an endorsement, and as always, other ELN vendors are available! If you are thinking of implementing an ELN, this section will give you some idea of the capabilities available, and whether they may be suitable for your particular needs, but you should also take a look at what other options are currently offered. Many websites provide lists of available ELNs, although many of these become out of date as time passes, so some homework will be required to keep up to date with the latest offerings. Many of the larger scientific software development organisations are targeting both large-scale industrial and academic laboratories with their offerings, whilst many of the smaller organisation are focused on products for individuals or small teams. In Chapter 7, we will also discuss what you need to consider when selecting an ELN provider.

The primary purpose of the ELN is to capture the record of the experiment. For the majority of researchers this means at the very least that the ELN should capture notes about the experiment as it is being run, but for many it also means to capture the design of the experiment. The design of the experiment may be based upon academic literature or previous experiments that have been recorded within the ELN and subsequently used as a reference. There may also be links to other information and materials provided by the ELN. For example, data from experiment analysis may be retrieved from instruments and stored in the ELN, or there may be access to information about sample storage and chemical inventories either from within the ELN or a connected system that can be accessed and linked to from the ELN notebooks. The majority of laboratory scientists use similar processes when they plan and conduct experiments, analyse their results, and record their experimental outcomes. As a result, the majority of ELNs facilitate these processes in similar ways. A challenge for ELN developers is to manage both the similarities and differences between the different disciplines.[16] ELNs have other capabilities beyond the experiment recording that may be more or less important for a particular audience. For example, support for specific disciplines may include integration with subject-specific libraries, dictionaries, data sources, and services; there may be varying levels of security available for the protection of data; and different options provided for the storage and management of data. We will talk more about these different additional capabilities when we discuss considerations for choosing an ELN in Chapter 7.

ELN providers may develop a desktop version of their ELN, a Web-based version, or both. Desktop versions of ELNs tend to be tied to particular operating system platforms, for example, an ELN may work only on Windows, and be installed on a local machine. These ELNs may have a server component that is installed on a local server or remotely. Web-based ELNs, in contrast, are accessed *via* a browser and therefore do not need to be installed on the machine directly, and can usually be used on any operating system. Web-based ELNs are also often cloud-based, so that the data is stored remotely, but this is not always the case.

The first ELN examples we will look at come from major scientific software developer PerkinElmer Informatics. PerkinElmer Informatics have developed both a desktop and Web-based ELN, together with a range of tools for scientific informatics, including the popular ChemDraw application and other scientific productivity tools, in addition to well-known Laboratory Information Systems (LIMS) and Laboratory Execution Systems (LES).[17] PerkinElmer Informatics produce ELNs tailored for both biology and chemistry with discipline specific modules and tools for each. For example, for biology, tools for assay data management, plate management, and instrument integration are provided. For chemistry, functions are provided to enable chemists to create reaction schemes, to automatically generate stoichiometry tables, and the tools also contain facilities that understand chemical terminology. The ELNs also enable users to perform calculations and process and perform analysis of data, as well as capturing data directly from instruments. Integration is provided to enable import and export to everyday tools such as Microsoft Office, PDF, and image files, but as might be expected, there is also integration with their own tools, including a sample registry and material inventory. For analytical laboratories, an optional bioanalytical module, complete with a library of templates, is available for workflow management for SOP and Good Laboratory Practices (GLP) compliance. Collaboration enables users to be linked to specific projects, and biology and chemistry users can exist together within the same ELN. Results from multiple experiments can be analysed together enabling project data to be managed in one place. Electronic signature witnessing is provided, and audit trails and regulatory compliance is claimed for the ELN. The ELNs are domain specific with modules provided for the target disciplines, but additional functionality is not provided that would overlap with the existing tools, such as LIMS, LES, sample tracking, and inventory management. There is a 30-day trial available for the Web-based 'Signals' ELN, and this appears to be targeted to users interested in cloud-based storage, improving productivity, and collaboration. The primary desktop 'E-Notebook' ELN appears to have no trial version, at the time of writing, and is targeted more towards the corporate industrial user interested in managing research, testing workflows, and protecting their IP. For large academic organisations that want an ELN that would work for both biologists and chemists, an ELN that enables both types of users to share storage and access, but still be able to use domain specific tools, could be an advantage.

BIOVIA is a similar organisation to PerkinElmer Informatics, providing ELNs that can be tailored for different disciplines through the use of third-party services and modules, and that also integrates with their other laboratory informatics offerings, including LIMS and LES.[18] Tailoring is available for many disciplines including analytical chemistry, process chemistry, and synthetic chemistry, as well biology, drug development, and pharmacology. Again, this provider does not mention a free or trial version, and appears aimed at the corporate industrial market, although there are different subscription rates for commercial and academic licenses. The ELN was previously known as 'Contur iLabber', before acquisition, and at the time of writing is called 'Biovia Workbook'.

An example of an organisation that targets research and development, and that pitches itself to both a commercial and academic market, is Benchling.[19] Benchling provides a discipline specific ELN for molecular biology and is an integrated solution with a Lab Notebook that integrates with a Molecular Biology Suite, Bioregistry, Sample Tracking, Workflow Management, and Request Management modules. The included modules and their available functions depend on the licensing option chosen, with a free version, a ten-person licence for a small team, and an enterprise version. The laboratory management modules, 'Sample Tracking', 'Workflow Management', and 'Request Management', are unsurprisingly only available in the corporate version. The Lab Notebook includes the kinds of features that are expected in an ELN, such as tables, sharable links, collaborative editing, data storage, and search, along with calendar entries for planning meetings and requests if included. The biology specific elements come from the Molecular Biology Suite and includes CRISPR (a genetic engineering tool), cloning and sequence design, sequence visualisation and tools for working with sequences, and the Bioregistry for tracking biological entities and linking between digital entities and physical samples. The tools are claimed to be regulation compliant. As with many of the larger software development organisations, added value is provided through support services, such as assistance with implementation, configuration, training, and on-going support. Often, these costs are included in the licence fee, and so usually do not apply to free or trial versions.

Another organisation with an ELN with different editions for industry and academia is Labfolder.[20] The ELN is the central tool, and therefore this is not an ELN designed to integrate with a LIMS or LES systems, although an Application Programming Interface (API) is provided for the ELN, so there is the potential for integration with software and hardware beyond the small number of built-in integrations. Instead, the ELN provides some of the functionality that might be expected in the other systems, such as a materials database and some task management. Labfolder provides both a cloud and on-site server install option, and the ELN provides the usual kinds of functionality that would be expected in an ELN such as integration of data, ability to customise the data format, and collaboration tools, including creating and sharing protocols, templates, and projects. The ELN claims compliance to a number of different regulatory requirements, and provides timestamping, digital signatures, and audit trails. Data can also be exported from the ELN in both PDF and XHTML format. The prices are comparatively low when compared to some of the larger development organisations, but the company still supplies additional support. Unusually, a free version is also available for industrial use as well as for academic use, although in both cases, this is a for a small number of users, and the advanced features are only available in the paid versions. More advanced features are to be found, as might be expected, in the industry version, especially for features related to compliance.

SciNote is one of the very few ELNs that mentions anything about utilising ELN contents for the purposes of generating publications, using a Manuscript add-on, a feature that is likely to be of most interest to an academic audience.[21] SciNote is similar to Labfolder in that the ELN is the main focus with some additional extensions. SciNote enables you to create projects, experiments, and tasks, and provides facilities for collaboration and generating reports. SciNote is an example of a generic style of ELN, with no functionality aimed at any particular discipline. There is a free version and a paid version that contains additional functionality, including extensions for project management and compliance, and an additional support service.

Another example of a generic style ELN is the LabTrove ELN, developed by the University of Southampton.[22] LabTrove is a Web-based ELN that utilises a blog style interface to capture digital notes, images, and data files. The ELN is targeted at academic audiences and is used by a number of academic groups from different disciplines around the world, including physical and synthetic chemists, biologists, and even physicists. LabTrove provides the facilities you would expect in an ELN for data management and collaboration, and also has a number of extensions to support researchers in the lab. These extensions include barcode generation to enable samples to be tracked in the lab, facilities for chemical drawing and chemical search within the interface, automatic term identification within the notes, and also the ELN has an API that enables content to be published to the ELN remotely. LabTrove has been used by a number of groups to automatically publish data to the ELN from instruments and sensors that can then be linked to the appropriate experiment records. LabTrove is an open source project, which means that other organisations can make changes to the code of the ELN to suit their own needs. Third-party plugins can also be integrated with the ELN to add additional functionality, for example, the Royal Society of Chemistry has worked with the development team to add chemistry specific functions through the use of plugins.[23] We will talk more about LabTrove later in this chapter as an example of how different groups make use of an ELN. There are a number of similar ELNs to LabTrove available, including many that are directly targeted at academia, but most have only limited functionality. For example, eLabFTW is a PHP-based online ELN, with templates, the ability to add 'steps' to an experiment, and basic editing features including tables. Data files, images, and code samples can be imported into the ELN.[24]

OpenWetWare is a wiki-based ELN aimed at biologists.[25] The focus of the ELN is collaboration with content shared on wiki pages to enable teams to share their research with other teams. Many groups use OpenWetWare, but the features of the wiki for an ELN are a little limited. Materials and protocol descriptions are shared on the platform. An individual wiki page is created for a selected date within the notebook. A basic template is provided that can be customised for an individual's needs. It is possible to include scientific specific information such as drawing chemical diagrams or graphs, but these require knowledge of specialised mark-up languages rather than being able to work with a user-friendly GUI interface.

There are also many examples of ELNs that have been developed within an academic environment that frequently have more experimental elements. For example, the BioTISCH system replaces the traditional workbench with a glass covered table top system that presents information on the bench's surface and includes some augmentation of objects in the form of reagent sensors.[26] The system allows the biologists to view plasmid charts from a database, display an experiment log that allows the user to see the steps in the experiment, and mark off each step as it is completed. A concentration calculator is used to perform conversions for the experiment, and the reaction sensors highlight the next reagent to be used. A similar system, the eLabBench is a digital table top system supporting biologists in the organisation of personal information, capture of experimental work for later access and management in the office, and providing access to a variety of computational resources directly at the lab bench.[27] The design of the system supports an open way of organising information rather than requiring a strict sequential or time-based organisation seen in workflow-based or notebook systems. It supports on the fly changes to the experiment, enables lightweight capture of information during execution, and enables computations to be made directly at the bench rather than needing to go back and forwards between the laboratory and the office. The system comprises an interactive table top with access to the kinds of tools that the biologists would have access to on their computers in the office, in addition to a Web-based notebook and an area for making rough notes. Interaction mechanisms include keyboard, mouse, pen, tagged physical objects such as racks and instruments, and buttons, for example, to activate an overhead camera.

The range of ELNs described above is a small sample of the different tools available and demonstrates both the similarity and diversity of such systems. In the following section, we will briefly look at some examples of ELN implementations and user experiences.

6.4 Experiences with Electronic Laboratory Notebooks

It is a reasonably common practice for potential users of ELN systems to try out an individual ELN, or perhaps a number of different ones, for a period of time before making a selection, but there are relatively few published accounts of these experiences. One such account comes from Guerrero *et al.*, who tested two electronic notebooks for three months with 28 scientists from a number of biomedical laboratories.[28] The chosen electronic notebooks were Microsoft OneNote and PerkinElmer Elements (now known as Signals). The two were chosen out of a list of many potential candidates and selected due to documented successes in academic research and biomedical laboratories. The candidates were also evaluated on criteria such as flexibility, security and regulatory compliance, searching, collaboration, and ability to work with smart devices and tablets. Guerrero *et al.* found that the vast majority of participants in the study agreed that electronic notebooks would be able to effectively replace their paper laboratory notebooks. Interestingly, however, all of the researchers preferred the generic electronic notebook, OneNote, over the formal scientific ELN. The reason for this unexpected finding was down to flexibility of the systems. OneNote was found to be more suitable for capturing the non-standardised assays with few experiment details conducted within the research laboratory, compared with the Elements ELN system that was designed to handle the highly standardised experimental settings and routine experiment workflows of testing and other industrial laboratories. The study also made use of a variety of tablet and smart devices, with the majority of users agreeing that these devices improved the use of the electronic notebooks and paper-based notes, for example, because they could be used in the lab to replace printed protocols, but the devices could also be used as convenient reference tools in meetings to discuss data and the results of research.

Another account of experiences with ELNs comes from the diverse community of users of the LabTrove ELN as reported by Badiola *et al.*[6] As mentioned previously, LabTrove is a blog-based generic ELN system that is used internationally across a wide range of disciplines, including analytical chemistry, medicinal and synthetic organic chemistry, physics, laser-based studies, biological evaluations, computational chemistry, process engineering, interdisciplinary teams, and university teaching. One of the advantages of the ELN for these different communities is the ability to be able to capture the record in a flexible manner. Entries created by different disciplines can be entirely different to reflect the needs of the different disciplines and the different groups of researchers. Badiola *et al.* detail how the ELN has been used by a number of these different groups, for example, the Open Source Malaria project uses the ELN as an alternative to paper notebooks to facilitate collaboration between the research team and colleagues in a local hospital, making use of facilities for repeating experiments, tracking samples, and making links between samples and the experiment record.[6,29,30] The version of the ELN used by the project is entirely public and makes use of many notebooks for different purposes, including recording synthetic chemistry experiments, biological evaluation of compounds, molecular modelling, and summary notebooks for the collection of materials for sharing and publication. The use of the ELN for this Open Science project is particularly interesting because of the use of collaborative features that would not be possible with paper notebooks, and also because of how the flexibility enables the team to create notebooks that can be used for many different purposes of value to the project. We will examine the Open Source Malaria project in more depth in Chapter 12 when we talk about Open Science.

A contrasting use of the LabTrove ELN is by the Ultrafast X-ray group, which uses the ELN for internal collaboration and as the centre of discussions in team meetings. Members of the team do not use individual notebooks, but each contributes regularly to a team notebook to discuss plans, equipment, experiments, and outcomes. Other materials such as images, spreadsheets, computations, and literature sources are stored in the ELN for use by the team. The team have also made use of the LabTrove API to automatically capture data from sensors in the labs and new data is automatically added to a new notebook entry whenever data is generated by the X-ray instruments, enabling new data to be reviewed rapidly. Badiola *et al.* also provide other examples, including more conventional uses that reflect a substitution of the ELN for the paper notebook, but with the additional capabilities of the ELN supporting the work of the different groups. For example, making it easier to review notes incorporated in notebooks compared with paper for meeting regulatory requirements and for linking together experiment records with both data and results.

These examples of ELN use in practice demonstrate that the different capabilities of ELNs can make them powerful tools for a wide variety of different groups and different kinds of scientific work. Different organisations and disciplines have different needs, and as we have mentioned with the different ELN examples available, some ELNs focus on the specific needs of one discipline or organisation type, whereas others try to provide a more generic service that could be used by a wider audience. The needs of an analytical laboratory are quite different to the needs of the research laboratory, for example. The analytical laboratory is much more likely to be focused on routine sample processing and analyses in accordance with standard operating procedures (SOPs). The data from these labs are likely to be highly structured and quantitative. Sample tracking, ensuring adherence to the proper procedures, and the accurate and reliable capture of data is important, and these needs are likely to influence the designs and capabilities provided by an ELN system suited to that environment. There is also likely to be more integration with other systems such as LIMS or project management systems, as well as automation of the analytical tests and procedures themselves. Data is often captured automatically but may also be directly entered through form-based interfaces to ensure everything required is recorded. Some environments may also have a focus on improvement of processes, validating methods, and compliance to regulations, and the technological functions that support these activities will better support workers in those environments. The scientists within these labs may have a variety of roles, and those responsible for preparation of samples, or setting up the experiments and running them may not be the same person who performs or interprets the analysis of the data. Roles may also be focused more on the lab and instrument maintenance, management of the lab, project management, quality control, and adherence to regulations and procedures, and the more commercial ELNs tend to provide access to more advanced functions that support the users and organisations for whom these tasks are important. Access to records is likely to be confined to a single site, even if an organisation is comprised of multiple laboratories, and access control and security are likely to be of primary importance to protect sensitive data and IP.

In academic and research laboratory environments, different functions that are provided by ELNs may be more useful. In these environments, there is likely to be a greater focus on the capture of the procedure and actual running of the experiment, with a less rigid structure required to capture data. Procedures are likely to be more fluid and are likely to change between experiments or tests. For small-scale laboratories, the rigidity of the industrial ELNs is likely to be a hindrance as exploratory investigation is unlikely to match previously defined protocols and data structures, and the nature of both are likely to change over time in unanticipated ways. Data may be captured directly through instruments, although instruments are perhaps less likely to be automated in the same way as a large-scale laboratory. Data may need to be manually uploaded or transferred to the ELN, or a manual link may need to be made. It is also possible that data may be manually entered into a spreadsheet or other tabular form. Observations may be captured in a variety of formats, such as drawings, diagrams, photographs, and even video. There is more likely to be a need for collaboration in academic and research laboratories, as related work may take place at multiple locations, and potentially across multiple time-zones. ELNs that provide discipline specific functionality, such as functionality that enables searches for information about chemical and biological materials, are likely to be very useful within these specific environments. In Chapter 8, we will take a closer look at how different user roles and their goals can influence the requirements and designs of digital tools, such as ELNs, for capturing the research record.

6.5 Barriers to Adoption of Electronic Laboratory Notebooks

Despite all the benefits of ELNs described previously, and a rapidly growing market, their use is by no means ubiquitous, and is still highly limited within academia in particular.[15,31–34] Within academia, awareness levels of ELNs are relatively low, and there is also a perception that they are designed for the needs of corporate organisations.[34] Even within commercial and industrial organisations, ELN uptake is not necessarily as high as might be expected because of the high level of comfort researchers have with paper-notebooks and the perceived complexity of ELNs. This anxiety about the perceived relative complexity of ELNs is a major factor in users' willingness or not to make use of them. There are a variety of factors that influence the adoption of ELNs within an organisation, and we will examine these factors, including ease of use below.

Cost is always an important factor to weigh against benefits when selecting any software. Costs are known to be one of the largest concerns in the use of ELNs in particular, in part because commercial options can be extremely expensive, but also because of the additional costs involved; costs associated with software inevitably extend beyond the initial off-the shelf cost of an item.[15,31,32] Software developed for use in corporate environments typically requires the purchase of many licences. For example, a licence may be required for each user of the software or for each device where the software is installed, or even per instrument. Often, the purchase of a corporate license for a premium price is a more cost-effective option for a large institution or company with many staff. If licences are required, these may be a one-off purchase, or an on-going expenditure, if they need to renewed on an annual basis for example. Licensing costs can become very expensive. For corporations, funding is likely to be available on an on-going basis, but the same may not be true for start-up companies or academic organisations, where funding is more likely to be available only for a limited time. There is a risk that access to stored data could be lost if the license period expires, especially if the data is stored off site.

ELNs often have a server-client model where the 'back-end' of the software, which might include a database and administrative components, is installed on a server machine and the 'front-end' interface to the software is installed on individual personal computers or other devices. The client version of the software then needs to be able to connect to the main server to work, or at least to be able to fetch previously stored data and make updates to the records. Additional hardware costs may come from purchasing new server machines to run the software and for additional storage space. However, other hardware costs may include improvements to the networking infrastructure to ensure that machines can connect to the ELN from within their intended environments of use. For example, more Wi-Fi routers or network connection points may be needed to facilitate access to an ELN within laboratories, which typically have less infrastructure than offices. Other infrastructure costs include hardware for reliability and safety, such as load-balancing, data storage virtualisation, and uninterruptable power supplies. There may also be costs associated with personal computers to consider. For example, new machines may be needed for access to the ELN within the laboratory or office, or existing machines may need to be upgraded to meet hardware or operating system requirements for the ELN. Around twice as many ELN vendors support Windows compared with Mac OS, and the ability to support the current infrastructure availability may determine the choice of ELN by an organisation.[15]

There are also likely to be personnel costs in terms of the original set-up of the ELN, on-going maintenance and back-up of data, and costs associated with upgrading and migrating the software in the future. There is also likely to be an on-going need to have staff that are responsible for the administration of the ELN and any associated databases. For example, an ELN administrator may be required for adding new users and assigning permissions, but also troubleshooting, configuration, tailoring of the ELN to meet organisational requirements, and managing back-ups. Costs are also associated with storing media used to create back-ups and data archives to store the data long term.

An alternative to the local server-client set-up is the use of a cloud-based service, which many ELN vendors provide. Instead of having a local server with a back-end for the ELN installed, the back-end is a service managed by the vendor or third-party storage provider and accessed over the Internet. The use of a cloud service may mitigate some of the costs by removing the need for additional hardware and on-going maintenance costs, although there may be an additional subscription fee to pay and these may rise with increasing use of storage space over time. The front-end to the cloud-based service is typically a browser, and this usually enables the ELN to be accessed on any platform and may widen the choice of ELNs that are compatible with an existing infrastructure for many organisations.

When ELNs are provided as a cloud-based service, significant concerns about security of data can be raised because the data is stored elsewhere, and users are therefore reliant upon the security controls of the cloud-based service provider. Whilst data stored within the organisation can be physically protected *via* lock and key and the use of a firewall, there is a perception that data held externally is more at risk. The nature of cloud architectures means that data from multiple organisations reside on the same servers outside of the administrative and physical control of the owner organisation. This 'multi-tenancy' leads to a risk of data visibility to other collocated users, and malicious code stored on the same server could potentially breach security measures in order to expose or corrupt data. The complexities of running cloud storage and processing systems can make data more vulnerable than in a traditional IT setup, and therefore greater security controls are required. Some cloud providers can accommodate a private cloud, where only users and data from a single organisation reside on a particular server, reducing some of the potential vulnerabilities. A lack of physical access to a server stored off site can potentially make sensitive data more vulnerable to insider attack, although access to the physical server is not needed to steal or corrupt data for users with the appropriate access.[28]

Another concern with some cloud-based services is whether the provider has the right to access the information and whether the uploader of the data still owns the data once it has been uploaded. In the earlier days of cloud-based services, the terms and conditions sometimes stated that the provider of the service owned the content that was uploaded to the service, although this is now unusual. However, the terms and conditions may still state that data and private information stored can be accessed by the supplier or third parties under certain circumstances.[15] A well-known case involved the Evernote electronic notebook, where a change in the terms and conditions of the software would enable Evernote's staff to access the private content of users' notebooks in order to improve machine learning algorithms. Evernote was forced to reverse their changes to the terms, asking users to opt in rather than opt out, but their staff still have the right to access the notes under other circumstances.[35]

The collaborative nature of ELNs compared with the private nature of paper notebooks often leads to concerns about the security and data protection of such systems. It can be argued whether a software system is more or less secure than a paper notebook. The paper notebook that never leaves the laboratory and is kept under lock and key when not in use, is probably as secure as paper can get, whereas the notebook that is moved from location to location may in fact be quite at risk from being forgotten on a bus, stolen from a bag, or dropped in a river, as previously mentioned. It seems that many people have the misconception that because sharing in an ELN is possible, that means that the content is always shared. However, the vast majority of ELNs are designed with security, data protection, and access control in mind, and all implement at least basic authentication and role-based access control. Those ELNs concerned with IP as a priority provide additional security measures including the use of certificates, public-key authentication, encryption, and hashing. These measures not only ensure that the users of the system are who they claim to be, but also that the content of the ELN has not been tampered with.[8] Where intellectual property is a concern, choosing an ELN that provides the appropriate reassurance about the protection of the contents of the notebook and any associated data is paramount.

For many users, especially in larger laboratories, ELNs may need to be integrated with a variety of other software systems, in addition to connecting to instruments for the purpose of data acquisition and experiment control. Integration with laboratory informatics and content management systems such as LIMS, Scientific Data Management Systems (SDMS), instrument data systems, and corporate information management systems such as Enterprise Resource Planning (ERP) software, supports the direct sharing of data between systems and provides support for planning and monitoring activities within the laboratory. There may also be value for some teams in integration with external repositories and services, for example, repositories of published literature, content-hosting sites, scientific data repositories such as the Protein Data Bank, Genbank, and services such as Flickr, YouTube, and Slideshare.[22] However, there can be a number of problems connecting an ELN to both hardware and software systems. A lack of standards for both protocols and APIs make it hard for ELNs to connect and integrate with these systems, especially when attempting to connect between new and legacy systems.[36] It is possible to make use of other tools to act as 'middleware' between these systems including the use of bespoke software, workflows, or Web services.[1] Adding this middleware for processing data sharing between software and hardware can also provide opportunities to add other new functionalities. The middleware is responsible for handling the communication between the different components in the system and may provide extended functions such as transforming data between different applications, managing business processes, and handling complex events. A key aspect of creating a middleware system to connect software and hardware is to ensure that it is flexible and 'loosely coupled' to enable better support and adaptability.[37] A loosely coupled system means that only one interface is needed to enable communication each time any new software or hardware components are added. There is likely to be significant additional work in planning and implementing a middleware solution in-house or with the help of an ELN supplier, even with the purchase of middleware software designed for the purpose.

The lack of accessibility of ELNs, in terms of availability both when and where the researcher needs it, are also a concern, particularly with the common requirement of an access to a network connection.[37] Laboratories are often less well supported than offices in terms of IT infrastructure, and a lack of connection to a network is common in the place where the ELN is needed most. For field scientists, connection to an online electronic notebook may be even more problematic because of limited cellular connectivity in rural environments. Many electronic notebooks have the ability to capture content while offline and upload it later, whilst previously captured content is often stored locally as well as remotely, but there are limitations on how much data can be stored on a device. Any information that has not been synchronised with server storage can be also be irrecoverable if the device is damaged or lost. Computers are also not suitable for use in all environments. In the field, they are heavy, bulky, and usually not robust enough to cope with wet, muddy, and rugged conditions; and in addition, they can be a temptation to thieves.[38] Even within the laboratory it may not be practical or possible to bring computers because of the risk of contamination by chemicals and other materials, interference with sensitive equipment, or magnetic equipment such as Nuclear Magnetic Resonance (NMR) spectroscopy machines, that can damage the hard drive of laptops, as well as other magnetic media, watches, and mobile devices. These problems all put at risk the precious notes and data stored within, especially in field environments where it may be difficult to regularly back-up the data safely.

Software is relatively volatile. New versions of software are relentlessly developed, not only ELNs, but also the operating systems and other prerequisite software on which they depend, changing the landscape, and often making previous software versions out of date and even obsolete within a few years. Companies developing the software often come into and out of existence in similarly short times; many smaller products are consumed by the bigger developers in the food chain or simply cease to exist. Open source projects may have initial momentum, but gradually fade as interest wanes or contributors move on. Scientists on the other hand need their research and their data to have a long lifespan. Our paper notebooks, as we know, can last for centuries to inform future readers about our endeavours. Even without that ambition, data for many organisations, both industrial and academic, are subject to retention requirements for several years, and sometimes even longer, after the end of funding or final publication for a project.[39] Data and notes in an ELN may need to be accessible for decades, an enormously long time in the world of software. There are a number of concerns with the storage of digital data in software. The software company may go out of business or be bought out, and the ELN software may cease to be supported. If this is a cloud-based service, not only may the ELN no longer be usable, but the data itself may entirely disappear as the software is 'sunsetted' by the developers. The concern is then whether the data can be exported in a format that enables it to retain its usefulness, ideally by being able to be imported into a new ELN. However, many ELNs use a proprietary data format, meaning that data cannot be extracted, and even if it could, it would be next to useless. Many ELN providers mitigate this risk by providing an export function into a different format, but often this format is PDF. The ELN content can be printed and archived, but it is not a particularly convenient format for use, and many important functions such as the ability to be able to conduct complex searches of the notes and access linked data will be lost. Because many ELN providers use a propriety data format, even if the current ELN enables export in a standard or self-describing format such as XML, there is no guarantee the data can be usefully and meaningfully imported into a new ELN. Even if the ELN developer does not go out of business, there may be a desire within an organisation to change to a new ELN provider in order to gain new capabilities, at which point similar problems are likely to be experienced. Fortunately, however, there has been much more cooperation in developing standard data formats for building solutions recently, even between organisations who had previously developed proprietary data formats for competitive advantage.[40]

For some research, using a computer may also be perceived as distracting and as interrupting the flow of normal activities. Using a computer often involves taking attention away from the task at hand to a greater extent than writing in a paper notebook. Electronic notebook designers need to address these concerns and take account of the affordances that paper provide so that they are not a revolutionary change from the day-to-day experience that the scientists are used to.[41]

For many potential users of ELNs, the largest barrier to overcome is the relative 'ease of use' of paper notebooks compared with electronic systems. Compared with electronic notebooks, paper notebooks are easier to use, easier to read, easier to transport, inexpensive, flexible, annotatable, persistent, readily available, they require no training, do not need batteries, and require minimal IT support.[2,32,42,43] In comparison, ELN software for taking notes is more difficult to use, takes more time, and is less flexible. Despite the fact that the vast majority of scientists are using computers and other technology in their day-to-day activities, there are still anxieties about the stability of computers. There are fears about the loss of precious data through hard drive failures or power surges, but also as stated by one researcher 'computers crash much more than books'. Computers can behave in unexpected ways for a variety of reasons, and just when you thought you understood how to use a certain piece of software, the developers release an update and sometimes it no longer works how it did before. Ease of use is very important to users and has a big influence on how users choose ELN software.[15]

Researchers sometimes appear to struggle to work out how to use an ELN because of what I call the 'blank canvas' effect.[44] When presented with an empty electronic notebook, they are unsure of how or where to start or what is expected. They are unsure what is needed to be recorded in an electronic notebook because they do not perceive it as an equivalent to the paper notebook. In user studies, even the presence of exemplar notebooks does not necessarily help with this, because users tended not to navigate away from their own notebook, and therefore gain no additional contextual information to work from. User studies demonstrate that researchers do not necessarily understand the capabilities of an ELN or see why they should use them because the beneficial functions in the ELN are not functions they would use in a paper notebook. For example, when encouraged to add links to related information from another notebook into their own as part of designing a new experiment, researchers declined to link to anything, even though the information was directly relevant to their own work and would have saved them effort in manually duplicating the method. Users were also put off by requirements that the system tried to impose upon them, which would not be needed in a paper notebook. For example, the need to save the notebook entry, to add metadata, and to create information externally in order to import into the notebook, such as diagrams and tabular data. ELNs are much more likely to suffer from the blank canvas effect than generic electronic notebooks because of their additional functionality and complexity adding to the perception that they are different to a notebook. Generic electronic notebooks tend to behave more like a paper notebook and are therefore perhaps more familiar and less intimidating.

We are becoming more accustomed to easy, intuitive interfaces through the use of products such as smartphones and tablet computers; a trend that is also being noted by a number of ELN vendors leading more to focus on making ELN interfaces simpler to use.[45,46] While vendors are busy improving their interfaces, many groups are taking advantage of free or cheap, easy to use technologies such as generic electronic notebooks. Many of these applications perform a subset of modern ELN capabilities. For example, generic electronic notebooks such as Microsoft OneNote, Evernote, GoodNotes, and WritePad provide note-taking capabilities; cloud backend applications such as Dropbox and Box.net provide data management and collaboration capabilities; and image capture and barcode scanners provide input sources.[12] However, these technologies do not necessarily provide all the functions that scientists desire, and there is therefore still an opportunity to create a good electronic notebook that has both the flexibility and functionality that scientists require.[47] To survive, any tool must be usable as well as functional. While large corporations may be able to afford extensive training and IT support to help overcome complexities in the software, smaller organisations, teams, and individuals need tools that are intuitive, reliable, and provide value beyond the paper notebook. In Chapter 8, we will examine a variety of techniques that can be used to improve the usability of software and provide some examples of designing ELN software based upon the needs of real-world researchers.

An additional concern with the use of digital notebooks over paper notebooks is the effect that they may have on the actual recording of the scientific record. In Chapter 2, we discussed the significance of the paper notebook as an artefact in its own right; the physical object as a record of participation and a trophy to be stored with pride on the shelf. Computer-based notes are perceived as dull in comparison. This may in part be because of the lack of physical presence of a digital record; you cannot hold an ELN in your hand in the same way and flick through the pages to bring back the memory of 'doing the science'. Another reason may be that the preciseness and structure provided by the computer takes something away from the uniquely personal nature of hand-written notes. Some authors have observed that their note-taking behaviour has changed as a result of switching from paper to digital. Although some feel that they have become more organised as a result, others lament that they no longer seem to write down their observations and feelings like they once did, focusing instead on the collection of data rather than capturing the personal experience.[48] Many electronic notebooks, whether they are discipline specific solutions for the field or the laboratory, make use of templates or forms to capture information, in contrast to the blank page of the paper notebook. There is a real possibility that using forms or templates may change the information that is recorded compared with a paper notebook, and this could have a significant effect on the scientific record produced as a result. In Chapter 9, we will explore the benefits and consequences of template use and what can be done to improve the quality of the scientific record through the design of electronic notebooks.

6.6 Conclusion

There are many benefits of electronic notebooks for capturing the scientific record as discussed above. Moving from using paper notebooks to electronic notebook software for recording scientific research can be a daunting prospect, especially if there is anxiety about the perceived complexity of ELN software. Individual researchers experimenting with ELN use have control of how they use the software and for what purposes, but for teams and organisations, the move will inevitably have a significant impact and result in changes to working practices which can cause considerable resistance amongst staff. Despite anxieties, the majority of users, once they can see the benefits of them in practice, tend to view electronic notebooks as a suitable replacement for their paper notebooks. A variety of factors, however, determine the likely success or failure of an ELN implementation project and in Chapter 7, we will discuss factors that can affect the success of such a project. We will look in detail at planning the implementation of an ELN, including creating a business case, managing the project, and overcoming resistance from staff. We will also discuss factors that might influence the choice of ELN provider and other considerations to take into account before selecting an ELN.

References

1. H. K. Machina and D. J. Wild, *J. Lab. Autom.*, 2013, **18**, 126.
2. C. L. Bird, C. Willoughby and J. G. Frey, *Chem. Sci. Rev.*, 2013, **42**(20), 8157.

3. S. Borman, *Chem. Eng. News*, 1994, 72(21), 10.
4. F. L. Macrina, *Scientific Integrity: Text and Cases in Responsible Conduct of Research*, ASM Press, Washington DC, 4th edn, 2014.
5. J. D. Myers, E. S. Mendoza and B. L. Hoopes, in *Proceedings of the IASTED International Conference, Internet and Multimedia Systems and Applications*, ed. H. Hamza, Acta Press, Anaheim, 2001, p. 334.
6. K. A. Badiola *et al.*, *Chem. Sci.*, 2015, **6**, 1614.
7. H. Wilhelm, in *NIP 22: The 22nd International Conference on Digital Printing Technologies*, The Society for Imaging Science and Technology, 2006, p. 308.
8. J. D. Myers, in *Proceedings of the 2003 International Symposium on Collaborative Technologies and Systems: CTS 2003*, ed. W. McQuay and W. W. Smari, Simulation Council, Inc., 2003, p. 13.
9. K. Shankar, *Arch. Sci.*, 2006, **4**, 367.
10. M. Kim, *Duke Law Technol. Rev.*, 2002, **1**(1), Available at https://scholarship.law.duke.edu/dltr/vol1/iss1/43 (accessed Sep 2018).
11. A. Tabard, W. E. Mackay and E. Eastmond, in *Proceedings of the ACM 2008 Conference on Computer Supported Cooperative Work – CSCW '08*, ACM Press, New York, 2008, p. 569.
12. M. Elliott, *Scientific Computing*, 2004, Available at http://www.scientific-computing.com/articles/2004/08/its-not-about-paper? (accessed Apr 2016).
13. R. C. Hice, *Scientific Computing*, 2009, Available at https://www.rdmag.com/blog/2009/05/roadmap-clear-definition-eln (accessed Sep 2018).
14. P. Naskrecki, in *Field Notes on Science and Nature*, ed. M. R. Canfield, Harvard University Press, Cambridge, 2011, p. 201.
15. S. Kanza, C. Willoughby, N. Gibbins, R. Whitby, J. G. Frey, J. Erjavec, K. Zupančič, M Hren and K. Kovač, *J. Cheminf.*, 2017, **9**, 31.
16. K. Taylor, in *Collaborative Computational Technologies for Biomedical Research*, ed. S. Ekins, M. A. Hupcey and A. J. Williams, John Wiley & Sons, Inc., Hoboken, 2011, p. 303.
17. Available at https://www.perkinelmer.co.uk/category/notebook (accessed Oct 2018).
18. Available at http://accelrys.com/products/unified-lab-management/biovia-electronic-lab-notebooks/workbook/index.html (accessed Oct 2018).
19. Available at https://benchling.com (accessed Oct 2018).
20. Available at https://www.labfolder.com (accessed Oct 2018).
21. Available at https://scinote.net (accessed Oct 2018).
22. A. Milsted, J. Hale and C. Neylon, *PLoS ONE*, 2013, **8**(7), e67460.
23. A. E. Day, S. J. Coles, C. L. Bird, J. G. Frey, R. J. Whitby, V. E. Tkachenko and A. J. Williams, *J. Chem. Inf. Model.*, 2015, **55**, 501.
24. Available at https://www.elabftw.net (accessed Oct 2018).
25. Available at http://openwetware.org (accessed Sep 2018).
26. F. Echtler, M. Häussler and G. Klinker, in *Proceedings of the 28th International Conference: Extended Abstracts on Human factors in Computing Systems – CHI EA'10*, ACM Press, New York, 2010, p. 3439.

27. A. Tabard, J. Hincapié-Ramos and J. Bardram, in *Proceedings of the SIGCHI Conference on Human Factors in Computing Systems - CHI '12*, ACM Press, New York, 2012, p. 3051.
28. S. Guerrero, G. Dujardin, A. Cabrera-Andrade, C. Paz-y-Miño, A. Indacochea, M. Inglés-Ferrándiz, H. P. Nadimpalli, N. Collu, Y. Dublanche, I. De Mingo and D. Camargo, *PLoS ONE*, 2016, **11**(8), e0160428.
29. Available at http://opensourcemalaria.org (accessed Oct 2018).
30. M. N. Robertson, P. M. Ylioja, A. E. Williamson, M. Woelfle, M. Robins, K. A. Badiola, P. Willis, P. Olliaro, T. N. Wells and M. H. Todd, *Parasitology*, 2014, **141**, 148.
31. F. Rudolphi and L. J. Goossen, *J. Inf. Model.*, 2012, **52**(2), 293.
32. N. H. Goddard, R. Macneil and J. Ritchie, *Autom. Exp.*, 2009, **1**, 4.
33. K. Shankar, *J. Am. Soc. Inf. Sci. Technol.*, 2007, **58**(10), 1457.
34. C. Lepere-Douard and D. Allard, *Technological survey on LIMS and ELN in Life Sciences*, University of Nantes, 2012.
35. Available at http://www.wired.co.uk/article/evernote-cancels-privacy-policy-read-user-note-data (accessed Sep 2018).
36. A. Laugier, T. Steele and F. Falco, *Lab. Autom. News*, 1997, **2**(1), 22.
37. D. J. Drake, *Drug Discovery Today*, 2007, **12**(15-16), 647.
38. G. B. Schaller in *Field Notes on Science and Nature*, ed. M. R. Canfield, Harvard University Press, Cambridge, 2011, p. 19.
39. K. Briney, *How Long Should You Keep Data?*, 2013, Available at http://dataabinitio.com/?p=172 (accessed Sep 2018).
40. P. Van Eikeren, *Org. Process Res. Dev.*, 2004, **8**(6), 1015.
41. B. Lass, *Scientific Computing*, 2011, Available at http://www.scientific-computing.com/articles/2011/07/implementing-electronic-lab-note-books-part-3#.UguhpRZrHN4 (accessed Aug 2013).
42. W. E. Mackay, G. Pothier, C. Letondal, K. Bøegh and H. Sørensen, in *Proceedings of the 15th Annual ACM Symposium on User Interface Software and Technology*, ACM Press, Paris, 2002, p. 41.
43. P. Luff, C. Heath, M. Norrie, B. Signer and P. Herdman, in *Proceedings of the 2004 ACM conference on Computer supported cooperative work - CSCW '04*, ACM Press, New York, 2004, p. 523.
44. C. Willoughby, C. L. Bird, S. J. Coles and J. G. Frey, *J. Chem. Inf. Model.*, 2014, **54**(12), 3268.
45. E. Dahlstrom and J. Bischel, *ECAR Study of Undergraduate Students and Information Technology*, EDUCAUSE Center for Applied Research, Louisville, 2014.
46. A. King, *Chemistry World*, 2013, Available at https://www.chemistryworld.com/culture/notebooks-go-digital/6187.article (accessed Sep 2018).
47. J. Giles, *Nature*, 2012, **481**, 430.
48. J. L. Reveal, in *Field Notes on Science and Nature*, ed. M.R. Canfield, Harvard University Press, Cambridge, 2011, p. 187.

CHAPTER 7

Moving from Paper to Electronic Laboratory Notebooks

7.1 Introduction

In Chapter 6, we took an in-depth look at Electronic Laboratory Notebooks (ELNs) and the benefits that they can bring to individual scientists and organisations involved in scientific research. We discussed the benefits, but also some of the issues that may lead researchers and organisations to be reluctant to make the change from paper notebooks to ELNs.

Despite the reservations that may exist, ELNs can provide enormous value to laboratories, and for many people, the move from paper notebooks to ELNs is seen as inevitable. For large projects at least, there is an increasing need to be able share data with multiple groups across multiple sites using a single repository. Beyond sharing, there are many other benefits of ELNs as we have discussed, such as facilitating the long-term storage and preservation of experiment records; ensuring standard operating procedure compliance, providing interfaces to instrumentation, supporting Intellectual Property (IP) protection, and facilitating collaboration and open science. Perhaps most importantly from a management point of view, ELNs can give a unique insight into the operations of the laboratory, allowing experiments in progress to be tracked, tasks to be assigned, and efficiency to be assessed.[1] The question for many is one of when, rather than if, laboratory scientists will convert to ELNs.

The implementation of an ELN in any kind of laboratory is no simple task, even beyond any concerns about usability. Significant resources in terms of finance, time, and staff are required to ensure that the chosen

Recording Science in the Digital Era: From Paper to Electronic Notebooks and Other Digital Tools
By Cerys Willoughby

Published by the Royal Society of Chemistry, www.rsc.org

solution successfully meets the requirements of the organisation, but also successfully supports the staff in the laboratory. The move from paper to digital is a significant transformation and will inevitably change the way that scientists work. The change does not just affect the way that notes are taken and data collected, but is likely to transform workflows and interactions within the laboratory. Such transformations inevitably cause anxiety and resistance, and the implementation of an ELN project is in many respects the same as the implementation of any IT project within an organisation. The way that such projects are managed, and in particular the way that they are communicated to staff, can have a big impact on the success or failure of the project. In this chapter, we will discuss some of the issues that need to be considered and some best practices to follow when planning the implementation of an ELN.

7.2 Making a Business Case for an ELN

A first step in the plan for implementation is to make a business case for the ELN, weighing the potential benefits of an ELN system with the likely costs. As mentioned in Chapter 6, depending on the scale of the project, there can be some significant costs involved to acquire the appropriate software licenses and any additional hardware and infrastructure if it is required. There are also other costs associated with the implementation, including staff time away from other tasks, the costs of consultancy if any tailoring is required that cannot be done in-house, and any costs associated with training at the end of the process. Cloud-based systems can lower some costs by reducing the need for specialist support in-house, because the installation and configuration is performed externally. Hardware costs can also be reduced because cloud-based systems store data and back-ups externally. User interfaces for cloud-based systems are also often Web-based, enabling users to run the software on their existing hardware or devices, so reducing the amount of new equipment and software that may be required, such as servers and operating systems.

Calculating what the savings benefits of implementing an ELN will be over time can be difficult, but there are usually savings to be made in both money and time. Financial savings can come from reducing the amount of resources used, in particular, paper and associated materials, for example, not just paper notebooks and pens, but also reams of printer paper and toner used to print out data, and the costs of storing these materials. Time savings can come from improved project efficiency. Time can be saved on transcribing data between the notebook, instruments, and analysis software, reducing errors, and making it easier to troubleshoot problems. ELNs can reduce the amount of duplicated effort by enabling members of a team to easily search for similar experiments that have been attempted before, and improved knowledge management can reduce significant time in research and reporting. In researching and experiment planning, an ELN can make it easier to find and make use of information from previous experiments. For reporting, it is quicker to collect information that can be used for decision making, regulatory requirements, publication, and IP submissions. For quality assurance environments, enforcing compliance to standard operating procedures can save duplication and improve record quality because users are prompted to perform instrument calibrations and record information at the appropriate time. Time can also be saved in the review and witnessing process, because reviews can take place remotely and at times that are convenient for the reviewers. The use of an ELN with other systems such as Laboratory Information Systems (LIMS), Laboratory Execution Systems (LES), and inventories can enable extra value to be gained by better utilising those systems.

7.3 Overcoming Resistance

There may be a variety of reasons why a particular laboratory chooses to implement ELNs. ELNs may be chosen to reduce the amount of paper used and to combat the expenditure of notebooks and storage. They may be chosen because they will make work in the lab more efficient or more effective by improving the quality of note-taking, automating data capture, or ensuring compliance to procedures. They may also be chosen to improve knowledge and management sharing across a team, an organisation, or to facilitate interactions with collaborators. They may be adopted to improve education and learning within a university. There may be other reasons in mind why an ELN is the right step for any particular group, but it is important to have a common understanding in an organisation of what the reasons are for choosing an ELN. What value will the ELN bring not only to the organisation, but also for the individuals within the organisation? Communication of the capabilities and benefits of an ELN is extremely important for overcoming resistance. Staff do not necessarily know or understand the capabilities or benefits of an ELN before the implementation of the project, and it is important to be able to effectively articulate the benefits of the ELN for both the organisation and the individual. An effective way to communicate the benefits and capabilities of ELNs is to use an active demonstration of the system in action, especially if the examples provided are relevant or familiar. Using active demonstrations also gives staff the chance to raise questions and concerns that can be considered and addressed at an early stage in the project.

Resistance to new information systems is common, and frequently leads to under-use of the system, or even complete avoidance of the system.[2] A major reason for the failure of information systems implementation and acceptance has been an inability to understand and manage the expectations of staff. In addition to staff not understanding the benefits of the system, they may view it as threat if they feel that it may increase their workload, cause a loss of control or hinder some aspect of their work, or be more difficult to use than the current way of working. Thought needs to be given to the impact that the technology will have on staff, both in terms of how it will benefit them, but also how it will change the way that they work. Will it change their working practices, will they need to spend time learning the new software, and will that take them away from other tasks? The perceived ease of use of an ELN and the associated work practices are important here because software that is perceived as easier to use is also perceived as more useful. Software that is more complicated will be perceived as less useful and more of a waste of time, leading to more resistance. It has been found that the perceived complexity or rigidity of work routines is even more important than the perceived usability of the system itself, and it is therefore important that any work routines that change as a result of the system still support the perceived goals of the user. If work routines are likely to change significantly, then consulting with staff about the changes gives the opportunity for staff to provide feedback on what is likely to work and not work, and for potential problems to be ironed out before the implementation.

The adoption of a technological solution requires a cultural change within an organisation.[3] It can be a significant challenge to win the hearts and minds of users with a new system. If the members of the organisation are not willing to make the change, then the technology will not be effectively utilised and the proposed benefits are unlikely to be seen. As an example, one of the big differences in the way of thinking between paper notebooks and ELNs is the ownership and access to notebooks. As we discussed in Chapter 3, researchers in particular are used to the idea of the notebook as a personal object within which they control what and how they record. Hostility towards ELNs can not only arise because of anxiety about others being able to read and 'judge' the content of the notebook, but also because they may need to change the way they work in order to accommodate a different way of recording. The cultural change required is a transformation from thinking about 'my data' to 'our data', and an understanding of how that can be a benefit and not a frightening loss of control. A challenge of implementing and adopting ELNs is that most of the benefits come after some time using the ELN, when data from previous experiments can be searched and compared.[4] Initially, users may not only find it difficult to see the advantages of an ELN, but may also perceive it as unfamiliar and complex. An example comes from the X-Ray Group at Southampton University, where the team were reluctant to use an ELN because of fears about the complexity of the system and that their mistakes would be visible to all. However, very quickly, once they began to add content to the ELN, it became a hub of collaboration. The advantages readily became apparent, not only in terms of accessing data but also in reducing their original inhibitions and leading to the capture and sharing of more information than was possible in a paper notebook.[5] This reluctance to expose the 'raw' notes to others is a commonly recurring theme when discussing resistance to ELNs, but the experience of many that do use ELNs is that this fear is quickly overcome by the benefits of being able to share notes and gain rapid feedback. Sharing is usually reciprocal, so the researcher is also able to benefit in the same way from the shared notes and feedback of others.

Achieving cultural change and acceptance of technology is easier if affected individuals perceive problems with the current working practices. If problems genuinely exist with the current ways of working, then presenting a solution for the problems becomes a much easier task than trying to enforce a solution to a problem that is not perceived to exist. As an example, the witness and review process with paper notebooks can often be inefficient, causing frustration for both researchers and reviewers. It may be difficult to arrange reviews of the paper notebook because the reviewer is located at a different site, and the delay in review causes the researchers to make mistakes that are unchecked or to take the research down a different path than intended, leading to wasted time. In some cases, researchers may not be able to make progress because they cannot proceed without the appropriate review. It is also the case that researchers are frequently duplicating failed experiments already performed by others because they have no way of knowing what has been attempted before. Sharing of notebooks through an ELN provides solutions to these genuine problems that can easily be articulated. The reviewer will be able to view the notebooks whenever they need to, and not wait until they are on site. This means that feedback can be provided much more quickly, saving the researcher work and wasted effort because they can quickly be directed onto the right path and feedback can be provided directly in the notes to provide closer support if they run into problems. Reassurance can be provided that the notebook is private to the author and only those individuals who have a right to access the information. The researcher is also able to avoid duplicating effort by reading the notebooks of others in order to compare results and check work that has previously been done. Eliciting information from staff about problems and inefficiencies in the laboratory, and then effectively communicating how the proposed solution will solve the problem, supports both the organisation and the individual, and makes it more likely cultural change can be achieved. Gathering feedback on perceived problems can also help to identify key functionalities that need to be sought from proposed ELN providers.

If there are no apparent problems that need to be solved, then the proposed ELN solution must provide significant new benefits to be accepted by users. Even if the current workflows seem to work well, there are always opportunities to provide benefits with new functions that may never have been considered before. There are many things that are just not possible to do in a paper notebook that can have significant benefits, such as linking directly to data or integration with other information systems, such as sample control and biological or chemical data repositories. A discussion about what staff would want to be able to do, that they currently cannot do, can both highlight the benefits of an ELN solution and also provide requirements for capabilities that can help with planning the implementation and selecting an ELN provider. Any desired functionality should be backed up with an understanding of why it is desirable, and whether it is generally useful. Functionality that is valuable to many users can be given a higher priority to implement than functionality that would be used by fewer staff, but there may also be cases where functionality that would be used by only a small number of staff may still provide significant efficiencies or benefits for the organisation as a whole. It is also important to understand the feasibility of the desired features, if something is very complex to implement, it could potentially come with a high cost in terms of personnel and time. It can be easy to produce a longlist of desired functions, but comparative cost and value of additional capabilities for the organisation need to be weighed to ensure feasibility and determine the appropriate priorities.

7.4 Factors Impacting Success and Managing the Project

A lot of work is required up front to make an ELN successful and there will inevitably be teething problems because software systems are not perfect and can often be difficult to use. It is important to have a clear understanding of what is desired from the system, together with clear sponsorship of the project. A lack of vision and sponsorship often lead to project failure. Projects with inadequate resourcing and without proper support are also unlikely to succeed. A project may be more likely to succeed in some teams than others. For example, it can be easier to demonstrate the value of an ELN to teams who produce large amounts of data from individual experiments or where hundreds of experiments with slightly tweaked conditions are run, compared with teams where little digital data is produced and each experiment is unique. ELNs also have greater benefits for some individuals than others. For example, digital systems can be easier to use and facilitate better organisation for users with special needs. As examples, users with dyslexia or osteoporosis may find it easier to use a computer to create legible notes than they do using paper. Users who are already comfortable using computer systems to organise their work are likely to adapt faster and get more value from an ELN system, and will require less reassurance than users who are anxious about technology.

The success of any IT system implementation is dependent upon the commitment and knowledge of staff.[6] The goals of the implementation are the improvements that will be achieved by using the software and not the installation of the software itself. Staff must have a clear vision of what is intended to be achieved through the project, and for large projects, visible support from senior management throughout the project is essential. For smaller projects, a project champion may be sufficient, but for larger projects, a separate project manager and steering committee of stakeholders may be needed to manage resources, make decisions, and deal with problems. Ideally, the project champion should be a recognised expert within the organisation with an enthusiasm for the technology;

users are more likely to raise concerns and ideas with a trusted colleague and accept the need for change.[7] The implementation may result in permanent changes of roles for very large implementation challenges. For example, it may be useful to create the role of ELN manager for large implementations or a continuation of the project champion role to oversee on-going use, updates, and maintenance of the system. For large implementations, as mentioned previously, there are on-going administration tasks that will require continued IT support.

An ELN project extends beyond the initial installation, and further work is needed to configure the ELN to meet the requirements of both the users and the organisation. More effort is then needed to maintain it and keep it useful. Software maintenance is a lot like doing housework, it is never finished! There are everyday administrative tasks to manage, such as managing user access, performing back-ups, and curating the data, but there will always be updates to fix bugs and new versions to add new features. Both fixes and updates need to be tested and installed, and there may be additional tasks that are needed to migrate ELN contents to a new version. If the ELN integrates with other software or hardware, then work will be needed if new or updated software or hardware is introduced to the system. If the ELN makes use of an enterprise database solution, then a dedicated expert resource may be required to manage the day-to-day administration of the database. For an individual running an ELN on their own machine, these tasks may be relatively trivial annoyances, but for a large organisation, these day-to-day tasks may require a number of expert staff to keep everything running smoothly. As mentioned previously, some of these burdens, such as applying fixes and managing the database, can be taken away by making use of a cloud-based ELN if that is suitable for the organisation.

It may be possible to 'stage' the rollout of an ELN system if all staff do not need to adopt the system at the same time or if not all of the functionality needs to be implemented immediately. For example, if there is no reason to have all staff use the ELN at the same time, then a pilot can be run with a subset of users who may be volunteers. These volunteers can be used to test and refine the new system and act as champions for their teams or departments to encourage other users to adopt the system. They can also be used to help demonstrate the benefits of the system to other users and help provide education as new users are added. The system can then be scaled up to add new users over time. If not all of the functionality needs to be implemented at once, then some basic tasks can be implemented initially, leaving more complex integration and customisation until a later stage when requirements and capabilities are better understood. Staging is also an effective way to manage the development of capabilities. The high priority items can be implemented first, and new capabilities added over time as resources become available or important new features identified.

For any project, it is important to set a realistic timeline to achieve the desired goals. Milestones should be set for the accomplishment of significant steps in the project. Breaking down the implementation of large-scale projects into smaller pieces can make the process simpler and reduce the risks.[7] In the initial stages of the project, the appropriate skills for the project team needs to be identified together with the reporting processes that will be followed to keep management and other stakeholders apprised of progress. The first task of the project team should be to analyse the current laboratory processes and interactions with other groups or departments, so that current workflows and data inputs can be mapped onto the functionality of the ELN. These workflows include the creation and movement of information in the laboratory and between other groups, but also include physical assets such as samples and materials. Where appropriate, investigations should be made into how the ELN can be integrated with instruments and other software systems, including repositories and LIMS used in the laboratory or organisation, and whether any parts of the workflows may benefit from automation. Design of the proposed system including prototyping can then proceed, followed by any installation, tailoring, and configuration for the server component, along with testing with real data, before user testing to ensure the system meets the user requirements. Prototyping can be useful to demonstrate the capabilities of the system and to elicit feedback from users to fine-tune the design and configuration of the desired solution. The final phase includes installation of the front-end components and providing user training and support.

It is important to include mitigation within the schedule for problems that may be encountered or changes to requirements that come up during implementation. A risk register can be created for risk management in the project to identify potential problems that may cause delays in the project, for example, staff holidays or sickness, issues with acquiring hardware or configuring software, or changes that may be needed as a result of problems found during testing. There will always be unexpected events that can impact on a project, but considering potential problems in advance, however unlikely they may seem, can at least provide a back-up plan or alternative options if things do go wrong. For example, is there a disaster recovery plan for a major failure of hardware or loss of equipment due to a fire, do staff have the appropriate knowledge to continue the project if a senior technical member of the project was to change jobs, and can the project continue if the ELN provider goes out of business?

Without a fixed deadline and milestones for a project, it can be difficult to assess progress and the end date can slip backwards over time. An effective way to both assess and communicate progress is to demonstrate the progress at each milestone using walkthroughs with stakeholders and users. Walkthroughs also provide the opportunity to receive feedback on the project design and implementation and to reflect on lessons learned since the previous milestone. If a technical project is not appropriately managed, then the priorities and organisational requirements can sometimes get lost amongst the technical challenges. The use of walkthroughs also helps to ensure that the project is overseen and that progress is being made in the right direction.

7.5 Understanding the Requirements of the Project

The broad vision of the project needs to be understood up front, but the specifics of the implementation are not necessary to know in the beginning. The objectives for the project need to be agreed and understood. For example, what are the limitations and challenges of the current system, and how can the ELN system help to improve them? In addition, it is important to understand what is valuable about current ways of working to make sure that the benefits of those ways of working are not lost when the project is implemented. In order to have a complete vision, it is important to be able to measure what the benefits are when the system is implemented. For example, an ELN may be able to increase the number of experiments that can be run in a certain time or may help make work processes more efficient. There may be opportunities to integrate with instruments, add automation, or connect with other systems that could improve the efficiency of the laboratory by saving time in data entry or improving reporting. Resistance to the new system can be reduced if the ELN reflects the familiar workflows that users are currently following, and ideally, an ELN should mirror the user's workflow.[3] For example, there may be standard operating procedures or paper forms that can be converted into templates in the new system to facilitate consistency between the current work practices and the new environment.

It is important to define the requirements for the organisation and individuals before attempting selection of an ELN. Having the requirements defined up front makes it more likely that the chosen ELN will meet the needs of the organisation. Defining the requirements also makes it easier to hold conversations with different vendors, especially if you expect to make use of an external service provider for implementation or customisation of the ELN. To define the requirements for the system, you need to identify the user roles of those who will be using the ELN. The users include both those inside of the laboratory creating the information content and those outside of it needing to access the information within the ELN. The needs, goals, and work practices of different users are likely to be quite different from each other, and it is therefore important to make sure they are all considered. An ELN that is excellent at capturing the experiment data but rubbish at generating reports needed by management or exporting data in a format suitable for the IP lawyers will be unsuitable for organisations where some staff are reliant on those functions. It is important to understand what the information requirements of different groups are, so that any shared information can be captured and shared in a way that facilitates both groups. If there is a need to standardise the way information is captured in order for sharing to be effective, then the potential consequences of this need to be considered, for example, staff may need additional training, and there may be potential resistance due to changes from current practices.

To generate requirements, it is also necessary to be able to understand the goals and work practices of each of the user roles that will use the ELN.

One way to do this is to carry out a task analysis to identify the goals and workflows that are followed by staff when everyday tasks are completed. A task analysis can be conducted by observing the routines of staff and asking them to describe what they are doing and why, but alternatively, it is also possible to interview staff and ask them to talk through their work practices. It is important to investigate what data sources they use, what data and notes are recorded before and during the experiment (or other work activity), and how the data is analysed and stored after the results have been obtained. In laboratories where activities stem from work requests or new samples, the information flow is likely to be centred around these artefacts as the work request or samples move between different parts of the laboratory or between different personnel. It can be useful to understand whether each experiment or activity is documented independently or whether multiple users may need to contribute to a single record. It is also useful to know whether the records vary in content between each experiment or activity, or whether the same information is always captured. If experiments or analyses are usually similar, then the use of standard templates or cloning tools to replicate the details of experiments or analysis parameters can save time and improve efficiency compared with the paper-based approach.

It is also useful to record what interactions the user has with others, in terms of collaboration with other team members, reporting to managers, and interaction with other groups outside the laboratory. Investigating the flow of information between teams and departments can help to identify data that needs to be shared and search queries and reports that are likely to be useful to particular areas of the organisation. Understanding what information is used and for what purpose can ensure that all the required information is captured. If there are reports or queries that are currently not possible, but which would be useful for the organisation, then there is the opportunity to add the capture of the missing data through the use of templates or new workflows. Defining common terminology and vocabularies can promote consistency in the use of the ELN and facilitate information sharing across different teams. It is also worth thinking about the potential benefits of sharing experiment protocols, results, and outcomes that are not currently shared to reduce duplication and help with planning new experiments. Access to content that will become available in the ELN may change some working practices and influence the requirements for the ELN.

It is also useful to consider where the data is currently being stored and whether all elements of raw and processed data should be captured in the ELN, including graphs or processed images. If some data is currently stored in repositories or other information systems, then it is important to decide whether all data should reside in the ELN or remain in the current locations and how linking between the two will be managed. The location of the ELN back-end may be important to these questions in terms of who will manage the data in storage and whether there are associated costs with keeping content in one system over another. In the long term, retention of the data and associated records needs to be considered. Content from active projects may be better stored in the ELN, but perhaps transferred to a different system for archiving. It may be possible for the ELN system to automatically archive records of a certain age or a manual process may be needed to ensure that data is stored in the appropriate location. For some organisations, there may be no repositories for digital data already present, and these may need to be set-up and managed as part of the ELN project, or alternative external repositories identified. New procedures and training may need to be introduced for data management and the long-term preservation of laboratory records.

Another consideration with an ELN set-up is where the users are located and where they will use the ELN. If the users are all located in a single room in a single building, then a much simpler hardware set-up is likely to be needed compared with a large organisation where individuals are located across many rooms in many buildings, and even potentially across multiple sites in multiple countries. Machine access and network connections can vary significantly depending on the location of use. As previously mentioned, the network connections and hardware availability are often better within offices than within the laboratory, but if the ELN will be used mostly within the laboratory, then additional hardware and network infrastructure may need to be purchased, installed, and configured before the ELN can be set-up. Any restrictions such as unsuitable environmental conditions or a lack of power outlets also need to be considered, together with connectivity requirements for instruments and other devices within the laboratory. Making use of mobile devices, such as laptops and tablets, can enable ELN access in whatever locations the user needs. Access locations may also influence whether the backend servers for the ELN may be better located on-site or in the cloud.

It is important to consider what the potential interactions may be between the ELN system and any instruments. Interactions may include calibration of equipment or recording of data, but there may also be opportunities for automation, for example, by enabling the ELN to directly control the device or to automatically upload data into a notebook. Integration with other systems and instruments can be a complicated aspect of ELN implementation and identifying whether pre-existing systems use particular standard formats or provide an API can inform which ELN providers may facilitate the integration, either through native support for the standards or by providing a way to programmatically connect to the hardware and software. One of the difficulties with ELNs which have no standard APIs, is that any functionality to connect components needs to be written for that specific API, and the architectures can change over time with updates to the software requiring rewrites for the code each time it changes.[8] ELNs that implement the standard Laboratory Equipment Control Interface Specification or other standard protocols are likely to be easier to connect to instruments without additional configuration or programming.[9] The ability to connect to existing data sources can also be helpful in terms of prepopulating information and avoiding transcription errors. For example, if a LIMS system is used for sample management, information about registered samples could automatically be uploaded into the ELN. Integrating with a materials inventory could enable information about batch numbers and expiry dates for materials to be accessed from the ELN. Integration can reduce errors by enabling the ELN to give the user warnings if stock is low or the batch has expired or does not meet quality control requirements.[10] If a direct connection between an information system and the ELN is unfeasible, then other means could be considered for creating links between data and records, for example, barcodes could be scanned to import sample information to an ELN to reduce the risk of transcription errors and maintain consistent IDs across the different systems.

There are also likely to be non-functional requirements for the ELN system that need to be considered. For example, the organisation may require certain levels of security and data protection to be met, which may influence whether the back-end server must be stored on-site or if cloud-based storage is acceptable with encryption. Ensuring that data and records are safe is critical, and an important requirement to consider is how back-up and disaster recovery will be managed. Specific software, hardware, and staffing may be required to implement a solution for the protection of data and resources in an ELN, including long-term archiving. As mentioned previously, electronic records stored in ELNs may need to be kept for many years for the purpose of IP protection or compliance with legal requirements. It is necessary not just for the record to be safely stored and retrievable in an appropriate format during this time, but also that it is intact, unchanged, and verifiable through the use of record authentication and digital signatures. The provenance of the record and an audit trail of changes must persist alongside the content of the record for authentication and verification purposes. Many ELNs provide a mechanism to export content from the ELN for archiving and storage purposes including PDFs and XML that can store the content of the record in a retrievable and human-readable form, so even if the software itself is not in use, a reliable format is still available. It is important to consider both the physical storage and the data management processes that will be used to store the electronic records for IP and regulatory purposes. If the organisation, or any part of it, needs to adhere to any particular regulations or standards, then these will need to be taken into account in selection of an ELN provider to ensure appropriate compliance. Adherence to regulatory compliance and data protection practices are usually specified by the ELN provider and relate both to ELN software capabilities and the business practices of the supplier themselves. Common standards, regulations, and guidelines that ELN suppliers claim compliance often include the following:

- Compliance with Good Laboratory Practices (GLP) from The Organisation for Economic Co-operation and Development (OECD).[11]
- EudraLex: The Rules Governing Medicinal Products in the European Union, Volume 4, Good Manufacturing Practice Medicinal Products for Human and Veterinary Use, Annex 11: Computerised Systems.[12]
- Title 21 CFR Part 11.[13]
- ISO 9001: Quality management systems.[14]
- ISO 13485: Medical devices.[15]
- ISO 15189: Medical laboratories.[16]
- ISO 17025: General requirements for the competence of testing and calibration laboratories.[17]

EudraLex is a set of European guidelines primarily for pharmaceutical and biotechnological companies and claimed compliance is usually found in ELNs that have been designed specifically for these industries.[18] Annex 11 of the guidelines provides guidance for the use of software systems as part of the manufacturing process in the pharmaceutical and biotechnology industries, and includes guidelines on security, data management, digital signatures, and policies for validation and operating IT systems.[12] The OECD GLP specifies good practices for data management for software systems including authentication, data security, integrity, confidentiality, and archiving.[19] The OECD GLP also specifies that the computerised systems used for managing electronic records must be from a recognised supplier and have been validated for adherence with their Principles of Good Laboratory Practice.[20] The organisation acquiring the software is responsible for undertaking and documenting acceptance tests to ensure the software meets the requirements of GLP. Under the GLP, management are responsible for making sure the system is suitable for the intended purpose and for defining policies and procedures to ensure that systems are developed, operated, and maintained in accordance with GLP principles.[21] The principles also cover practices beyond the storage of notes and data, for example, the handling and storage of samples, standard operating procedures, and study plans that can help with recording requirements for environments that need to manage these elements. Even if compliance to the OECD GLP is not required by an organisation, it contains a lot of useful guidance about best practices that can help elicit potential requirements for any organisation.

Title 21 of the Code of Federal Regulations Part 11 contains the regulations concerning electronic records and electronic signatures as formulated by the US Food and Drug Administration (FDA).[13] Compliance to the title is required by organisations producing pharmaceuticals, medical devices, and other biotechnology that are regulated by the FDA and relates to ensuring that electronic records are as secure and reliable as paper records. The regulations provide requirements for the use of electronic signatures, authentication, data integrity, audit trails, data management procedures, access to records in human-readable form, record retention, and archiving. The requirements for digital signatures and maintaining integrity of the electronic records are extensive and mandate that electronic signatures must be as reliable as hand-signed signatures.

There are a variety of ISO standards that relate to laboratories. ISO standards provide requirements for documentation and process management associated with laboratory and other projects, including standard operating procedures, standard formats for documents, and data management, including access, audit trails, retention, and review. Many ELN suppliers themselves have implemented ISO 9001 to indicate their adherence to quality standards and to assist with the supplier-validation processes required by regulations such as CFR Title 21 and the OECD GLP.

There are many potential ELN solutions to choose from and understanding the needs of the organisation are key for determining the most suitable alternative. If the organisation has very specific needs that cannot be meet by an existing ELN in the marketplace, then it may be necessary to build a bespoke ELN solution in-house or to outsource the development of such a solution.[3] Such a venture is likely to require significant levels of skill and resources, and an excellent understanding of the requirements for the organisation so that an appropriate system is developed. For the majority of organisations selecting from one of the many available options, with potentially some additional tailoring or customisation, is a sensible approach. There are likely to be a number of systems that provide a reasonable fit for any organisation and each should be evaluated against the established requirements for the organisation and its users. Selecting the ELN that most closely matches the requirements is likely to speed up adoption and simplify the process of implementation, but it may also be possible to tailor a solution or pay a commercial provider to add important functional requirements that are missing from a preferred solution. It is necessary to weigh the costs of customisation with the potential long-term benefits. If additional development or customisation is required, then it is important to develop a good working relationship with the vendor or development team to ensure that specified requirements can and will be met by the ELN solution in the appropriate timescale. Specific ELNs tailored to particular scientific disciplines or industrial activities are likely to more closely match specific requirements for those areas than generic ELNs. However, some generic ELNs also have the option of adding modules or plug-ins to facilitate more specific use and may provide similar functionality at a lower cost.[8]

7.6 Selecting an ELN Provider

There are many factors to consider when selecting an ELN to support a particular research team or organisation. The needs of an individual or a small team will be quite different to those of a large, multi-laboratory organisation. Budget, scientific discipline, laboratory activities, and the defined requirements for a system will all have an impact on selection.

Large ELN implementation projects, where funding is not an issue, are likely to be best supported by developing a close relationship with a suitable ELN vendor in order to get extra support and customisation. Many of the higher-end ELNs provide additional functionality, such as those commonly associated with a LIMS system, that may benefit organisations that do not already have these systems in place, or where the system needs upgrading. For smaller projects, selecting a simpler 'off the shelf' option may be more appropriate.

There are many discipline and industry specific ELNs that are more likely to suit the needs of departments or organisations that match these environments, for example, chemistry, biology, and pharmaceuticals. For highly specialised work, an ELN with a high level of customisation and extendibility may be most beneficial so that specific needs of the organisation can be met by developing a custom solution rather than trying to work within the limitations of the ELN. For organisations or teams where the research is more varied and where more flexibility may be required, a simpler and more generic system may be more desirable.

The requirements for a specialised system need to be well defined, in particular so that potential suppliers and the implementing organisation both have the same understanding of the proposed solution and how it will meet their needs. The basic functionality provided by different ELN providers is very similar, but the extended functionality varies widely across the market. The ELN providers can be narrowed down by matching the desired functionality with the functionality provided by the available systems to identify a shortlist of 'best fit' options. If further extended or customised functionality is required, then the availability of additional modules, customisation, or APIs can inform whether the desired capability is practical. Customisation can be managed in a number of ways, for example, through scripting and programming, through plug-ins or modules, and through the use of templates. Customisation may need to be performed by the supplier themselves, particularly for proprietary systems, or it may be possible to utilise skills in-house or purchase third-party services. Sooner or later, if customisations or extensions are required, direct interaction will be needed with potential suppliers. It may be possible to invite a shortlist of suppliers to demonstrate their ELN solutions and to discuss what customisation or extensions are possible. Such a meeting or discussion is an opportunity to assess the commitment of the supplier to the project and the practicalities of working with a supplier team, as well as an opportunity for the team to ask questions of the suppliers. Requirements, expectations, and timelines for the project can also be discussed. If a supplier is responsible for the entire implementation and customisation, it may be difficult to track and monitor the project, so it is important that regular progress meetings can be held with demonstrations of the progress made to instil confidence and ensure that any misunderstandings are identified and fixed early in the process.[3] If customisation is to be done in-house, it is important to verify that the intended development is actually achievable within the selected

ELN. This may involve reading the product documentation to investigate the functionality of any APIs or modules, or to contact the ELN support team directly.

Guerrero *et al.* suggest a number of features of ELNs that can be used for evaluation that may correlate with user perceptions of the usefulness of the ELN.[18] These include flexibility, ease of searching, collaboration, external resources, and security and legality. Flexibility and ease of use are important considerations for an ELN. In a quality control environment, flexibility may be less desirable, but in many research and development environments, the ability for users to create their own workflow designs, to capture the information in the formats that they find most natural (such as drawings and tables), and to be able organise their own work can help improve adoption and effectiveness of the ELN implementation. The ability to easily find information in an ELN is a key task for both content creators and other users of the ELN. Different search options can be useful for different users and for different kinds of information stored in the ELN. For example, some users may want to use methods other than text for search, such as matching DNA sequences or graphic search for chemical structures. Depending on how data is captured or analysed, it may be useful if search works beyond the notebook contents and also works within data files, attached documents, and images. In addition, the ability to group or filter notebook entries by topic, date, and users can also be very useful to help find particular information within a system. The usability of search within a system is important for all users, but it can be especially challenging for staff outside the laboratory who have not been formally trained in the ELN system. It may, in fact, be simpler for staff who are external to the organisation, such as IP lawyers, inspectors, and auditors, to locate information for reconstructing studies using paper notebooks compared with complex ELNs.[10] This demonstrates just how important it is to consider the usability of the system for all user roles.

Collaboration and sharing have previously been discussed, but there are different levels of sharing and collaboration features available to consider. At the most basic level, an ELN can allow read-only access to notes captured in the ELN to authorised users. Many ELNs enable viewers of the content to add comments to the notes, and some will allow editing, in addition to features for witnessing and review. Other ELNs have features suitable for task management and sending messages, whilst some provide project management and team management facilities such as presentation modes for meetings. ELNs can also be evaluated based on their security capabilities and compliance to certain regulations or best practices such as EudraLex and Title 21, as previously mentioned. Aspects to consider include how access to the records and data in the ELN are restricted, how users are authenticated and their activities tracked, and how records are protected, not only from malicious tampering but also from physical loss and corruption. If the procuring organisation needs to produce audit trails for regulatory or quality purposes, then the capabilities of the ELN need to match the requirements of the organisation, and at the appropriate level of detail, such as per experiment, notebook, or individual. Other considerations for the protection of data include whether data is automatically saved and whether version control is implemented so that changes made to the record can be tracked and explained. Requiring a reason for a change to a record is useful when a mistake or addition is found after the creation of the record and an explanation is desirable. However, asking the user for a reason for their change after every update whilst they are still creating the record can be annoying and frequently causes the users to work around the annoyance by entering minimal information or nonsense. For some organisations, it may be desirable or even a requirement to capture the reason for a change every time, but an ELN designed for flexibility should enable such a feature to be turned off.

Another aspect to consider in the evaluation of ELNs is the type of data formats used. As mentioned previously, the use of proprietary formats by ELN suppliers is of concern to many ELN users because of the potential for loss of data if an ELN provider goes out of business and the difficulty in transferring data to new systems in the future. Data formats that are open, well documented, and accepted by the community are less likely to become obsolete and more likely to be supported by commercial tools for migration and emulation, making it easier to transfer and transform data between different ELNs and other software.[22] Many proprietary formats are opaque, meaning that it is impossible to access the contents. For example, if the file is opened in a text editor, the content will not be human-readable. After the software that created the files is obsolete, it is usually impossible to build tools to extract the content, unless export to a different format is possible within the original software. Often, the format for export within ELNs is PDF format, which is human-readable, and suitable for printing or archiving, but does not usually enable the content to be imported into a new ELN. Data formats that have a simpler structure are easier to read in different software, and it is also easier to develop tools that enable transformation and migration, especially if they are in a self-documenting structure such as XML, where context is associated with the content. In some ELNs, the metadata associated with text-based content may be stored in separate files, making it more likely that the context of the original content is lost, including potentially important provenance information such as dates, author names, and versioning information. If the data from the ELN is stored in a database, it may be difficult to retrieve it without specific tools to do so, and strategies may be needed for extracting content for the purposes of back-up and archiving.

There are many different ELN suppliers as previously mentioned. These suppliers represent a variety of organisations from small groups of open source developers to large scientific information system developers. In choosing an ELN provider. it is sensible to check the future viability of the supplier as far as possible.[18] For example, if an ELN provider goes out of business or gets bought by another company, then the ELN may cease to have bug fixes and upgrades released, and there may be no support organisation to help with technical issues. If the ELN is cloud-based, there may even be a risk that the data stored in the ELN becomes unavailable. If a supplier is expected to provide assistance with the implementation, training, long-term support for the project, or data storage, then it is important to identify any risks and mitigations for changes in the supplier's status. For example, is it possible to take a regular snapshot of the data in the ELN in a format that can later be migrated to a new ELN or different provider if required?

7.7 Next Steps

Once an ELN provider has been selected, it is possible to plan the deployment in greater detail, including where software needs to be installed and in what order. If additional hardware and software infrastructure are required as prerequisites to the ELN deployment, then these need to be acquired, installed, configured, and tested first. For any scale project, it is appropriate to install and configure the ELN on a small number of machines for testing before a complete rollout. It is helpful to plan the testing before deployment begins to include functional testing of both the ELN and the associated infrastructure. The ELN may also need to be tested with other organisational software or instruments with which it is intended to be used. User tests should also be planned to include representatives from different user roles and to include testing within the laboratory and any other locations where the ELN is planned to be used. For example, if the ELN is to be accessible both in the field and in the laboratory, then it needs to be tested in different locations and with different network connectivity to ensure that access is still practical when lower rates of data transfer are experienced, such as with cellular data in the field. For staged deployments, prototyping can be used for testing the implemented functionality for a particular stage. Any software or hardware that needs to be integrated with the ELN system should be tested individually as each new component is added. A log should be kept of any issues encountered so that defects can be raised with the ELN supplier if necessary, and a record of planned resolutions and current status can be kept and monitored.

Once the ELN has been installed and configured, whether for a pilot or prototyping, the process of transferring from paper to digital notebook will begin. It is possible for users to try to replicate their paper notebooks within the ELN and try to follow the same workflows as before. The benefits of this approach are familiarity and a reduction in new learning, but such an approach may not gain the most benefit from the new capabilities provided by the ELN system. For research and development environments, freedom in the recording process can enable the recording to reflect the haphazard and uncertain nature of the experimentation.[23] For more constrained disciplines and activities such as quality control and analytical testing, a more rigid recording process may be required. Providing training and setting up templates can help to ease the transition, communicate new capabilities, and facilitate best practices for all users. Best practices may include sharing new and changed procedures such as managing records, recording experiments, and acquiring electronic signatures. Training, and the presence of expert users within the user community, can also help to ensure rapid adoption and use, as well as compliance with important procedures to facilitate good laboratory and record-keeping practices.[3] Procedures should also be developed for the long-term running and maintenance of the ELN to include policies for the storing and archive of electronic records, back-up and disaster recovery, and security of systems and data.

Many users of ELN systems spend some amount of time duplicating the record in both the ELN and paper-based notebooks as reassurance that all the information is being captured, and perhaps clinging onto the familiar paper notebook. If they feel frustration with the ELN system during this time, it can be very easy to go back and simply return to using the paper-based notebook and no research has been lost. It may be a useful exercise as part of the training with the ELN system (or at an even earlier stage in the project) to ask participants to bring along their paper notebooks and attempt to replicate their last (or a representative) experiment record in the ELN. This is likely to highlight any areas of difficulty and potential frustration in the tool so that appropriate support and guidance can be provided, but it is also potentially an opportunity to demonstrate useful functions of the tool that can add value to the user.

7.8 Conclusion

Implementing any kind of new information system presents a variety of both technical and human challenges. For an ELN implementation to be successful, time, resources, and leadership need to be put into the project. There needs to be considerable thought put into what capabilities the system needs to support for both the organisation and the individual members of staff, and how the chosen ELN matches those requirements. Resistance to new systems is common and can, if not handled appropriately, lead to the failure of the project. However, resistance can be overcome if communication about the benefits of the system are matched to user requirements, staff are engaged in the implementation process, and if the impacts of the change on working practices are considered and discussed.

ELNs are complex, and as mentioned, this anxiety contributes to resistance to using them and resisting the move from paper to digital. We can help users to overcome some anxiety by providing training in new systems and personal support, but ideally, the software itself will be intuitive to use. One of the reasons why so many researchers choose the generic electronic notebooks over the more complex software is that they are more familiar and easier to use. ELNs are often complex, but capable software does not have to be difficult to use. In Chapter 8, we will examine a variety of techniques that can be used to improve the usability of software. We will explore some examples of the different roles that typically use ELNs and some examples of how capabilities and designs of interfaces for ELNs can be developed using knowledge based upon the needs of real-world researchers. Following on, in Chapter 9, we will take a look at the design of interfaces for capturing the experiment, and in particular the positive and negative impacts of using templates for recording the scientific record.

References

1. H. K. Machina and D. J. Wild, *J. Lab. Autom.*, 2013, **18**, 126.
2. S. Laumer, C. Maier, A. Eckhardt and T. Weitzel, *Euro. J. Inf. Sys.*, 2016, **25**, 317.
3. D. J. Drake, *Drug Discovery Today*, 2007, **12**(15-16), 647.
4. T. Dickens, 2017, personal communication.
5. K. A. Badiola *et al.*, *Chem. Sci.*, 2015, **6**, 1614.
6. G. S. Streitberg, P. T. Bwititi, L. Angel and K. A. Sikaris, *J. Assoc. Lab. Autom.*, 2009, **14**(2), 94.
7. A. Parr and G. Shanks, *J. Inf. Technol.*, 2000, **15**, 289.
8. A. E. Day, S. J. Coles, C. L. Bird, J. G. Frey, R. J. Whitby, V. E. Tkachenko and A. J. Williams, *J. Chem. Inf. Model.*, 2015, **55**(3), 501.
9. Available at https://www.omg.org/spec/LECIS/About-LECIS/ (accessed Oct 2018).
10. J. Sydor and G. Kim, *Bioanalysis*, 2017, **9**(16), 1203.
11. Available at http://www.oecd.org/chemicalsafety/testing/good-laboratory-practiceglp.htm (accessed Oct 2018).
12. Available at https://ec.europa.eu/health/sites/health/files/files/eudralex/vol-4/annex11_01-2011_en.pdf (accessed Oct 2018).
13. Available at https://www.accessdata.fda.gov/scripts/cdrh/cfdocs/cfcfr/CFRSearch.cfm?CFRPart=11 (accessed Oct 2018).
14. Available at https://www.iso.org/iso-9001-quality-management.html (accessed Oct 2018).
15. Available at https://www.iso.org/iso-13485-medical-devices.html (accessed Oct 2018).
16. Available at https://www.iso.org/standard/56115.html (accessed Oct 2018).
17. Available at https://www.iso.org/standard/66912.html (accessed Oct 2018).
18. S. Guerrero, G. Dujardin, A. Cabrera-Andrade, C. Paz-y-Miño, A. Indacochea, M. Inglés-Ferrándiz, H. P. Nadimpalli, N. Collu, Y. Dublanche, I. De Mingo and D. Camargo, *PLoS ONE*, 2016, **11**(8), e0160428.
19. OECD, *OECD Principles on Good Laboratory Practice*, 1997.
20. OECD, *Compliance of Laboratory Suppliers with GLP Principles*, 2000.
21. OECD, *The Application of the Principles of GLP to Computerised Systems*, 1995.
22. Available at https://www.loc.gov/preservation/digital/formats/sustain/sustain.shtml (accessed Oct 2018).
23. A. Milsted, J. Hale and C. Neylon, *PLoS ONE*, 2013, **8**(7), e67460.

CHAPTER 8

Designing the User Experience

8.1 Introduction

Software tools can be extremely powerful and enable us to perform tasks that would not be possible with just pen and paper alone. Computers and their software should be making our lives easier. However, poorly designed software can actually hinder our progress and leave us feeling frustrated or even make us feel stupid. In this chapter, we explore a variety of user experience design techniques that are used to help make software useful and usable for the intended audience. We investigate how we can make use of our understanding of researcher needs and behaviours to generate requirements for electronic notebooks using these techniques, together with examples. We also examine some best practices for designing interfaces with examples of good and bad interfaces to illustrate the issues. We also take a look at how interface designs can be checked for usability. Although this chapter is focused on examples of user interfaces for Electronic Laboratory Notebooks (ELNs), the same principles can be applied to the development of any digital tool.

Although good visual design can often help with usability by making things clear and easy to understand, and they can often be perceived as easier to use than less aesthetic designs, swishy graphics and a pretty interface alone do not make software usable.[1] There are many examples of websites that have attractive graphics, animations, sounds, and stylish navigation schemes, but are completely unusable for a variety of reasons. The elements that make them attractive also often tend to make them very slow to load because the pages contain large numbers of images, videos, or animation. The slow speed of loading makes it less likely that users actually see the page in the first place. A number of studies have shown that half of us will abandon viewing a web page if it takes more than three seconds to

Recording Science in the Digital Era: From Paper to Electronic Notebooks and Other Digital Tools
By Cerys Willoughby

Published by the Royal Society of Chemistry, www.rsc.org

load. The design of such websites can also make it unclear what elements are interactive on the page, often forcing the user to move the mouse randomly around the screen to find out what is clickable. Unexpected behaviours may occur when the user moves the mouse over objects on the page, and the resulting behaviour is often inconsistent, sometimes taking you to a new page, sometimes playing a sound, or opening a window overlay on the screen. These websites show off the talents of the visual artists or the programmer, but often provide limited value for the user. For example, there are many marketing websites for products that make use of these visually impressive websites to showcase their company's vision or concepts, but the websites actually provide little information to the consumer about what the product might do for them. Usability is about understanding user goals. If you do not understand what users need or what they are trying to achieve, then you cannot produce a usable interface.

So, what is usability? A good definition of usability comes from a set of International Standards related to usability for human–computer interfaces and ergonomics. ISO 9241-11:2018 defines usability as the extent to which a system, product, or service can be used by specified users to achieve specified goals with effectiveness, efficiency, and satisfaction in a specified context of use.[2] In other words, users must be able to achieve their goals without much difficulty and in an acceptable period of time. Can the user easily use a product to achieve their goals without being shown how? Users need to be able to achieve their goals in the 'context of use', in other words, the environment in which they intend to use a product. For example, using a PDF bus timetable to choose a bus journey may be easy to do using a laptop, but may be much more difficult on a mobile phone late at night when you are trying to get home. Considering the context of use is particularly important in designing applications for science. What may work well for capturing or looking up information on a desktop in the office may be insufficient for use with gloves in the laboratory or in a muddy field in the rain.

As we have already discussed in Chapter 6, poor usability causes frustration and prevents researchers from doing their research effectively and reliably. The prime goal of usability is user satisfaction. An unhappy user will choose not to use a product or will struggle and waste time making mistakes or fixing problems. In contrast, if a product is usable, the user is able to quickly get value from it through increased user-efficiency and productivity. The easier something is to use, the quicker the user can do something useful with it, the less training they need, and the higher the chance that they will use it again. The user is more likely to buy a product from the same developer again and also to recommend it to others. Thinking about usability up front in the software development cycle is also much cheaper for an organisation than realising they have a usability problem late in the cycle and having to fix the problem at a late stage. If an organisation finds out there is a significant problem with the product after it has been released to the public, it is very expensive to fix that problem. There is also a significant risk that losses will be made as a result of the damage to the reputation of the company or product as a consequence of releasing a product with such a significant problem.

Poor usability often arises for a number of different reasons. Software is not intrinsically hard to use, but if the right processes are not used to create the software, then the result can be a difficult to use product. There are often conflicts between the needs of developers and the needs of users that can lead to problems with usability, particularly where there is a rush to develop product 'functions' before considering how users might want to use the software. Software designers often try to guess at what users might want or develop features that they themselves desire, leading to software that does not meet the goals of users. Software engineering is very complex and often this complexity is passed onto the user through difficult interactions even though a thoughtful design could have avoided these problems.[3] The underlying complexity of software development and unrealistic perceptions of user skill levels are often used as excuses to make software more difficult to use than it needs to be. I have even known some developers to admit deliberately exposing the complexity of their code to users in order to show off their technical prowess! There can often be unrealistic expectations about the knowledge and capabilities of users. There is often the perception that the user has, or should have, the same skills and knowledge as the developer themselves. Not only that, but often the user is expected to have a clairvoyant knowledge of how the technology works and therefore how to use it! In reality, users are often much less skilled than software developers expect, with only around 5% of the population in developed countries having strong computer skills; and even those skills are still much less advanced than those of a typical software engineer.[4] Regular use of technology does not necessarily lead to the high-level skills in computing and information literacy that we might expect, even for disciplines that are perceived to be more technical, such as the physical sciences.[5–7]

Some software engineers, and even computing 'power users', defend the complexity and difficulty of using software because they personally enjoy the challenge. Unfortunately, I have been in numerous meetings with some software engineers who have openly mocked users who have struggled with their software along with members of the development team who try to advocate for them, dismissing the pain and frustration of so-called 'dumb users'. Some expect that users should have to adapt to the technology rather than make the technology work better for the users. Should we change the technology or change the user? Some would argue that what we need to solve the problem is to make humans understand technology better by improving computer literacy and programming skills, in other words, change the humans to make it easier for the computers. Human behaviour is undoubtedly changed by our interactions with computers, and over time, we adapt with more experience. For example, we learn how to type more efficiency, moving from carefully selecting and pushing the keys on the keyboard with a single finger at a time, to using multiple fingers on both hands without looking; or we may progress to seamlessly using our thumbs for texting and swiping. However, the fact that our behaviours change to adapt to some new ways of working is not the same as forcing users to learn software that is overly complex and which is not intuitive to use. Smartphones and social networks have a large impact on how we communicate, how we interact with others, and how we spread news. In fact, it could be argued that the success of these technologies is in part related to their ease of use. As mentioned in Chapter 7, the perceived usefulness and ease of use of software has a major effect on whether users are willing to accept it. Many smartphones are in fact quite simple to use compared to earlier generations of phones. The most successful smartphone applications are the ones that are very simple and require little time to learn. The benefits of ease of use for adoption and on-going use of technology is plain. Is it really reasonable then to expect users to undergo complex training in order to more effectively complete their tasks? Do users really need to have to understand the inner workings of a computer system in order to perform some task related to their work, even if the work itself is complex? Although undoubtedly many researchers develop greater computer literacy and skills such as programming over time, having a high-level of technical skills should not be a prerequisite to using software that should not require them. Users have different knowledge, backgrounds, and goals, and it is necessary to understand what they are trying to achieve, what their capabilities are, and the environment that they are working in, in order to design software that is usable.

At the end of this chapter, we consider some examples of good and bad practices in designing for usability, but first we consider what processes can be used to improve the usefulness and usability of software. Interaction design is an umbrella term describing a variety of processes and activities associated with designing user experiences of software, interfaces, and products.[8] In this chapter, we will explore some interaction design techniques to assist in the design of tools for keeping and managing records. In order to create software that has good usability, it is important to make sure that there is a focus on user needs from the very first stages of developing the software, in particular during the requirements gathering and design stages. This makes it much more likely that the software will be both usable and useful for the intended audience. Including user-focused activities such as creating prototypes and undertaking user testing throughout the software development lifecycle also ensures that the original requirements are still valid and that the designs generated as a result are usable. The examples provided in this chapter illustrate how the techniques could be used to develop digital tools to support scientists. The examples provided are based on my own experiences with different research teams and illustrate how there are similarities and differences between their needs and their priorities for tools. Some requirements and roles are likely to be similar across different disciplines, types of organisations, and different kinds of research, but undoubtedly there will be many differences between them. There are a vast range of practices in science and the way that scientists conduct their research, how they choose to record their research, and how they go on to make use of the information that is recorded may be quite different, even within the same research team. There is no 'one size fits all' solution, and that is why making use of interaction design techniques to elicit user needs can be so useful.

8.2 Requirements Gathering

In order to design tools for researchers we first need to have an understanding of our users, their goals, and their needs. It is rarely possible to talk to every potential customer of a software product, and therefore it is necessary to find alternative ways of eliciting information about what users may want or need. We can learn a lot about user behaviours and attitudes through techniques such as surveys, interviews, task analysis, and observation. Replacing a primarily paper-based system with a computer-based system has been a common problem to solve in the history of the development of computers and software, and traditionally, as a part of 'systems analysis', users have been observed in their place of work to identify the business processes in use and artefacts, such as paper forms, that need to be replicated. This is still a valuable technique to help understand how future users of a system currently work and interact, but it is also valuable to observe the areas of work where users are currently having problems and where computer systems can add functionality to overcome problems or improve efficiency, rather than just trying to replicate exactly the manual system that exists. It is also useful to elicit ideas for new features from the users themselves, for example, what would the users like to be able to do that they currently cannot? Observing users and eliciting feedback about the design of potential future systems can also be used when a current computer system is being updated. This is a good opportunity to get feedback on current problems with the system and where enhancements might be made.

There are also various techniques in interaction design that enable us to create proxies for users that can help us to better understand their needs, goals, and the environments in which they are working. We will discuss these techniques in more detail below, but, for example, we can create a 'persona' to represent a potential user of a product describing their skills, goals, and concerns. The persona can be used by the development team to understand what a user might want to do and how they might expect to use a product to achieve their goals. These techniques are particularly useful for the generation of new tools or where a tool is being developed to extend functions beyond those that are currently available. The requirements that are generated using these techniques progress into a design stage, so that the user interaction with these capabilities can be more precisely defined before coding begins.

8.3 Conceptual Models and Metaphors

Interfaces for digital tools need to be understandable and familiar for users. A common cause of complexity in software is that the tasks and actions presented to the user do not match their goals or the workflows with which they are familiar. One way to reduce the complexity is to present the user with familiar concepts and metaphors within the software. Conceptual models are easy to understand paradigms and metaphors for the objects and activities that we encounter in everyday life when applied to computer-based interfaces. For example, the majority of computer file systems are based on the conceptual model of the office filing cabinet, where documents are arranged in folders and sub-folders. The majority of shopping websites provide an approximation of the experience of visiting a supermarket by enabling us to browse items for sale, add them to our basket, and then go to the checkout to make a payment. Metaphors and patterns like these can be used to conceptualise what we are doing (*e.g.* recording an experiment), to construct the interface we are using (*e.g.* the notebook), and to visualise an action (*e.g.* take a photograph). Conceptual models can be developed from the analysis of the activities and behaviours of users in our intended audience. These conceptual models can then be used to identify tasks that the researchers might want to perform using the software, together with appropriate terminology that can be used in the interfaces.

In Chapter 3, we explored how researchers capture and manage their research. The actions and activities involved in recording and making use of their research can be used to generate concepts and metaphors that can support the design of interfaces in digital tools for capturing and managing the research. For example, we can consider how researchers plan their experiments, record the execution of their experiments and the observations they make, how the results are analysed, and other ways that the materials generated might be utilised. For example: sharing results with colleagues and collaborators, turning analyses into graphs for publication or presentation, and making use of the data to design new experiments, refine a technique, solve a problem, or look for patterns within the results.

In the physical world, researchers perform some of the activities in the following list for developing plans and capturing experiments using paper notebooks. The **verbs** and *nouns* in these activities have been highlighted to demonstrate how familiar terminology could be translated to a virtual environment to create a meaningful vocabulary that can be refined and used in the interfaces to represent particular actions:

- **Search** for and **locate** *information*
- **Extract**, **refine**, and **write** *information* to make *plans*
- **Identify** and **write down** *materials*, *measurements*, and *hazards*
- **Design** the *experiment protocol* or *instructions*
- **Read** *information* and **follow instructions** from the *plan*
- **Write** *notes* about *ideas* and **record** *observations*

- **Draw** *sketches* and **take** *photographs*
- **Make** *annotations* and **highlight** *important information*
- **Draw** *tables* and *graphs*, and **record** *results*
- **Allocate** *IDs* and **label** *materials*, *samples*, and other *resources*
- **Record** *IDs* in the *notebook*
- **Paste** printed *data* or *photographs* into the notebook
- **Flick** through a *notebook* to **locate** *information*
- **Explore** *information* in a *notebook* or a *specific experiment record*
- **Compare** *information* and *results* from different *experiments*
- **Extract** *information* from the *notebook* and **transcribe** into different *formats*
- **Share** the *notebook* and **discuss** the *content* with *collaborators*

Other subtasks could be identified within this process to increase the level of detail. For example, for 'Write notes' we might also include subtasks and other decisions that the user might make, including:

- **Start** a *new page*
- **Choose** a *pen colour*
- **Record** the *date*
- **Start** a *new line*
- **Write** a *bulleted list*
- **Draw** a *sketch*

It may or may not be appropriate to include all of these actions in the design of a digital tool to replace a paper notebook, but the advantage of capturing them in this way is that familiar actions can be presented to the user and to avoid potentially important capabilities from being missed out. It is particularly valuable to use the observations and discussions with users to derive these various actions, rather than to try and guess at what tasks users may do. For example, the ability to compare information and results from different experiments is an extremely useful function that may not have been considered if only the 'functions' of capturing an experiment in a notebook were considered. This can be tricky to do in a paper notebook, so there is also the opportunity to add value by making this more straightforward in the digital version.

There are other tasks that are impossible to do in the real world with a paper-notebook, or at least impractical. When we asked researchers to identify requirements that they would have for an Electronic Laboratory Notebook system, many desired to add functions for actions that they already do in the real world on paper. For example, functions were requested such as completing forms for risk assessments and acquiring sign-off, providing a table of contents, and performing a structure search. However, there were also a number of requests for functionality that would not be at all possible with an actual notebook, for example, spell checking, 'liking' experiments, automatic tweeting of new experiment entries, different ways of organising their information, and identifying future collaborators. It is important that the tools provided to support researchers are able to perform the tasks that researchers already do in the real world, but digital tools can also add immense value to researchers by including these kinds of extended capabilities. Identifying these extended functions can be difficult. Users themselves can generate novel ideas and suggestions for useful functionality beyond what they can currently already do, but they can also be limited by their own individual preferences and experiences. Software engineers sometimes have an erroneous expectation that users are able to tell them exactly what they want, but users are both limited by their own experiences and by their lack of knowledge of what is actually technically possible. Making use of further interaction design techniques can help to verify the individual desires to a broader audience and to elicit less obvious but potentially very valuable requirements for the design of digital tools.

8.4 Research Roles

For any given product, there will be at least one user role who will make use of the product, but often there is more than one to consider. Within an organisation or a team, there are usually different roles. For example, in a typical analytical laboratory, there may be laboratory assistants who perform the day-to-day analysis and recording of results, supervisors who monitor the work of the analysts and the results generated by the laboratory, managers who handle strategy, and technical support workers and their managers who take care of the various laboratory and IT systems. Each user role will have particular tasks they perform or goals that they want to achieve. For any given system, all these users may need to interact with it, but the aspects they use may be quite different. For example, the technical support workers need to install the system, configure it, manage user access, and maintain it, but the chances are they are not ever going to use the core features that are designed to be used by the laboratory assistants and their supervisors. When we talk about a user role, there is not necessarily a one-to-one match between people and roles. It is entirely possible for one person to perform multiple roles, and for more than one person to be needed to fulfil a single role. For example, within a large organisation, multiple users may have the same role such as analyst or technical support, whereas for a small team or an individual researcher, a single user may be fulfilling the roles of system administrator, researcher, and supervisor, managing the system, their own research, and monitoring that of their students.

Different digital tools have different user roles. For example, in designing software for a car, we might only need to consider the roles of driver and engineer; each role has entirely different goals and needs from the system. In this section, we will look at a number of example user roles that would be appropriate to consider during the development of an ELN system.

These roles have been chosen to represent different kinds of users common within an academic research environment, and alternatives are likely to be more appropriate in a commercial or industrial environment, for example, test analyst, laboratory technician, and manager.

8.4.1 Administrator

In some organisations, an individual may choose to install an ELN on their own machine without the support of anyone else. However, a more likely scenario for most research organisations and groups is that the ELN will be shared by multiple users and installed on an internal server or remote location such as a cloud-based service. In such a shared ELN scenario, an important (and often forgotten) user role is the administrator of the system. The administrator is likely to be the person responsible for installing, configuring, maintaining, and updating the system. Day-to-day administration activities may include user management, security, backup, database administration, training, and other support tasks. The administrator role may be held by a member of the research team, an internal IT-support team, or even an external contractor, depending on where the ELN system is hosted and who is responsible for its set up and maintenance. The concerns and needs of the administrator role are likely to be quite different to those of the other user roles, and quite different interfaces may be provided. Although this role might be considered a 'power user', holders of this role could have a range of technical ability, and in academia in particular, it may be a member of the research team without any specialist skills and only a part-time responsibility for helping others with the system.

8.4.2 Supervisor/Principle Investigator

The second role identified for the ELN is the role of principal investigator or supervisor, depending on the nature of the research group. The 'supervisor role' is responsible for monitoring the research activities of the team and providing feedback to individuals. They may also be interested in exploring the research of their team, including comparing experiments and results, and looking for patterns. The supervisor may want to be able to extract information from notebooks and other resources and then use that information to hold discussions, create presentations, and include it within other publications. The supervisor is likely to be involved in coordinating and developing relationships with external collaborators. Some supervisors may be interacting with their researchers remotely. When we investigated how ELNs were being used, we found that a large percentage of notebooks were used by supervisors for project and team management purposes, acting as a tool for organising meetings, distributing information to research teams, and for capturing personal thoughts and activities. The prevalence of this use for the ELN suggests that project and team management are also important goals for the supervisor role.

8.4.3 Researcher

The third role identified for the ELN is the role of researcher, acting as the author of the research information that is to be captured and utilised in the system. The tasks performed by the researcher have already been mentioned previously, but we will also examine this role in more detail through the use of personas in the next section.

8.4.4 Collaborator

The fourth role of the ELN to be considered in this chapter is the role of collaborator. Collaborators want to be able to access the research of staff using the ELN and to be able to collaborate in a variety of different ways. Collaborators may be other researchers, second supervisors, collaborators from other disciplines, or active members of the wider research community. Although in Open Science projects the notebooks may be visible to the entire scientific community, an individual is not counted as a collaborator unless there is a two-way dialog between them and members of the research team. Open research and the potential for public interaction through an ELN system understandably raises concerns about individuals engaging in nefarious activities, with the possible abuse of facilities provided. With any system accessible outside a firewall, the potential exists for spamming, trolling, denial of service, data theft, and corruption. These are legitimate concerns and mechanisms may be required to ensure security of data, enable moderation of comments, blocking of users, and blocking of external inputs to systems. Here we will focus only on 'happy day scenarios' where everything works as expected; errors, exceptional conditions, and misuse or abuses of the system will not be considered, but in any system where there is interaction between different parties, particularly if these are anonymised, then an additional 'moderator' role for a tool may be required.

8.4.5 Observer

An individual who merely observes the research without engaging in dialog is a fifth role 'observer'. For example, interested members of the scientific community and members of funding bodies may observe the research process and the results obtained without directly contributing to the conversation or the research effort. Intellectual Property (IP) lawyers may belong to this role if they have access to the system to extract content, but do not actively engage with the researchers using the system. The Observer role may also describe auditors who need to access the research record to ensure that appropriate regulations are being adhered to. An IP lawyer or auditor may also belong to the collaborator role, or require a role of their own, if they do provide active feedback in the system, for example, by adding comments, asking questions, or assigning tasks.

8.4.6 Teacher

The final roles of the ELN to be considered here are the roles of teacher and student. In academic environments, ELNs can be used as tools for both teaching science and for teaching data management skills that will later be useful knowledge for roles in industrial laboratories. Teaching can also be used to encourage the adoption of ELNs by researchers. Teachers need to be able to plan activities, create and share materials, monitor student activities and experiment records, and to provide feedback to students.

8.4.7 Student

The student role is likely to be similar to the researcher role, although the information that they capture and how they interact with other materials and other students may be different.

8.5 Personas

Personas are used as a representation of a potential user of a product, and are presented together with some personal information, but primarily information about their skills, goals, and concerns. Personas are created as part of the design stage of a product and are then used throughout the development process to remind members of the development team about the needs of the users, but they also act as a basis for asking questions about the design and how it might affect particular users. Persona use throughout the software development process helps to ensure that the original requirements are still valid at different points in the process. Personas are usually fictitious users with imaginary personal details, but they are meant to be realistic archetypes of actual users based upon research about and discussions with actual future users of the system. Personas include details of identity, goals, skills, and motivations that build up a picture of a person that can be used to facilitate a common understanding of what the user wants to achieve and what they are able to do. Only a small number of personas need to be developed for most projects, and each contains precise descriptions that represent a realistic person, but not someone who has the unusual quirks or preferences of any one individual. Developers often struggle with understanding who their audience is and what they are trying to achieve, and personas can help to make users more real and overcome the instinct to assume the user has the same knowledge as the developer. Having a small number of personas with precise descriptions helps to ensure that all members of the development team have the same understanding of who that person is. It is important to remember that the personas should represent 'users' of the software of the product rather than buyers of the product, who may be quite different. Personas of the buyers of products are appropriate for use by marketing teams, but are unlikely to help develop a usable product unless the buyer and user are the same person, although this might be the case for specialist software and small teams or individuals.

A persona typically contains precise details of the individual including a picture, name, their role or job description, their goals or motivations, and skills. The persona may also contain other details such as their age, their frustrations, personality traits, day-to-day activities, and other biographic information. Goals represent the purpose of the persona: what is the persona trying to achieve? For example, goals for the user of an electronic notebook may be to record their research, remind themselves how to perform a particular procedure, investigate a problem, look up results, or to improve a process. It is important to remember that users want to achieve these goals in a way that enables them to work efficiency without feeling frustrated. User goals do not include learning to use the software, reading the manual, configuring settings, fixing bugs, or using a particular technology or platform.

In the following section, a number of example personas are presented based upon the different user roles we discussed above, and the activities of a variety of different research teams to illustrate how personas can be developed. The personas have been developed based on the activities and challenges faced by the individuals within these research teams. We will also examine more about the activities of these different research groups and how we can make use of the personas to elicit requirements when we discuss the related scenarios in the following section.

8.6 Scenarios

A scenario is a narrative description of how one or more personas might interact with and use a tool to achieve their goals. Scenarios can be used to elicit information about the requirements for digital tools and inform ideas about the kinds of interfaces that would be most effective in the contexts and environments of use. Scenarios ideally represent the day-to-day use of a tool, rather than unusual situations. Scenarios can help to understand what the user's motivations are to use a tool and how they might go about using it. They can also begin to refine the tasks that the user needs to complete in order to achieve their goals. To demonstrate the use of scenarios, four everyday scenarios have been developed based on discussions with a number of teams in different academic organisations. For each scenario, a brief narrative is provided with the roles and identities of the personas that are used to represent them. As with the personas, the scenarios are fictional, but they are based on the current working practices of real scientific research teams and organisations that have been interviewed and observed in their working environments.

8.6.1 Scenario 1: Closed Research Including Interaction with a Funding Body

At the Mountain University, Professor Felix runs an environmental and analytical chemistry research group including eight research students. Felix spends a large amount of his time at a different location, making it a challenge to manage the research and interact with his team at Mountain University. Much of the research is funded by external funding bodies who enforce detailed quality standards for research documentation. Periodically, the funding body needs to have access to the notes kept by the research students to check that the documentation matches the standards and that the research is progressing as planned. Felix's students, once trained, are able to work on their experiments autonomously, but need to have sign-off from Felix before they start every new experiment. The research students also like to receive feedback from Felix to reassure them that they are on the right track. The experiments that they perform involve intensive activities across different buildings on the campus, meaning that they need to take notes in several different locations. Their experiments also typically take several weeks or months to perform, and they are often working on multiple experiments at the same time. The team are interested in using an ELN to help Felix with managing the research project and to provide support to the researchers remotely, and also to enable the researchers to work more effectively with their funding bodies. In Scenario 1, Professor Felix is a supervisor, his student Alejandra is a researcher, and auditor Alvin, from the funding body 'Central Environmental Services', is an observer on the project. Figures 8.1–8.3 show the persona details for Felix, Alejandra, and Alvin.

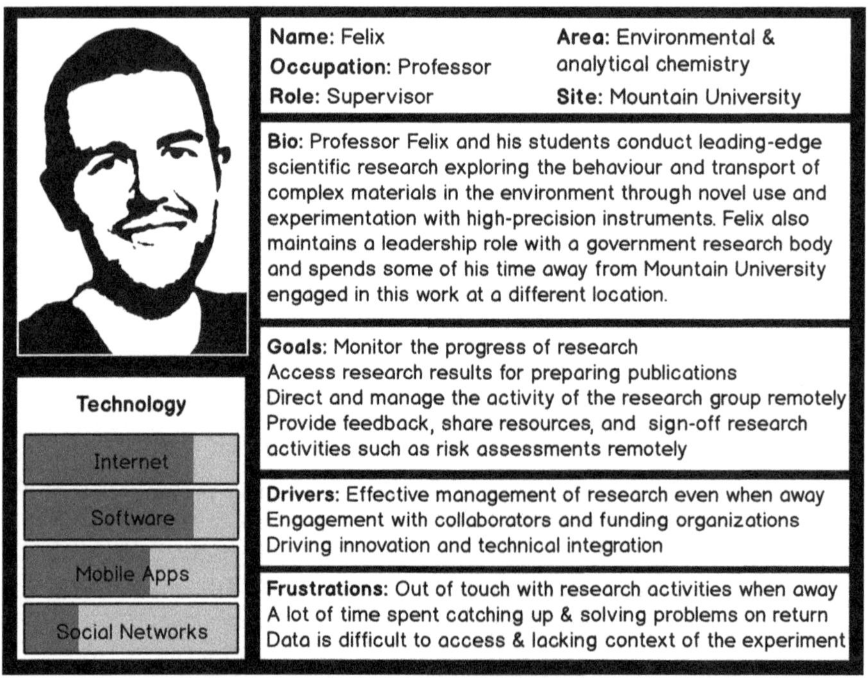

Figure 8.1 Supervisor persona 'Felix'.

Name: Alejandra
Occupation: Researcher
Role: Researcher
Area: Environmental & analytical chemistry
Site: Mountain University

Bio: Alejandra is a research student in Felix's research team at Mountain University. She is in the third year of an externally funded PhD project investigating properties of materials in the environment using NMR analysis. Her experiments involve working in several different buildings and take a long time to run, so she can be running several experiments at a time. She helps Felix manage the project and team while he is away.

Goals: Successfully complete experiments for her research
Produce results for publication
Get feedback, direction, and sign-off from Felix
Share findings and work with other researchers in Felix's team
Document & complete activities in line with funder standards

Drivers: Produce high quality research for PhD &future career
Meet the requirements of the funding body
Make writing of PhD thesis less painful

Frustrations: Funding body requirements means every page and lot number has to be signed & funder wants to check
Harder to get feedback and approval when Felix is away

Figure 8.2 Researcher persona 'Alejandra'.

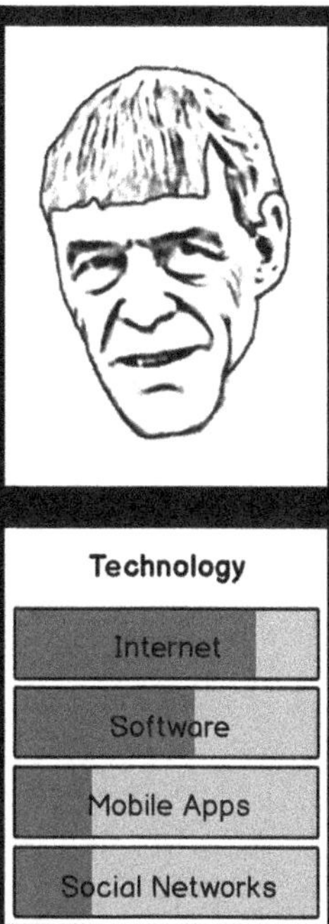

Name: Alvin
Occupation: Auditor
Role: Observer
Area: Funding Body
Site: Central Environmental Services

Bio: Alvin is an auditor for Central Environmental Services, one of the funding bodies for Felix's research activities. He is responsible for checking that groups funded by the body are following the standards that are set for documenting their research & recording their results, and that the research is going to schedule. This involves visiting research locations & checking documentation for compliance & research progress.

Goals: Verify team is adhering to prescribed standards
Examine documentation produced by researchers
Examine results to determine if research is on schedule
Communicate and resolve any issues with the project lead and other funding body members

Drivers: Ensure research is conducted to high standards
Ensure research is progressing on schedule
Effective communication with stakeholders

Frustrations: Access to research data and documentation is restricted by time and geographic constraints
Compliance checking is manual and time-consuming

Figure 8.3 Observer persona 'Alvin'.

8.6.2 Scenario 2: Fully Open Research Including Interaction with Global Scientific Collaborators

At the City University, Professor Guy runs an organic chemistry research group including seven research students and two post-doctorate researchers. Guy is an evangelist for Open Science and works with a variety of international collaborators on projects to improve techniques for building organic molecules for use in the treatment of disease. Managing the interaction with the collaborators and trying to develop new relationships with relevant research groups is a challenge. The nature of the experiments in Guy's research group means that the students are repeating very similar processes, and they save duplication by not recording information that they have already learnt. This makes it a challenge for new students to the group to understand the working practices of the team and to be able to take on the work of experimenters that have moved on. Guy's students find the risk assessment process to be onerous and involve unnecessary duplication: information has to be manually extracted from databases or records, transcribed, and then countersigned by Guy. The students would like to streamline this process to make it easier and less error prone. Students need to work with collaborators, but find it difficult to keep up to date with their activities and to find specific information in their records. The team are interested in using an ELN to help with interacting with collaborators, to make some processes easier, and to reduce duplication. They also want to use an ELN to provide a platform to encourage open science and generate new collaborations. In Scenario 2, Professor Guy is a supervisor, his students Ollie and Lukas are researchers, and Professor Ben from the Island University is a collaborator from the other side of the world. Figures 8.4–8.7 show the persona details for Guy, Ollie, Lukas, and Ben.

8.6.3 Scenario 3: Closed Research in a Local Research Group

Dr Pal is a postdoctoral research student in the Synthetic Chemistry research group at Castle University. Pal's research is funded by an industrial funder and involves the development of novel techniques for synthesising molecules. Pal's research has significant commercial value for the funder, and the resulting intellectual property and results must be protected from competitors. As a result, Pal works independently and shares his results once a quarter with his managers who are located elsewhere. Pal is also responsible for supervising a small group of research students who are also working independently on their own projects. Pal finds it difficult to keep a track of what the students are doing and making time to check their work. To check their work, he needs to sit with them in order to manually go through their notebooks and data, and to provide feedback. Pal would like to find a way to improve how he can manage both his own research and that of his students, so that both he and the

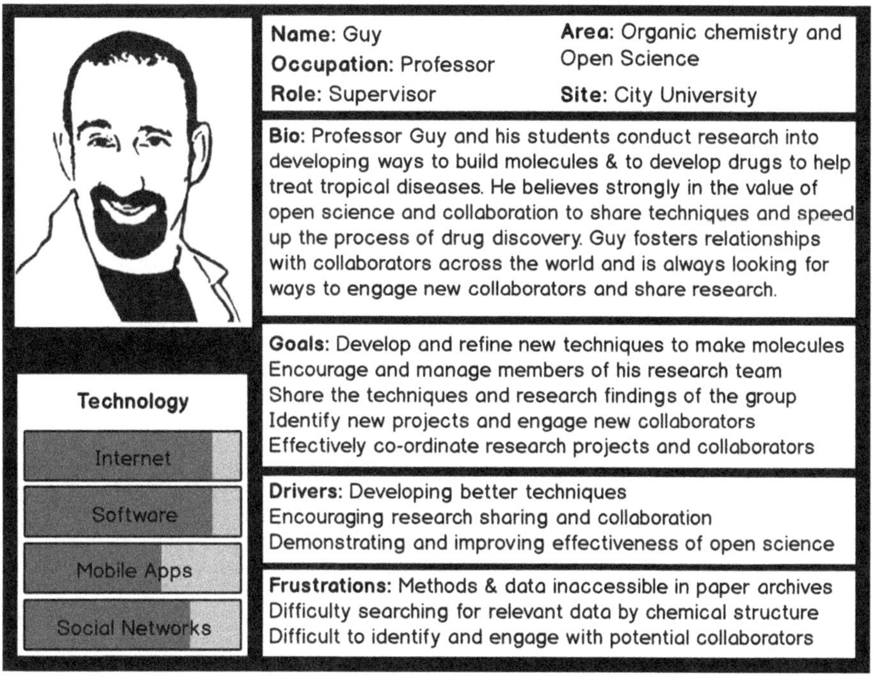

Figure 8.4 Supervisor persona 'Guy'.

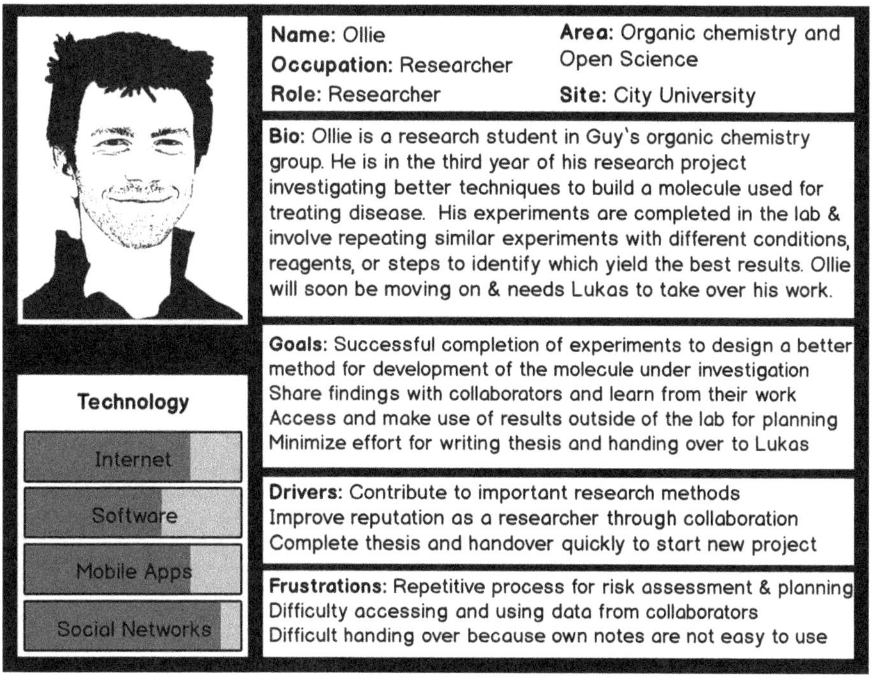

Figure 8.5 Researcher persona 'Ollie'.

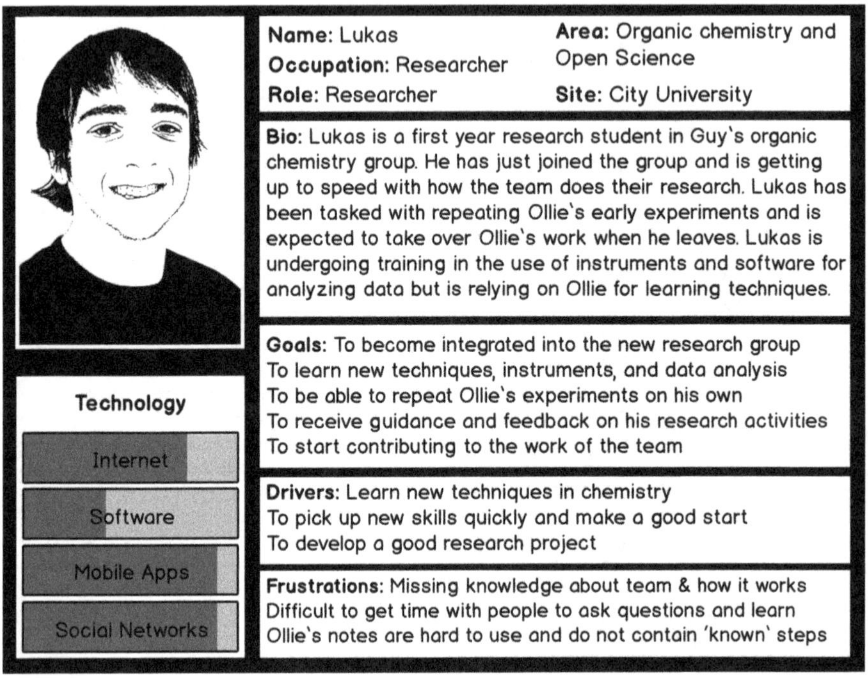

Figure 8.6 Researcher persona 'Lukas'.

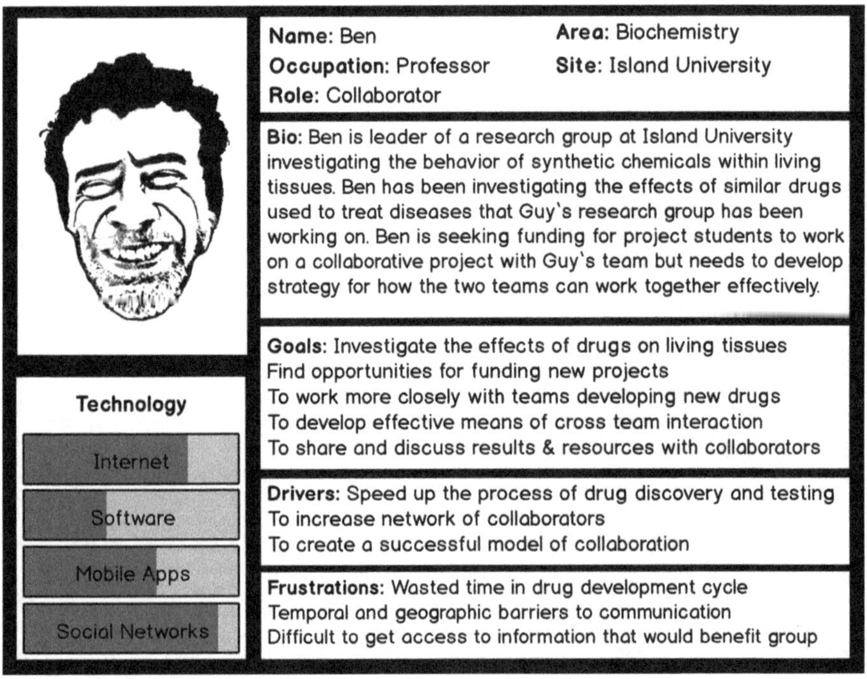

Figure 8.7 Collaborator persona 'Ben'.

students can get more regular and faster feedback on their work. An ELN has been suggested to Pal as a way that he might be able to improve the management of his own and his students' research. Although he is suspicious of using an ELN for his sensitive data, he likes the idea of being able to instantly access information about his students' activities. Kat is one of the Pal's research students. She is new to the team and is struggling with the new techniques she has to learn, and with repeating experiments from the literature. The more experienced members of the research team have all gone through the experiments she is trying to replicate when they first joined, but she finds it hard to get time with them to ask questions and learn from their mistakes. Kat spends a lot of time trying different things to solve her problems and feels frustrated because she does not know if she is proceeding in the right direction. She does not want to waste Pal's time by asking him simple questions. In her undergraduate course at Highway University, Kat liked using a Virtual Learning Environment (VLE) system with an area where the students could ask their lecturers questions and get quick responses on their practical courses. Kat would like something similar so that she and the other new researchers could help each other with the experiments and learn from more experienced staff. In Scenario 3, Dr Pal is both a supervisor and researcher, and his research student Kat is a researcher. Figure 8.8 shows the persona details for Pal and Figure 8.9 shows the persona details for researcher Kat.

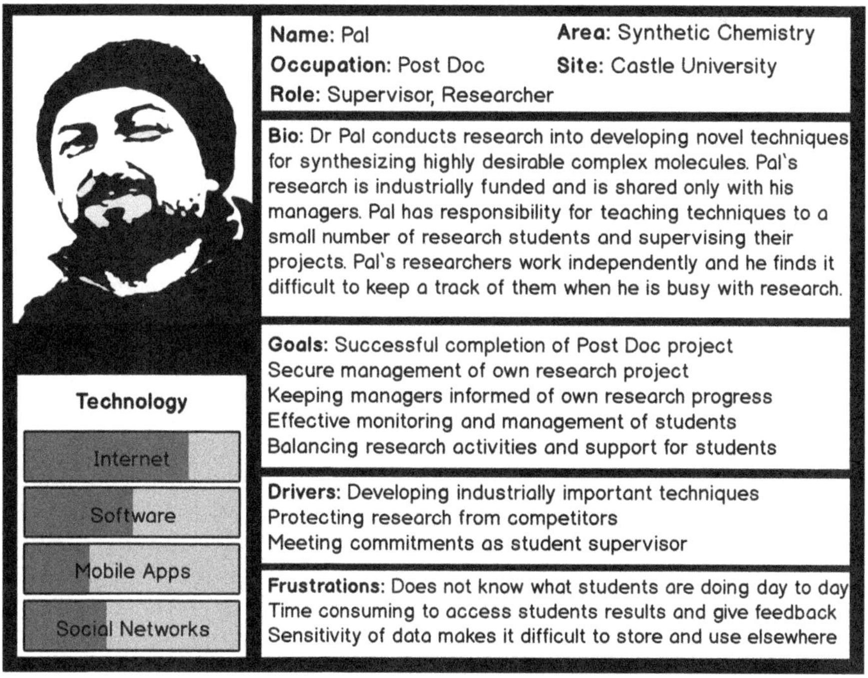

Figure 8.8 Supervisor persona 'Pal'.

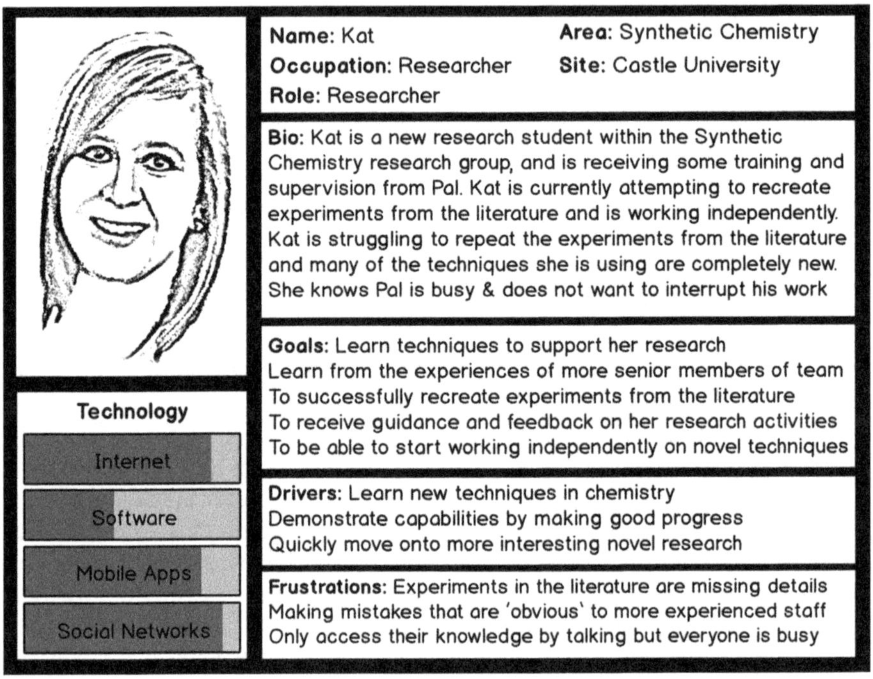

Figure 8.9 Researcher persona 'Kat'.

8.6.4 Scenario 4: Undergraduate Practical Course

Dr Jenny is a lecturer in the Physical and Analytical Chemistry group at Seaside University. Jenny is a technology enthusiast and is actively involved in projects to assess and implement technologies applicable to chemistry within the department, together with other collaborators outside the university. Jenny is keen to try out new things to help support her students and teach them analytical techniques. Although Seaside University has trialled ELNs with researchers, Jenny is keen to see the undergraduates use an ELN in their courses to teach them about data management and prepare them for future research or careers in industry. Jenny also hopes that an ELN system might streamline the student marking process, which is currently manual and very time consuming. Dan is one of the 90 students on Jenny's Advanced Analytical Techniques course. Dan is a bright student and although he enjoys the practical work, he gets frustrated having to wait for the other students to finish before he can move on with other tasks. He also gets frustrated with the feedback process, which takes a very long time and involves handing in his notebook after the practical sessions. He would prefer to get immediate feedback and to be able to use his notebook between practical sessions to remind himself about the previous experiment. In Scenario 4, Dr Jenny has the teacher role and undergraduate student Dan has the student role. Figure 8.10 shows the persona details for Jenny and Figure 8.11 shows the persona details for student Dan.

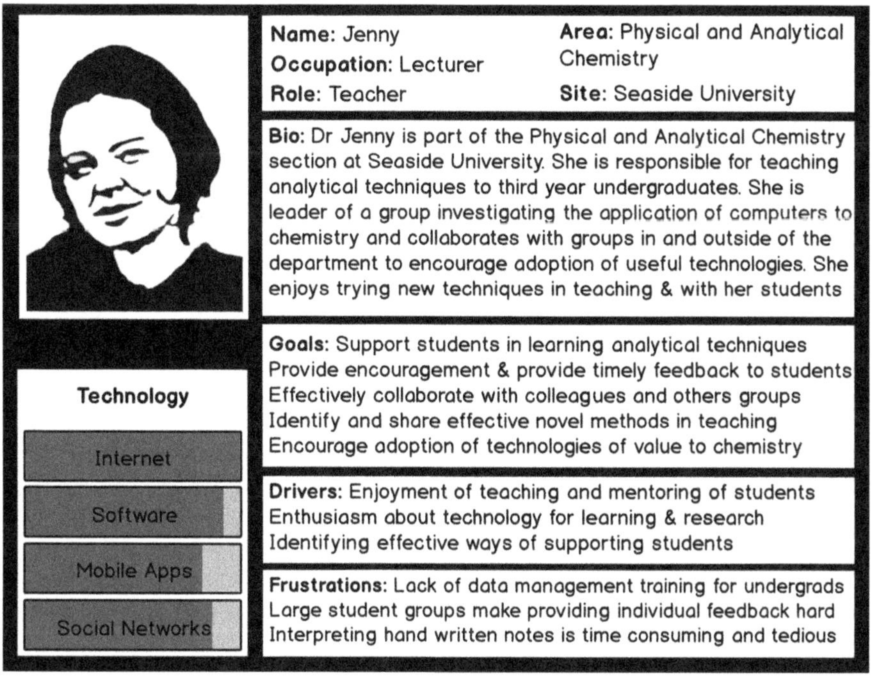

Figure 8.10 Teacher persona 'Jenny'.

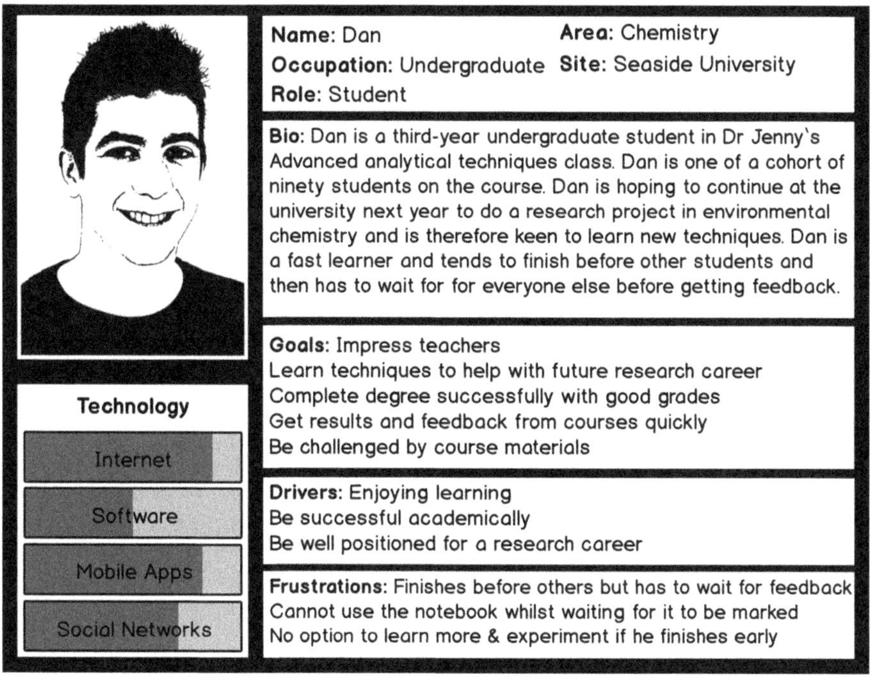

Figure 8.11 Student persona 'Dan'.

8.7 Storyboards

Storyboarding is a useful way to transform scenarios and personas into a format that can help to develop specifications for new tools by providing examples of how tools might be used and what they might be used to accomplish. Storyboards provide a narrative of an everyday interaction between the users and the tool based on the personas and scenarios that have been developed for a product. Storyboards can help to highlight features that would provide a significant benefit to users, including those that may not have previously been considered, in addition to handling user expectations about more obvious tasks. Storyboards can be useful to validate requirements based on what we know about the anticipated users of a digital tool, but they can also be useful for finding unexpected requirements that would add value for the users. Such requirements would be less likely to have been guessed by the designers thinking about features for the tool or suggested by the users when interviewed about the features they might like to see in the tool. In this section, example storyboards have been created for each of the scenarios described in the previous section, together with the personas that belong to the scenarios. For each storyboard, examples are also given of unexpected requirements that could be drawn from the storyboard that would enhance the value of the tools for the researchers in those different scenarios.

Figure 8.12 shows the first storyboard for Scenario 1, a closed research project that interacts with funders. The storyboard illustrates how the supervisor persona Felix might use a digital tool to interact with his team and approve experiment plans remotely.

Figure 8.13 shows the second storyboard for Scenario 1 and illustrates how a remote auditor or funder could use the same digital tool to monitor the progress of a research project. In the storyboard, auditor persona Alvin is able to provide feedback and guidance to the project team on adhering to guidelines.

There are a number of unexpected features that could be derived from these storyboards for Scenario 1. For example, experiment plans and their approval are an important part of the communication between Felix and his researchers to ensure that they can work autonomously. Highlighting that an experiment plan has been derived from a plan that Felix has already approved, what the differences are, and who approved it, makes it easier for Felix to determine whether the new plan is suitable for approval. It might also be useful for auditors, supervisors, and researchers to be able to see whether a notebook has been assessed, to see when it was last assessed, and for the system to send out a notification when a new assessment is due. These two functions would provide capabilities that would also be useful for provenance and auditing purposes. Standard templates can be useful for recording the experiment and results, particularly when a specific format or specific information needs to be captured, as mentioned in Chapter 6, and as we will discuss in more detail in Chapter 9. In addition

Figure 8.12 Felix and Alejandra storyboard – managing a team remotely.

Figure 8.13 Alvin, Felix, and Alejandra storyboard – monitoring and auditing a project remotely.

to providing tools that enable templates to be created and edited, it could also be useful to provide a method of approval for new or changed templates. The auditor or supervisor can then approve the template for use, having assessed it for compliance to standards or local best practices.

Figure 8.14 shows the first storyboard for Scenario 2, a laboratory working on open research. The storyboard illustrates an everyday situation where researcher persona Ollie plans an experiment, accesses information to complete a risk assessment, and gains approval from his manager, supervisor persona Guy.

Figure 8.15 shows the second storyboard for Scenario 2, illustrating how a digital tool could be used to help a new team member integrate with the research team. The new team member, researcher persona Lukas, uses the tool to find out about current members of the team and details of the project. Lukas uses team notebooks and shared plans as a reference to replicate an old experiment and also for comparing with his own results. When he finds a discrepancy, he uses the tool to ask researcher persona Ollie, who performed the original experiment, to review his new experiment and provide feedback for his work.

A third storyboard for Scenario 1, shown in Figure 8.16, illustrates how a digital tool could help collaborator persona Ben to find and engage with Guy's project. The storyboard also shows how the tool could be used to facilitate communication and sharing between the two collaborating research groups located in different places and time zones.

These three storyboards provide a number of examples of potential features that may not have originally been considered for the design of a digital tool for capturing the research record. The potential features focus on interaction and communication between researchers in a team and with their collaborators. For example, enabling experiment records to be 'buddy checked' or peer reviewed provides an opportunity to help new members of the team to be reassured about their work and to enable others to provide guidance. Providing gamification elements, such as trophies, badges, and ratings could encourage positive interactions within the tool. These interactions could not only encourage better quality in the recording of experiments, but also in the design of plans and templates that can be shared across groups. Providing a way to hold public and private discussions *via* the tool could also make use of gamification elements to encourage helpfulness by rewarding interaction and answers to questions. Discussions would also be useful for communication about the project and for facilitating interaction between different groups collaborating. Project updates that can be shared as news both privately and publicly, for example, through social media, could help to drive connections to other scientists working on the same problems and to link to potential collaborators.

Figure 8.17 shows the storyboard for Scenario 3, a closed local research group. The storyboard illustrates researcher persona Kat trying to replicate an old experiment. Kat makes use of our digital tool to get information to

Figure 8.14 Ollie and Guy storyboard – planning an experiment.

Figure 8.15 Lukas and Ollie storyboard – integrating a new team member.

Figure 8.16 Ben and Guy storyboard – research collaboration and project management.

Figure 8.17 Kat and Pal storyboard – running an experiment.

help her with the experiment set-up and to provide a way to seek help when something unexpected happens.

The storyboard shown in Figure 8.17 suggests a number of unexpected features that are quite similar to those in Scenario 2. If there was a facility to ask questions, then users would not only be able to ask a question related to their problem, but the answers to questions would also provide a repository of knowledge. Questions marked as 'solved' could be searched and used to provide tips to others, whilst enabling users to add an urgency or importance associated with a question could help to make them more visible to other users using the tool at the time, so the poster of the question could get help more quickly. Again, gamification elements could be used to encourage helpfulness and to rate the suggested answers so that only the best or most relevant are used by future users with the same issues or questions. Such facilities are obviously not substitutes for appropriate interactions with supervisors and colleagues, but they could provide alternative ways to get assistance when others are inaccessible.

The storyboard for Scenario 4, an undergraduate laboratory practical course, shown in Figure 8.18, illustrates how a digital tool could be used by teacher persona Jenny to design a practical experiment for her students and provide them with feedback. The storyboard also suggests how student persona Dan might use the tool to help him run and record the experiment, and also how a discussion feature might facilitate learning and sharing of best practices between the teacher and students.

The storyboard for Scenario 4 captures many of the same interactive elements as highlighted in the other storyboards and may be even more useful in an educational context. For example, enabling reviewing and rating of experiments can enhance learning and provide exemplars for students. Facilities for discussion and asking questions can encourage students to help each other, whilst gamification elements can also encourage them to work together. Private discussions could be used for feedback or to facilitate group working. Templates could be used in a slightly different way in an educational context, not only ensuring the correct information is recorded, but also to ask specific questions to assess student learning and understanding. Such a tool could also have a facility to automate reminders and send messages to the students. An additional unexpected requirement from the storyboard is the requirement for a capability to enable linking between experiments, annotations, and the various project discussions and resources, for example, to highlight an example of a problem in an individual experiment for the benefit of others.

8.8 Next Steps in Designing Digital Tools

Once requirements have been gathered, further activities within interaction design are focused on narrowing down specific requirements. For example, 'use cases' are a more formal method of describing the idealised requirements of a software system without the constraints of any particular

Figure 8.18 Jenny and Dan storyboard – undergraduate practical.

technology for implementation. In order to evaluate potential interface and system designs, prototypes can be constructed before coding begins, and test cases can be developed to validate the system functions as intended once coding begins. These techniques can be used by members of the development team to solidify the design and verify its suitability, but can also be used at both design and test stages with real or representative users even more effectively.

A use case typically describes the interaction between a user (often known as an actor) and a software system, although, sometimes, it can describe an interaction between an external system and the system under development. For example, if the system retrieves data from another system or service. The use case describes the sequence of actions that the designer believes will enable the user to achieve their goal. Use cases may make assumptions about the technology that is being used and include details about particular interface elements that the user will interact with, such as buttons, menu choices, and navigation. However, use cases do not need to be that specific, and during the design of a new tool, it can be advantageous in working towards the final design to be less rigid. For example, recording in the use case that the user will authenticate themselves to the system, without being specific about how, enables greater flexibility over the implementation of the function for authentication. The use case could be specific and record that the user enters a password at a particular point in the sequence, but being less specific, would enable the design of the implementation to consider other possible alternatives that might be appropriate, such as scanning a fingerprint or entering a code sent to their phone.

The focus of a use case is the so-called 'happy day scenario', where everything goes according to plan, and the user is able to provide all of the information that the tool requires and, in turn, the tool does everything the user expects. It is important to consider what issues the user might encounter, so other alternate sequences of events, such as actions involved when an error condition is encountered, are also usually recorded. Use cases may be recorded as a narrative in text or in a graphical form, often using Unified Modelling Language (UML). Use cases usually contain a title for the use case representing the goal or task of the actor, the role of the user, the basic sequence, and extensions such as the flow of actions under an error condition. The use case can also contain other sections depending on the needs of the project. For example, it might be useful to include a background to the task, the status of the system after completion of the tasks, and prerequisite conditions that need to be satisfied before the task can be started. It may also be useful to record circumstances or actions that might trigger the use case, for example, the user has decided to create a new entry in a notebook or the user is responding to a notification sent by the system. An advantage of use cases is that they can be used to form the basis of prototypes for validating the design of interfaces and to assist in designing test cases that can be used as the development gets under way. Table 8.1

Table 8.1 Example text-based use case for uploading a photograph to Flickr on a smartphone.

Use case name	Uploading a photo to Flickr	
Short description	User selects a photograph, configures settings, and uploads to Flickr using smartphone app	
Triggering event	User has taken a photograph they want to share	
Actor	Smartphone user	
Pre-conditions	User has taken and edited photograph they wish to upload Flickr app is already installed User is already authenticated to Flickr using the app	
Post-conditions	The selected photograph is uploaded to Flickr with chosen settings	
Normal flow	*Actor*	*System*
	User navigates to Photos on smartphone	System displays photographs
	User navigates to correct folder	System displays selected folder
	User selects photograph	
	User chooses 'share' option	System displays apps that can be selected
	User selects Flickr app	System opens Flickr app
	User selects filter from options	App applies selected filter
	User selects 'next'	App prompts user to add caption
	User adds a caption	
	User selects 'Share'	App uploads photograph with caption to Flickr website
		App exits and system returns to photographs
Alternate flow	Photograph is not available User does not change filter User does not add caption Network connection to Flickr website is not available	

shows an example use case for a user uploading a photograph to Flickr using a smartphone. When multiple actors are involved in a use case, it can be useful to use UML to illustrate the actions and flow of the use case. For example, Figure 8.19 shows a basic use case example for an online ticket ordering service where the customer represents the primary actor, but the system also talks to a second actor in the form of a third-party credit payment service required to handle the payment of the order.

The most effective way to check the usability of any product is to get others to try and use the product, or a prototype of the design, without your assistance. Testing is an essential component of ensuring usability of software products. Prototypes and testing can be used to validate the design of interfaces from the design stage of the project to final release of the product, but the earlier in the process the design is tested, the cheaper and easier it is to fix any problems identified. A prototype provides a way for both the development team and users to interact with a potential design of

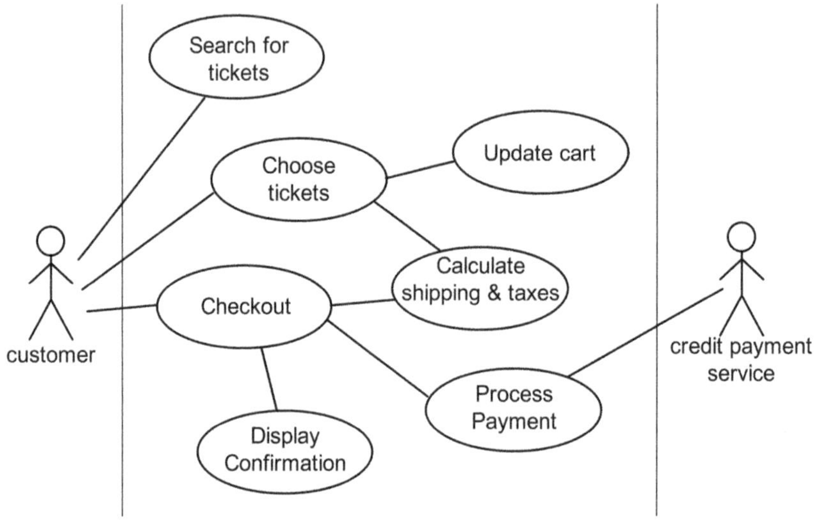

Figure 8.19 An example use case for a ticket purchasing system using UML format.

an interface, enabling them to explore its suitability and to provide feedback on the design. Different prototypes can be used to enable designers to work with users to try out different designs and choose between alternatives. Prototypes can vary in complexity, from low-fidelity prototypes such as simple paper mock-ups used during the design process, to almost complete versions of the software under development used for more formal usability testing in later stages. In either case, the user is asked to complete a task or series of tasks and is asked to talk through their actions, explaining to the observer what they are doing and what they expect to happen as they progress through the task. Observers take a note of where the users struggle, get lost, or make mistakes. If a low-fidelity prototype is used, changes can be made to the design immediately and the test continued. Later in the development cycle, when it is harder to change the interface or adapt prototypes, a formal defect raising process is more likely to be used to request changes.

Low-fidelity prototypes are ideal for evaluating potential designs in the early stages of the project because they are cheap and fast to produce, and easy to change. Alternatives can cheaply and easily be presented, for example, to enable two different designs to be tested side by side. Designs can also be rapidly altered during a testing session if any issues or new ideas are encountered. The simplest form of low-fidelity prototypes are paper mock-ups where a designer or developer draws a basic representation of a screen or web page on a sequence of index cards or pieces of paper. As the user interacts with buttons or navigation on the screens, the card can be replaced with the next appropriate card. If a change is found to be

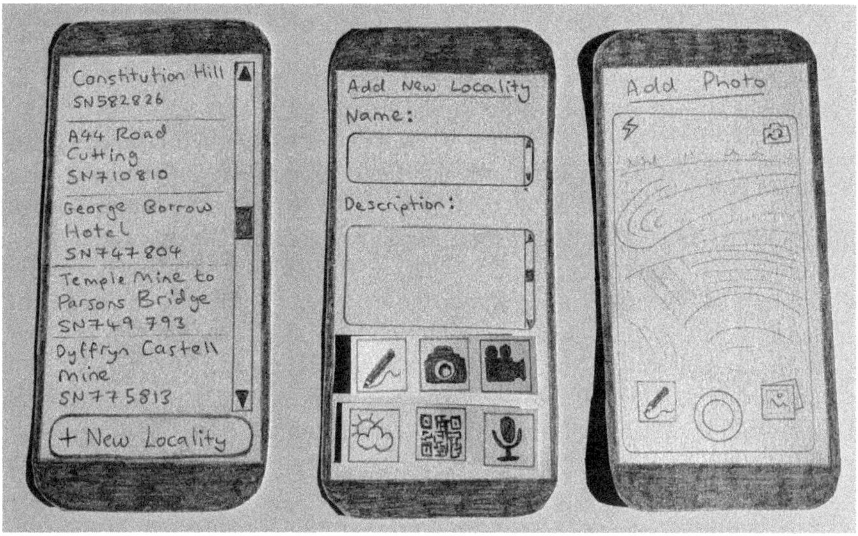

Figure 8.20 An example of a paper prototype.

needed, because the user is confused by a part of the design or an expected action is missing, a new or edited design can quickly be made so that their progress is not hindered. Figure 8.20 shows an example of a paper prototype of some screen designs for a mobile application that could be used to capture information and media about a locality for scientific fieldwork. As the example demonstrates, prototypes do not need to be sophisticated or artistically drawn to convey the essentials of the interface design.

Some people have an anxiety about producing paper prototypes because of their own inhibitions about drawing, but there are many reusable design resources available on the Web, such as images and icons, that can be printed out and included in interface designs. To create slightly more sophisticated prototypes, often known as wireframes, there are many software tools available to help design the layout of interfaces, especially for web pages and applications for smartphones and tablets. Many of these tools not only display the interface designs, but can also be used to connect different screens and pages together, so that when the user clicks on an interface element, they are automatically taken to the correct screen or page in the sequence. Such wireframes can be created using specialist software, but they can also easily be created in software commonly used by researchers such as PowerPoint and Keynote, where slides can be linked together to create an interactive interface. Wireframes are kept simple, with the emphasis on the flow of actions rather than the aesthetics of the interface. Although at first it may seem difficult to imagine using an interface drawn on paper or a very simple software design, keeping the design simple enables users to focus on the design of the interface for accomplishing a task. If the interface designs are too complete looking, users can become distracted by features of the design that are less important in the early stages, such as colours and font choice.

High-fidelity prototypes are usually created using the software of the system under development or specialist software and are usually much closer to the final look and feel of the interface. High-fidelity prototypes are most appropriate for later stages in the development process, including usability testing or during actual system development, because they are more expensive to create and more difficult to change. Using high-fidelity prototyping can be helpful during iterative software development, where the design of the actual interface of the system is evaluated at the end of a cycle of development and can then be updated based on feedback.

Many large software development organisations use a formal usability laboratory to conduct their usability testing. These rooms often have a one-way mirror, so that the user can be observed unobtrusively as they use the software, and cameras and microphones are used to record their reactions and interactions with the software. Sometimes the users are also provided with a telephone so that they can 'phone support' if they get stuck whilst they use the product. Behind the one-way mirror, many members of the development team, including programmers, testers, usability practitioners, and managers, can observe first-hand how the users interact with the product. However, such sophisticated set-ups are not necessary to do effective usability testing, and they can be quite intimidating for the users. For desktop software, a desk with a computer in a quiet location is all that is necessary, and for a product that is meant to be used on a mobile device such as a phone or tablet, it is even better to test it in the environment in which it is expected to be used, such as in the laboratory or out in the field. For less formal settings, two observers are sufficient, one to run the test and support the user, whilst the other records the observations. Software can also be tested remotely, for example, the test can be conducted over the telephone or Web. Another alternative is to ask users to keep a diary of their independent use of a product. This enables testing to take place in a more realistic environment over a longer period of time, but it does rely on users remembering to keep the diary. The user needs to have access to the software or other product under test during the time they are testing, but this is a useful technique for testing in the real environments where the product will be used and how well it works over an extended time period.

For formal usability testing, it can be useful to run each test with four or five different users. This number is ideal for identifying common problem areas, whilst avoiding focusing on individual preferences. If real users are not available for testing, then volunteers from other parts of the development organisation can be used, but they should not be directly involved in the development of features under test. Asking the volunteers to represent a particular role or using a persona as an example can help to get the volunteer to think like the user they are representing. It is important to assess whether the test subject is behaving in the same way as a real user would under real conditions. Some test subjects will behave in an unrealistic way during a test, for example, being a 'good' user and diligently attempting to read a 60-page long readme before they attempt to install some new software, even though the vast majority of real users would never do that! Users expect to be able to install and use software without sitting down and reading all the documentation first, and the test subject only engages in this behaviour because they are being observed and are worried about being judged as a 'bad' user. Natural behaviour is more likely to highlight genuine usability problems that need to be fixed. It is important that test subjects in usability testing are reminded, before testing begins, and during it if necessary, that it is the software, not the user, that is being tested. The idea is to find problems with the software, so if something goes wrong and they are unable to progress, it is the fault of the software and not the test subject. Even so, difficult to use software makes users feel stupid and frustrated, and test subjects can often get upset during testing if they get stuck. Under such circumstances, reactions can range from bursting into tears to angrily storming out of the test. Although such reactions can be an eye-opening surprise to a development team, proper evaluation of the interface design as development progresses can lessen the likelihood that such extreme problems are encountered during formal usability testing. If users do get stuck during sessions, they can be prompted to talk about what they expected the behaviour of the software to be at that point, and how the interface could be improved to overcome the problems they encountered. One of the reasons for preferring a less formal style of testing environment is that the visible presence of the usability testers makes it easier for users to ask for help and they are less likely to get overwhelmed or embarrassed if they struggle with the testing task. The wider development team can be involved in the usability testing by being invited to help observe and design the usability tests. Taking part is sessions with real users provides team members with an insight into the usability testing process, but also the expectations of users and the problems and frustrations that they encounter.

Usability can also be tested formally by specialised test experts as part of the software testing regime. Heuristic evaluation is a method often used to evaluate the usability of a system. This technique can be used by non-experts by following a checklist of usability principles, or by using an expert that has a good understanding and experience of usability principles and design problems. The technique is relatively inexpensive and requires less time and resources than formal usability testing, but will not pick up more subtle problems with the interface design or interaction flow. It is often useful to use the technique earlier in the development process and then follow it with formal usability testing at a later stage.

8.9 Usability Principles

This section describes some important usability principles that can be adopted by software development teams to help make software more usable. These principles are applicable to any software including desktop applications, applications for mobile devices, and also web pages.

8.9.1 Keep It Simple!

Increased flexibility in software leads to reduced usability by making products more complex. Adding new functions to software is relatively inexpensive and it may therefore be tempting to give the user as many new functions as possible. Although this may seem like a good idea, additional features add to the complexity of the software and can overwhelm both the user and the interface. It is therefore important to only add additional functions to software for the right reasons. A classic example I use to demonstrate this problem is a software interface that enables a user to configure a set of properties on a server object they have created in their environment. The user is presented with a window with a list of 134 different properties to configure. These properties are presented in a single list in alphabetical order by property name. In the real world, the users only usually need to configure a dozen of these properties, and in fact, dozens more of these properties are internal to the software and are not even configurable by the user! The developers have chosen to expose all the possible options, even those that make no sense, leading to complexity and confusion for the user. The developers argued that they do not know what the user will need so therefore they provide everything, but this is a poor excuse.

A number of improvements can be made to make such an interface more usable. First, all of the settings that the user cannot configure should be removed. If these settings become configurable later, then they can be added back into the list. The remaining properties can be categorised and organised by the different kinds of properties. For example, the properties could be grouped by related function, such as network settings and security settings. An alternative way to present the properties is to show the user the most commonly used properties first and the less commonly used included in an 'Advanced' tab or panel in the interface. Instead of being overwhelmed by a huge list of properties that the user is unsure whether they can or need to configure, the user is instead presented with a restricted number of properties in a sensible order.

It is common that additional functions get added in new releases of software products, and an original and simple design can often become cluttered and confusing as new buttons, tabs, and other controls are added to the interface. Often, new functions are requested and used by a minority of users, making it difficult for the majority to work out what the functions are used for. This tendency to add new functions leading to a cluttered interface is particularly common for mobile interfaces where many functions may be presented on a single screen. In order to avoid over cluttering the main interface, it may be better to rethink the entire interface, or at least to add new screens for new functions, rather than trying to fit them all into the original interface.

8.9.2 Familiar Concepts and Metaphors

Using familiar concepts can help users to understand the software better and work out how to use it, as mentioned earlier in this chapter. For example, the metaphor of the office has long been used in the design of software and computer systems. Filing cabinets full of records or documents are used to represent databases and the filing systems of operating systems. Dragging items to the trash bin is an obvious metaphor for deleting files. Metaphors are used to create meaning in the abstract world of software and data, but they need to be used carefully to support user understanding without constraining the capabilities of the software to what can be achieved in the real world.[9] Using familiar concepts also extends to the use of terminology in the interface, and making use of simple and clear language and concepts makes software simpler and easier to understand. Concepts are often used in the design of icons and buttons for interfaces, and although not always possible, it is better to make use of meaningful symbols and designs than to generate new abstract icons that cannot easily be interpreted. It is also helpful for users to avoid using very similar icons because users can struggle to distinguish between them, and as a result, may be more likely to make mistakes. Figure 8.21 shows a variety of common icons, some of them are much more recognisable than others because they represent objects in the real world, for example the camera, calendar, and padlock. The user has a good chance at guessing what the icon represents and what selecting it might do. Other icons are less obvious and the user may have to look up the meaning.

It is also important to be consistent with the use of these concepts and metaphors across the software product. The same icon should not be used

Figure 8.21 Metaphor is commonly used in icon design.

in different parts of an interface to represent different actions. Consistency with other software and standards is also important because users expect familiar concepts from familiar software to be present in other software that they use. For example, one of the most recognised icons is the floppy disk symbol used to represent a 'Save' action in the majority of software, even though many people have never actually seen or used a floppy disk. In fact, many users misinterpret the design of familiar icons to match their own experience. For example, users who have never seen a floppy disk often think the icon represents a garage or loft 'where stuff gets stored'. Other users, familiar with the padlock icon when shopping online, often do not realise that the icon is meant be a padlock and instead interpret it as a purse or handbag! Even though the interpretations are wrong, they still work for those users because the concepts and associated actions are consistent. The familiarity with these standard icons demonstrates that consistency with what users already know is better than inventing new icons and symbols unnecessarily. There should be both consistency of the icons used and the functional behaviour. Providers of platforms for software development, such as Microsoft, Apple, and Google, often have their own style guidelines to help developers to create a better and more consistent experience for their users, including the use of icons.

8.9.3 Make Choices Visible and Let the User Correct Mistakes

A user often needs to make choices in software and so it is important that these choices are clearly visible, and that appropriate controls are used in the interface to support them. When choices are clear, users are able to learn and complete work tasks efficiently. There should never be any ambiguity in the labelling of interface elements or messages presented to the user. For example, buttons should be labelled meaningfully so that it is clear what the user is choosing to do. A common usability problem in software comes from the design of 'dialog boxes' where the user has to choose an option. Often the default choices are 'OK' and 'Cancel', or other choices that can be ambiguous and confusing for the user, causing them to make mistakes. For example, in the dialog box shown in Figure 8.22, the ambiguous choice of Delete or Proceed leaves the user uncertain as to what action they are performing and provides no way for them to change their mind. In contrast, the dialog box shown in Figure 8.23 is well designed, informing the user about the action they are choosing to take and making it clear which button will delete the photograph and which allows them to change their mind.

The design of an interface can also be constrained in such a way that the user cannot make mistakes. For example, if the user has an either-or choice such as turning a particular behaviour on or off, then making use of a switch control makes it very clear those are the options available and which of those options is currently selected. Alternatively, if the user can choose from a range of values, then a slider control indicates the choice available,

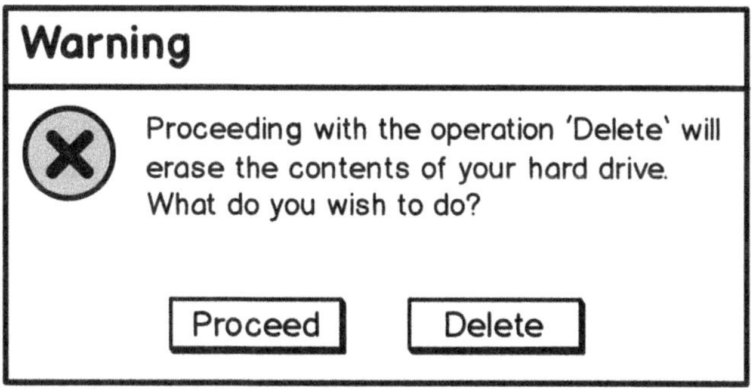

Figure 8.22 A badly designed dialog box, the options are ambiguous and confusing.

Figure 8.23 A well designed dialog box, the choices are clear and unambiguous.

and providing sensible maximum and minimum values on the slider enables the user to make a sensible choice.

Both new and experienced users can find themselves in situations where they want to change their mind, have make a mistake, or want to return their data or system to a previous state. New users, in particular, may click on various buttons or controls on the interface to see what happens. A user should be able to make mistakes without fear of causing any permanent damage, such as losing data or permanently changing settings, and it is therefore important to provide ways for users to get out of a situation. A number of mechanisms can be used to help the user avoid unintended actions and errors. Providing an undo option or a reset option enables the user to easily revert the data or system into the previous state if they realise they have made a mistake. Ideally, however, we want to try to prevent the user from making mistakes in the first place by providing the option to cancel an action if they change their mind or by asking the user for confirmation before they perform actions that may have significant consequences, such as overwriting files. Validating what the user enters at the point they enter it also enables problems to be immediately identified and corrected. For example, if a user adds text into a field that should only contain numbers, warning the user at that point they must only enter numbers in that field is much better than waiting until something goes wrong because the data is invalid.

8.9.4 Appropriate Feedback

Feedback is very important to users so that they are aware of what is going on. For example, if a user clicks on a button and there is a delay as the system carries out an action, the user will become confused or frustrated if no feedback is provided. The user may think that the button press was not recognised or that the software has crashed. The user may repeatedly press the button until something happens or give up waiting and shut down the software, both of which can cause unexpected problems to occur and potentially even cause a loss of data. Simple feedback can include messages that indicate the system is performing the requested action and indicators of passing or remaining time, for example, egg timers and progress bars.

Sometimes an error occurs when the user has performed an action in the software. These errors may be due to a mistake the user has made, a problem with a system the software is interacting with, or some internal problem with the software itself. Error messages in software are often poorly designed with a lack of thought about what would be useful for the user trying to solve a problem they have encountered. They are often ambiguous or provide meaningless internal error codes that tell the user nothing about what the problem is or how they can fix it. When something goes wrong, it is important to provide meaningful messages to help the user know how to respond and whether there are any consequences as a result of the problem. An error message should make it clear in non-technical terms what went wrong, what the consequences are (for example, if any data may be lost or if the software needs to shut down), and what the user needs to do next to fix the problem. For example, if the user has entered data in an incorrect format then examples of the correct format can be provided, or if a connection to an external service was lost then the user can be asked to check their connection to the network.

8.9.5 Providing User Assistance

From time to time all users need help with using software and most software comes with some form of technical documentation explaining what it can do and how to use it. However, help should be provided where the user needs it, rather than forcing them to try and find the answer within the documentation or on help forums. On software interfaces, providing sensible labels and a sentence that explains the interface element can make it more intuitive to use. Text in the interface can provide simple prompts to help a user to know where to start or what to enter into a field, or provide more in-depth guidance with decisions or with learning a new interface.

There is a balance to be met between meeting the needs of new users and experienced users, and as previously mentioned, an excess of text and other items in an interface can hinder usability. A concept known as 'progressive disclosure' can be used to restrict the content initially, but enable a user to see more information or more complexity if they want to. For example, the user could click a button to choose to display additional information about a particular control or action in the interface, or to bring up an overlay that shows a brief description of various buttons. Software can also guide new users with tutorials and walkthrough guides to help them get started. Some software even restricts the user to the more basic functions as they learn. More experienced users can choose to access an expert mode, where more advanced features are visible in the interface and less help is provided to keep the interface clearer when it is not needed.

8.9.6 Designing for Mobile Platforms

Mobile platforms such as smartphones and tablets present challenges for good design. Taking an interface or website designed for a desktop computer and shrinking it down to fit a tablet or smartphone screen creates a poor experience for users, with text difficult to read and controls too small to use effectively. Different devices have different sized screens, and these are often much smaller than a laptop or desktop computers, meaning that the space needs to be used more efficiently. For Web-based platforms, responsive and adaptive designs can be used to ensure that elements of the interface fit appropriately on different screen sizes. Mobile devices also often have different ways of interacting with the software and inputting data, for example, with a touchscreen, camera, microphone, and a fingerprint reader that may enhance some experiences but make others more difficult. For example, it can be more difficult to precisely select a small element of the interface or draw an accurate drawing with a finger compared with using a mouse. Touch controls can be effective, but it is important to keep them simple and intuitive, and not overload the user with too many controls. For example, some applications for the iPad use complex combinations of multiple finger touches and one, two, and three fingers all do something different. Added to these are stretching, pinching, and swiping actions, leading users to become confused and unsure what they did to initiate a particular action. If many different controls like this are needed to control an application, then it is likely that the interface needs to be rethought.

8.9.7 Designing for All

Digital tools can be a lifeline for users with special needs, but poorly designed software can be impossible to use for people with disabilities and other needs. More than one billion people worldwide have a disability, but we can all benefit from software that has been designed with universal usability in mind.[10] When we think about disabilities, we tend to think about permanent disabilities such as visual and hearing impairments or mobility problems, but there are also other conditions to consider. For example, less commonly considered impairments are visual problems such as colour blindness, cognitive problems such as poor memory and dyslexia, and temporary disabilities such as a broken arm or forgetting your glasses. There are also situational problems that can benefit from accessible design. For example, if software relies heavily on sound for communicating information to the user, it will be difficult or impossible to use in certain situations, such as when the user is working in a loud environment, or even a very quiet environment like a library where sound needs to be turned off to avoid disturbing others. Disability can affect all of us at one time or another. Not only does our eyesight, hearing, cognitive abilities, and movement degrade as we age, but we can also suffer from unexpected disability at any time, for example, following an accident or illness. After spending many years presenting to others about the importance of designing for accessibility in software, I have had first-hand experience of the challenges of using software with a cognitive impairment after receiving a brain injury in an accident.

Blind users generally make use of specialist screen readers to use software, but users with less severe visual impairments benefit from being able to change the contrast, size, and colours of characters on the screen. Not using colour as the only way to signal information to the user is important, not just for users with colour blindness, but also because our ability to distinguish colour and contrast diminishes as we age, making it harder to pick out the words and links on web pages, for example. Interfaces that are cluttered and contain many small buttons can be difficult to use, confusing, and even overwhelming for users with cognitive impairments. Including plenty of clear space and large buttons in the design can help. For users with limited mobility in their hands, providing large buttons and other controls with space between them makes it much easier to accurately select the buttons and click them. Interfaces that are difficult to use by users with special needs are also usually more difficult to use by everyone else as well, so universal design ultimately improves usability for all users.

8.10 Conclusion

Designing for usability is undoubtedly an essential part of the software development process. Poorly designed software is frustrating for users and reduces the chance that a product will get used. Digital tools for science are often developed by small teams that may not have formal software development training or the budget for specialist user experience professionals to help with development. However, the tools of interaction design and the principles of usability presented here can be used by anyone to design software that is more useful and usable for the community. As mentioned in the introduction, the design of interfaces in digital tools affects the way that users behave, and in Chapter 9, we will take a look at the positive and negative effects that different interface designs may have on recording the scientific record.

References

1. M. Kurosu and K. Kashimura, in *CHI'95 Conference Companion*, 1995, p. 292.
2. ISO 9241-11: Ergonomics of human-system interaction – Part 11: Usability: Definitions and concepts, 2018.
3. A. Cooper, *The Inmates Are Running the Asylum: Why High Tech Products Drive Us Crazy and How to Restore the Sanity*, Sams – Pearson Education, Indianapolis, 2nd edn, 2004.
4. J. Nielsen, 2016. Available at https://www.nngroup.com/articles/computer-skill-levels/ (accessed Sep 2018).
5. S. Smith, G. Salaway and J. Caruso, *The ECAR Study of Undergraduate Students and Information Technology*, EDUCAUSE Center for Applied Research, 2009.
6. I. Rowlands, D. Nicholas, P. Williams, P. Huntington, M. Fieldhouse, B. Gunter, R. Withey, H. Jamali, T. Dobrowolski and C. Tenopir, *Aslib Proc.*, 2008, **60**, 290.
7. J. Goodger and W. Worthington, *Research Data Management Training for the Whole Project Lifecycle*, 2013 Available at http://research-data-toolkit.herts.ac.uk/document/rdmtpa-final-report/ (accessed Sep 2018).
8. Y. Rogers, H. Sharp and J. Preece, *Interaction Design: Beyond Human-computer Interaction*, John Wiley & Sons, Chichester, 3rd edn, 2011.
9. J. Donath *The Social Machine: Designs for Living Online*, The MIT Press, Cambridge, MA, 2001.
10. *World Report on Disability*, World Health Organisation, Geneva, 2011.

CHAPTER 9

Interfaces for Capturing the Experiment Record

9.1 Introduction

The information we capture in our paper notebooks acts as a personal 'external memory' and as a way of enabling our research to be reproduced by others. As previously mentioned, these notes form an important part of the scientific record that gives important context to the data we collect and the findings we make. It is essential that what we record is accurate, complete, and understandable. The nature of what we record often differs by discipline and the environment we are working in. For example, in the field we are more likely to record the details of observations about a particular locality we are exploring, or a particular species we have encountered; whilst in the laboratory, the focus may be on the processes and procedure associated with an experiment. As discussed in Chapter 3, what we record is a combination of what we have learnt in the classroom, from formal guidance or training, from colleagues or mentors, and through our own trial and error. In our paper notebooks, we can choose what we record, in what ways, and how we structure and organise our records. However, in electronic notebooks we are more constrained. Although there may be new ways to capture our records, for example, by being able to include photographs, video, and even audio directly within the notebook, it may be more difficult to capture information in the ways that we would in a paper notebook. When we are asked to capture our research on a computer, there is a likelihood that the design of the interfaces of the software we use will have a strong influence on what we record.

Generic electronic notebooks, which we have previously discussed, such as Microsoft OneNote, Evernote, GoodNotes, and WritePad, often mimic

Recording Science in the Digital Era: From Paper to Electronic Notebooks and Other Digital Tools
By Cerys Willoughby

Published by the Royal Society of Chemistry, www.rsc.org

the paper notebook in their design. Although some generic electronic notebooks are restricted to textual entry, many also include the ability to add drawings, in addition to extended options for capturing content such as photographs and videos. In contrast, specialised electronic notebooks, and in particular, Electronic Laboratory Notebooks (ELNs), have diverse interfaces and encourage users to interact with them in different ways. As discussed in Chapter 6, there are both generic and highly specialised ELNs, and although the more generic ELNs may have simplistic interfaces that are more like the electronic notebooks, many ELNs make use of form-like interfaces to capture information about scientific experiments and their results. The advantage of form-like or 'template' interfaces is that they can ensure the capture of specific information, particularly useful in environments where standard information needs to be captured for every experiment or test in the laboratory. In such environments, there is a risk that a 'free-form' interface will result in the loss of information because the user forgets to record some of the information. For environments where the information to be recorded is much less prescribed, the greater flexibility of the free-form interface may be more beneficial, and tension may rise if the user is constrained in what and how they record by being required to complete information requested by a template.

Not only do these interfaces potentially have different benefits in different environments, they also have the potential to significantly influence what information is captured in the scientific record. It is therefore important that we understand what impact the designs of such interfaces have upon the scientific record. Interfaces for ELNs should encourage the collection of information that is accurate, complete, and understandable to facilitate the needs of both the creators of the information and those who may subsequently need to assess and reuse the information. For an ELN to be fit for purpose, it must be able to capture the experiment record at least as well as a paper notebook, if not better, and not have any negative impact on either the record that is captured or the research that is being conducted.

In this chapter, we examine the potential problems with interfaces for recording research, and in particular, looking at how the use of templates and forms may influence the information that is captured in the scientific record. We will also take a look at the role of mobile electronic notebooks in bridging the gap between paper notebooks and ELNs, including some examples of both generic electronic notebooks and ELNs for mobile devices. We will also briefly examine some alternative interfaces for capturing electronic records.

9.2 Potential Problems with Using Templates

Within the laboratory, capturing the record is more than just capturing the experiment procedure or the result of a standardised test. The record should also capture quality information about the context of the experiment such as the design and objectives, details of the materials used, observations made during the experiment, any deviations made from the original plan or procedure, and an evaluation of the results. Explanations included with these notes can provide essential details linking observations made to both successful and unexpected results, and also linking any problems that occurred with the theoretical background of the experiment. The additional context provides important information to help with the design of future experiments, in understanding the success or failure of the experiment, conditions needed to replicate and improve the experiment, and to back-up research claims. Providing a formal structure, such as templates, for capturing the experiment might provide an advantage over paper notebooks by encouraging consistency, promoting good record-keeping practices, and reducing the risk of mistakes.[1,2]

In fact, many ELNs provide template or form-based interfaces for entering information to capture both the design and execution of experiments. These templates act as cues or reminders to record the particular information that has been requested. As we discussed in Chapter 2, studies of memory show that cues can change what we recall, and cues are frequently used within questionnaire and form design to act as prompts to encourage the capture of specific information. Although the use of templates and cues are very likely to encourage the capture of desired information, it is also possible that there may be a negative impact from their use by potentially constraining or changing the information that is captured. Studies investigating the use of electronic medical records have suggested that there may be a number of possible consequences of different interfaces for capturing information using structured templates and free-form reports. Whilst a large number of ELNs provide a template format for capturing information about experiments and other laboratory activities and resources, many ELNs also provide a simple free-form interface for capturing information. In fact, the majority of electronic notebooks of all kinds do so. These free-form interfaces are usually in the style of a word processor document, allowing users to organise the information on the page how they want to and add additional formatting as desired. In a flexible interface where the user is provided with a simple text box, we might expect that researchers will record much the same information as they would record in their paper notebook. If the interface is instead a formal 'form' or template style interface, then we would expect different information to be recorded. A potential problem with the simple text box or word processor style is that the user does not know what to write. During usability studies with researchers who were new to ELNs, it became apparent that they struggled to know what to write in the ELN because they did not view the ELN as a replacement for the paper notebook, even though we might assume that they would. As previously mentioned, I call this the 'blank canvas' effect, because the empty page of the electronic notebook provides no helpful prompt to help a user get started. In contrast, using a form or template-based interface provides prompts that have value in helping the user to at least have some idea

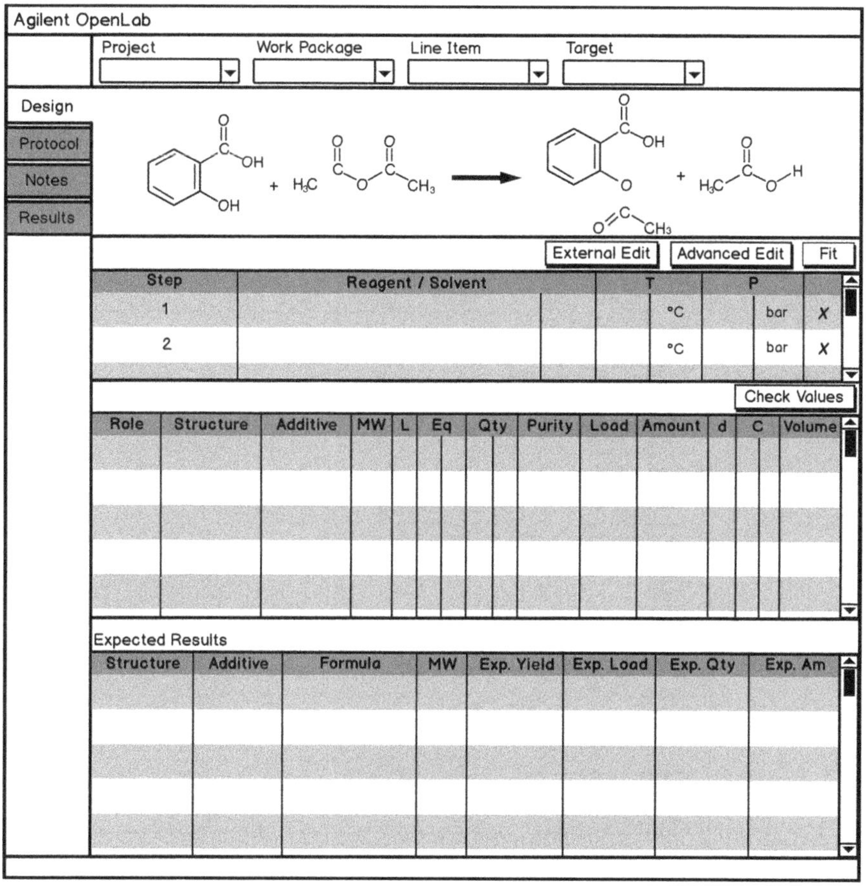

Figure 9.1 Example of an ELN template, based on Agilent's OpenLab ELN.

about what they should capture in an unfamiliar tool. Figures 9.1 and 9.2 show different styles of templates from two different ELNs. These examples are fairly typical of the types of templates found within ELNs, where users are prompted to record information in a table or in a 'free text' format with a prompt in the form of a title, such as the 'Methods' section in Figure 9.2. Users are also often able to capture additional content by importing other kinds of files such as images or spreadsheets.

A common complaint from users of structured templates is about the inflexibility of the template, where the sections in the template are not appropriate to a particular activity or record that needs to be captured.[3] Although many ELNs allow custom templates to be created, it can be difficult to design an interface or template that is suitable for all potential users, and even individual researchers working within the same discipline may find that they want to capture quite different information from each other. Even an individual researcher may need to capture different

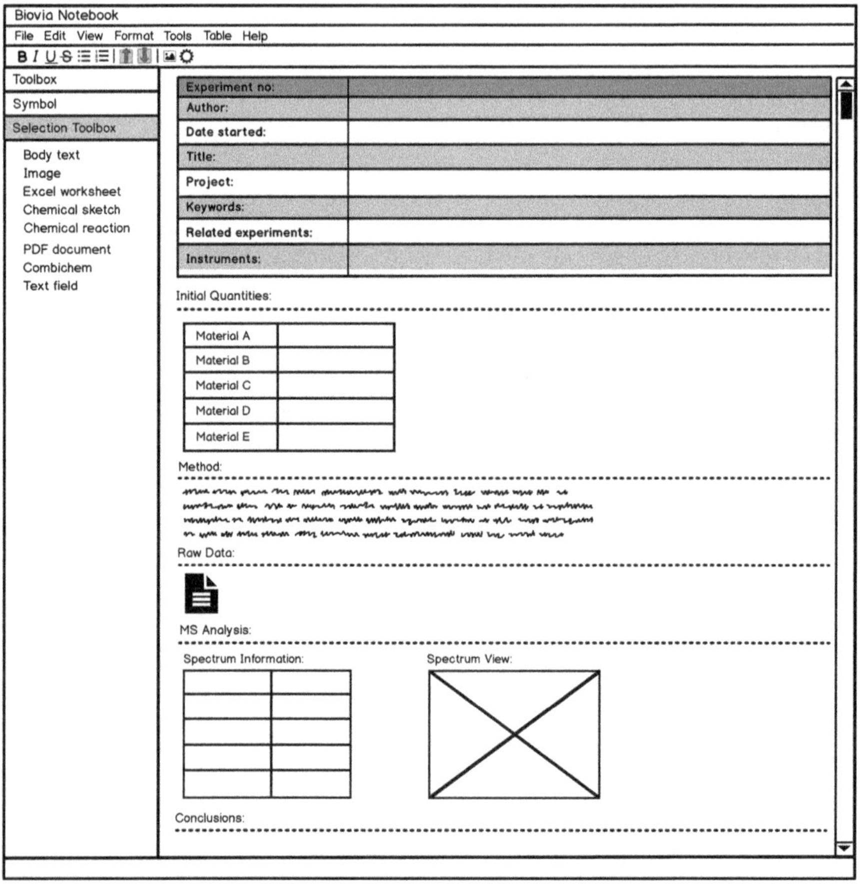

Figure 9.2 Example of an ELN template, based on Accelrys' Biovia Notebook ELN.

information between different experiments. Other researchers report that ELNs which enforce a rigid structure for capturing experiments were too cumbersome and inflexible to be used. In particular, that they are not suited to the less formal and more haphazard nature of academic research, with many feeling that templates are better suited to the more rigid (and consistent) requirements of industrial laboratories.[1] From my observations, inflexibility is often a reason why users who try out ELNs choose to later abandon them. It can be very frustrating and time-consuming to work around a system that does not enable you to do exactly what you want to do, as discussed in Chapter 7. Related to the inflexibility problem is an even more concerning issue. Although templates may act as a reminder to record particular information, if there is no space to record information about a specific topic within it, important information may not be captured at all. For example, a study looking at the use of computer templates for recording medical notes about chronic diseases discovered that the templates not only led to important health factors for patients not being recorded because there was no 'box' in which to record them, but in fact, the template entirely changed the way that the nurses interacted with the patients in health check sessions.[4] In particular, the templates had a minimum of space for recording free-text discussions and was instead focused on the collection of measurements. The change to the template changed the way that nurses interacted with patients, ensuring the capture of important clinical information, but also reducing the record to generalised data, rather than an accurate record of the patient's unique experience and specific symptoms. Such template design problems are not limited to health records, and similar problems can be observed when templates are designed for recording experiment records, as we shall discuss in more detail below. We will look at some of the effects of different interface designs and different ways to try to resolve some of the problems later.

9.3 Templates for Capturing the Experiment Procedure

Although Electronic Laboratory Notebooks can fulfil several purposes, the primary function of an ELN is to capture the experiment record. As previously mentioned, templates can be used to standardise the information that is captured within an ELN, but there is some evidence that the relative inflexibility and formality of templates may have a detrimental effect on the information that is captured. In the chemistry department at the University of Southampton, we were interested in the use of templates as part of understanding ELN use and improving the designs of ELNs. Based on the comments of researchers about their frustration with ELNs, we suspected that the inflexibility of templates may be a potential problem. However, it was unknown whether using templates would have a positive or negative impact on the recording of experiment-related information. We had expectations that using a template would cause the consistency of recording to be improved because we expected that the information we specifically asked for in the templates would be recorded. However, we also thought that the inflexibility of a template might cause some information to be lost. To investigate whether templates would have a significant impact on the information that was recorded for experiments in the laboratory, we ran a number of studies to compare the information recorded on a 'blank canvas' compared with information recorded using a structured template with undergraduate organic chemistry students.[5] In the first study we conducted, the students were asked to write a report of their experiments under one of two conditions: 'No Template' and 'Template'. The No Template condition was essentially a blank piece of paper on which the students could write free-form whatever they wanted to. In the Template condition, they were given a very simple template with the following high-level headings: Name of Experiment, Aim of Experiment, Balanced equation with relative molecular masses (RMM), Step-by-step experiment procedure, Results, Discussion, and Conclusions. These headings provided cues for information that we might be expect to be recorded in an experiment, based on the University of Southampton guidelines, with which the students would already have been familiar. As discussed in Chapter 2, cues provide a prompt for memory, and not only help to remind us to record information, but also prompt the recall of related information.

Because the headings in the template were based on what we already expected the students to be recording, we did not expect very dramatic differences between the two conditions, we expected that similar information would be recorded on both the template and the blank page. However, when we analysed the results, we were quite surprised to find out how different the results of the two conditions were. On the blank page, the majority of information recorded was about the procedure of the experiment. In contrast, much more information was captured on the template, with the majority of students providing the information that was requested by the cues, including information about what was learnt from the experiment, theoretical information about the experiment, discussions, and perceptions of success of the experiment. However, for the records captured on the blank piece of paper, there was much more information about the personal experience of carrying out the experiment, with observations and explanations provided about actions taken during the experiment that were frequently missing from the template version. Even more unexpectedly, a large number of the students switched style between the two templates. The majority of students used the past tense to describe their experiment on the blank piece of paper, but half of those students changed to a 'command' or instructional style in the Template condition. For example, on the blank piece of paper, a student wrote as part of the procedure 'A solution of CCl_4 and Br_2 was poured into a dropping funnel and the light was shone onto the rxn mixture', whilst a student who carried out the same experiment wrote 'while shining a light drop the bromine CCl_4 mixture drop by drop' on the template. On the blank piece of paper, the majority of students wrote a 'what I did' account of the experiment, including what they saw and what decisions they made and why, whereas on the template, the majority of them wrote a 'how to do it' account of the experiment, missing important personal elements of the experiment. The 'what I did' account is more like the kinds of notes that are typically recorded in the paper notebook, where not only the procedure followed is recorded, but also the thoughts, observations, and experiences of the scientist in carrying out the experiment.

The results suggested that a template could in fact have both positive and negative consequences for recording the experiment record, as we suspected. In a second study, we added a second template to try and recapture the information lost in the original template by asking using more personal questions as cues. Rather than use simple headings, the new 'Profile' template prompted the students for information in the form of questions, such as 'What were the aims of the experiment?' and 'What reactions were involved in the experiment?'. The Profile Template also attempted to prompt the recall and recording of more about the personal experience of the experiment by asking questions such as 'What did you do in the experiment?', 'What observations did you make in the experiment?', and 'What did you learn from the experiment?', as well as deliberately prompting for information that was missed before such as 'Did anything unexpected happen?'. Overall, the results of the second study backed up the important findings of the first, that using a template changes the information that is recorded compared with a blank piece of paper, and that using cues encourages the recall and recording of the prompted information. Interestingly, the new template that asked 'What did you do in the experiment?' had a significantly reduced amount of information about the procedure of the experiment. In fact, in many cases, the responses became just a list of techniques without any details of the specific steps taken!

From the studies, we learnt that templates can indeed elicit information from scientists that would otherwise fail to be recorded, but there is also the possibility that important information about what actually happened in the experiment could be lost. Getting the information back and in the appropriate context can be quite difficult. The nature of the cues, the specific prompts used, and even the combination of prompts makes a big difference to exactly what is recorded. We also ran another study along similar lines with academic researchers and staff working in various chemistry disciplines as participants. They were asked to complete an experiment using a Lego car powered by an Alka-Seltzer reaction, and asked to record their experiments using one of three Google Doc templates. Although this study was less formal than the first two, the activity provided a valuable opportunity to get feedback on the experience of using the different templates through a post-experiment discussion, where groups using different templates were brought together to discuss and share their experiences. This study used the same templates as before with some additions, including a 'results' table with appropriate column headings in the original titles-based template. There were a number of interesting outcomes from this study. The first was the genuine surprise expressed by the study participants when they realised how different what they recorded was between the different templates. They had expected that the records would be very similar to each other, but in fact they were very different. Much more information was recorded with the cues provided by the Profile template compared with the blank page where they could record whatever they wanted. Another surprising outcome of the study was that the addition of the results table in the title-based template had a significant undesirable effect that had not been anticipated. The presence of the table resulted in considerably less information being recorded about the experiment in this template compared with the other two. This was because the groups using this template had spent almost all their time diligently filling in the results table, but had neglected to enter other information. The lack of other information recorded in the template meant that little was actually captured

Table 9.1 Results table from the title-based template used for capturing the Lego Rocket Cars experiment.

Run Number ($\frac{1}{4}$ tablet)	Quantity of water ($\frac{1}{2}$, $\frac{1}{3}$, $\frac{1}{4}$)	Weight of car (g)	Aerodynamic car shape (Yes/No)	Time to launch (seconds)	Distance travelled (cm)	Comments
1						
2						
3						
4						
5						
6						
7						
8						

about why and how of the experiment, and certainly not enough for someone else to be able to repeat the experiment. In fact, these teams spent so much time filling in the table that they hardly ran the experiment! Not only did the change in the interface for recording the experiment change the information that was recorded, but it reduced the amount of research performed. The failure of the results table in the template demonstrates just how a poorly designed interface could actually change the behaviour of the researcher and even derail the research. Table 9.1 shows the results table that the researchers were asked to complete. Although the intention of the table was actually to be helpful, it clearly hindered their activities and reduced the quality of the record they kept. This effect is similar to that seen in the medical templates for chronic diseases described above. The actual nature of the activities performed by the nurses changed to be focused on making the measurements, and the same was true of the results table, in that the researchers chose to focus on the data at the expense of any kind of context information about the methods or other aspects of the experiment. Many templates within ELNs use tables in a similar format to record details about materials and samples used in an experiment, measurements, and results information. The finding of this study suggests that tables in templates need to be carefully considered to ensure that researchers do not rely on the contents of the tables alone to explain their experiments, and that they do not change the behaviour of researchers during the experiment in a negative way.

As we have mentioned before, researchers prefer flexibility when capturing their notes, and in the post-experiment discussion, the teams indicated that they liked the blank page for its flexibility and freedom, but they also recognised that important information had been 'forgotten' compared with what was captured in the Profile template. The teams found the Profile Template more laborious to complete, but they felt it was a good starting point for developing a template that would match their own requirements. The teams felt that they would like to be able to create templates based on personal need. These desires fit well with the options that many ELNs provide of being able to create and customise templates, and would help get around some of the problems of inflexibility.

The studies described in this chapter provide strong evidence that the design of interfaces for capturing experiments have an impact on the information recorded. The findings also indicate that relatively subtle differences in designs may not only have a large impact on the quality and value of the information that is recorded, but that some designs could have a detrimental impact on the way an experiment is actually performed. Providing a formal structure for recording experiments certainly has some benefits. The amount of information captured is greater and information that would otherwise be forgotten is recorded and therefore available for later use. The specific cues that are used do influence the information that is recorded, with the presence of a cue encouraging the recall and recording of that specific information, but also related information such as the related theory, lessons learned, and perceived success. The use of a template could improve the quality of the record, certainly in comparison with the 'blank canvas'. The experience of the researchers and the types of experiment being recorded may also affect what is captured within an ELN. For example, in the first two studies, the main information recorded was the personal memory of the procedure of the experiment, whereas the Lego Rocket Cars experiment was more repetitive and included making measurements, and consequently generated more 'self-structuring', with more information recorded about results and conclusions.

The use of cues does have some unexpected effects. Sometimes a simple prompt like the heading 'Aims' is all that is required to get the information requested, but the studies also demonstrated that the precise wording can make a significant difference to what is recorded and how. On the blank templates, all of the participants recorded information about the procedure of the experiment, and often the observations and explanations associated with the actions taken in the experiment. Often, no other information was included. In the first template, the cue 'Step-by-step experiment procedure' was used. This cue seemed to change the recording of the procedure for many of the participants from a personal record of what was done in the experiment to a more formal list of instructions of what to do. The result of this was a loss of detail about why actions were performed and observations made, together with a switch in style for the paper-based template. We had hoped that changing to a personal perspective with the questions in the Profile template would have led to a response more like the blank page, but the open nature of the question may have led to the even briefer responses, 'What did you do in the experiment?' is actually more open than 'Step-by-step experiment procedure'. Some studies have suggested that open-ended questions may lead to a more superficial search of memory, and therefore using closed questions can generate more results.[6] This may be because the more closed questions contain more specific cues, for example, 'What chemicals or other materials did you use in the experiment?', 'What instruments or equipment did you use in the experiment?', and 'What were the aims of the experiment?' produced sensible and specific responses from all of the participants in the Profile template condition. In contrast, the more open question, 'What reactions were involved in the experiment?' worked less effectively than the more specific 'Balanced equation with relative molecular masses (RMM)'. Care therefore needs to be taken with selecting the most appropriate cues to ensure the capture of relevant, useful, and accurate information.

The loss of explanations and observations, seen in the first study, were reinstated by asking specific personal questions such as 'What observations did you make in the experiment?' and 'What did you learn in the experiment?'. This increased the number of observations and explanations in the Profile template, and a similar effect was also seen in the original template, where more explanations and observations associated with the experiment were captured in the Results, Discussion, and Conclusions sections. A problem, which is not overcome by using these cues, is the separation of the observations and explanations from events and actions within the procedure of the experiment where they occurred. The separation between the experiment procedure and the associated observations and explanations leads to a loss of some of the temporal and contextual information from the experiment. Not only does the temporal and contextual information help the researcher to remember what happened in the experiment, but is also likely to be more useful for someone else trying to repeat the experiment.

There are benefits to all the styles of templates; although as we saw in the third study, including the wrong kind of cue in the template can have significant negative consequences. The inclusion of the results table in the template was not expected to generate negative effects, but the table had the result of constraining the information that was recorded by causing the researchers to diligently focus on completing the table at the expense of both completing the other fields in the template and even completing the experiment itself. Including a table of results as standard in a structured interface is problematic. Many ELNs include these kinds of tables, and even in ELN interfaces where tables are less prominent, such as the more generic LabTrove ELN, the majority of templates that are used and shared by research groups do include tables for results or for describing the properties of materials, samples, or instrument settings. Researchers find tables particularly useful for capturing information for repeating experiments. However, designing a standard table for an experiment is not practical, and even a customised table may cause problems. Providing tools for researchers to design their own tables and including them within templates may be more beneficial than enforcing a standard table structure. A cue to 'Create a results table', with perhaps a choice of some standard table templates as a starting point could be useful. It may even be possible to include some intelligence in the interface to suggest columns for a table based on the content of the experiment record and may be more useful than directly including a table in the template. For those environments where standard data is always recorded and does not change between experiments or test runs, then the use of standard tables make sense.

Study participants found using the Profile style template for recording their experiments to be more onerous than using the blank piece of paper. They also felt it provided less flexibility and freedom. A balance needs to be struck between capturing 'the right' information and making the process too much of a chore for the researchers. The question of 'What should be recorded for an experiment?' can probably only be 'it depends'. Requirements on what needs to be recorded are likely to be different between different disciplines, different kinds of experiments, and different kinds of organisations. As we saw in Chapter 3, different disciplines and sub-disciplines have their own guidelines and standard practices. What is recorded is also dependent upon the perspective of the potential audience for the record. For the author of record, they probably need a record that is sufficient to enable them to repeat the experiment, to view the results, to compare the results with other experiments, and to troubleshoot the experiment if something went wrong. For someone attempting to repeat the experiment, they need a full description of the methods and materials, together with explanations of why certain tasks were performed, why the protocol deviated from the plan, and whether anything unexpected happened. Someone repeating the experiment also needs to have a good record of the results, whether they matched the expectations of the author, and what processing or analysis may have been performed upon them. Other readers may want to know why the experimenter was carrying out the experiment, how it relates to previous work, and the theoretical basis of the experiment. For more exploratory work, more information is needed to be captured about the theory, rationale, and objectives for the research. More detailed procedure information is required with explanations when new protocols are being developed, compared with routine everyday repeating experiments or tests. Different templates had different benefits for capturing these different kinds of information, for example, the results table is ideal for capturing data where the measurements that are needed are known. For answering 'Why did you do the experiment?' the blank page was much less useful for capturing this information than either of the templates. For capturing information about what the researcher actually did in the experiment, the blank page produced the most useful information in narrative form. The original title-template also captured this information, although in a format that was perhaps less useful to someone other than the author of the record. There are also much more subtle questions that we might be trying to capture, that may or may not be enhanced by asking more questions or providing more cues. For example, we may get subtler and more useful answers by asking 'Why did you do this experiment?' rather than 'Have you done this before?', 'Are you following a recipe?', 'What do you expect to happen?' and 'What outcomes have you anticipated?'. However, it is difficult to know if the less specific question will result in a more or less meaningful result.

Providing researchers with ways to design, configure, and share their own templates with others may help to mitigate some of these problems. Providing example templates as a basis for user-generated templates, which can be customised for the needs of individual researchers and research groups, may be an appropriate strategy for achieving a balance between flexibility and capturing the right information. A benefit of templates is that they can be shared in a team or community. Sharing is beneficial for standardisation, but it is also useful for less experienced staff in order to help them learn good recording practices from more experienced staff and to give them a starting point for the customisation of their own templates.

9.4 Paper *Versus* Digital and the Role of Mobile Electronic Notebooks

As we discussed in Chapter 6, there are benefits and downsides to ELNs over paper notebooks. Where paper notebooks are simple to use, have great flexibility, and portability, ELNs, in contrast, can be complex, rigid, and be restricted on where they can be used. Even though, as previously discussed, there has been a resistance to the adoption of ELNs in research environments, scientists nonetheless have high expectations about their capabilities, including their functionality, portability, and ease of use.[7] The expectations for the capabilities and portability of ELNs often reflect the affordances of paper itself. Computers, or desktop computers at least, do not provide the portability of a paper notebook because they are static and require the user to move to their location to enter data or to read content, whereas portable devices such as phones, tablets, and laptops can be moved with the researcher to the locations where the research is carried out. Portable devices may therefore provide a possible bridge between the paper and electronic notebooks that could enhance the adoption of ELNs and other electronic notebooks by researchers. Native applications for mobile devices can also take advantage of the various capabilities of the device, in a way that Web-based tools are often not able to. For example, it is much easier to capture images, video, and audio directly through a native application, and make use of other capabilities of the device such as using the gyroscopes and other sensors.

Mobile device ownership has been increasing. In the 2014 ECAR Study of Undergraduate Students and Information Technology, it was found that 86% of undergraduates own a smartphone, rising to 97% by the 2017 study, with around 50% also owning a tablet, and 95% owning a laptop.[8,9] We are also becoming more accustomed to easy to use, intuitive interfaces through the use of smartphones and tablet computers that are typically designed to be used without training, outside of an office environment, and even on the move. These interfaces are typically simpler and more intuitive than desktop software, and focus on the essential tasks that need to be performed, rather than the complexities that are often inherent in desktop software design. This trend has been noted by a number of ELN vendors, leading to more focus on making ELN interfaces simpler to use.[10] Tablets provide the opportunity to both develop mobile applications that work in the way that researchers are familiar with and to design interfaces that can take advantage of the strengths of the platform.[7]

While many vendors are busy improving their interfaces and developing tablet and smartphone clients for their enterprise ELNs, many research groups are taking advantage of more generic electronic notebooks available for smartphones and tablets. These electronic notebooks are often free or relatively inexpensive and easy to use. Many of these applications provide a subset of modern ELN capabilities, and can be combined with other services to enable not only notetaking capabilities with applications such as Microsoft OneNote, Evernote, GoodNotes, and WritePad, but also data management and collaboration capabilities with cloud-based backend applications such as Dropbox and Box.net.[7] From discussions with a variety of researchers handling digital data, it is clear that Dropbox is widely used for data storage and also for collaboration, within the academic community at least.

Other applications available on such mobile devices can also provide useful tools such as image, sound and video capture, and applications for scanning barcodes and RFID tags that can link together notes with samples or tagged specimens. Integration with instruments and sensors is also possible with applications that provide an interface to instruments such as weigh stations.[11] Many applications make use of Bluetooth and Wi-Fi to connect to sensors and other devices, for example, wearable devices such as watches often use Bluetooth, as do biometric sensors for health and fitness. External devices can also be attached directly to mobile devices for control, as well as reading data. For example, telescope controllers, microphones and musical instruments, environmental sensors for use inside and outside the laboratory, weather stations, and even Geiger counters! There are also many medical and environmental testing instruments and applications for mobile devices, which we will talk more about in Chapter 12. Such technologies do not necessarily provide all the functions that researchers desire and there is therefore still an opportunity to create a good digital notebook that is flexible and useful enough for researchers, but to survive, any digital tool must be usable as well as functional.[12]

Mobile ELNs have the same advantages for researchers as desktop-based ELNs, but also include additional advantages that may help to mitigate some of the perceived drawbacks in comparison with paper notebooks. For example, tablet devices are relatively small in size and have long-lasting batteries, enabling the ELN to be carried and used wherever they are needed. Their use is not just restricted to the capture of experiment information in the laboratory or in the field, but can also be used to provide remote access to information that would not normally be contained within the paper notebook. This information in the laboratory, for example, could be the details of previous experiment results or instrument settings, research papers, notes from other researchers, or details of the experiment protocol. In the field, the information could include species identification guides, maps, and expedition notes from other researchers.

In a traditional laboratory environment, information is tied to physical objects such as notebooks and working documents, which support a limited set of interaction models and data types, making it difficult to both access and share information.[13] Digital capture methods, in particular, the advanced capabilities of smart mobile devices, enable recorded information to be captured in a way that they cannot be on paper or easily using a desktop computer. For example, it may be more effective and efficient to capture equipment set-ups or results data, such as TLC plates and gel electrophoresis by taking a photograph rather than drawing a diagram. Mobile devices can also capture free-hand sketches, audio files, and videos, enabling information to be captured in the most appropriate way for a particular experiment or piece of research.[13] Using these methods has the advantage that the data, such as a photograph, can easily be integrated into the electronic notebook and then later exported as a resource to be stored elsewhere or used in other materials, such as publications and presentations. Some researchers express the desire to be able to record sound and make use of voice commands, to both dictate notes and access information from an ELN system, to enable them to go 'hands free' in the laboratory.

A common complaint about existing ELN systems is their reliance on a network or Internet connection causing problems accessing the system in the laboratories, which often do not have network connections or wireless network infrastructure available. Researchers also want to able to access their research in other locations, for example, accessing their experiments at home or whilst travelling, or in remote locations for researchers in the field. Smartphones and many tablet devices can provide access through a cellular data network, in addition to Wi-Fi connectivity, enabling access to data stored in other locations even without a local network connection. Having a connection to data on a network greatly reduces the need to return to the office, enabling the researcher to spend more time in the laboratory or field, and providing access to much more information whilst carrying out the research. In terms of recording and sharing the research, another advantage of mobile platforms is that the data can be stored locally on the device and then uploaded to a server or other service, such as an ELN at a later time. This means that the data is accessible locally and information can be captured wherever needed, such as in the laboratory, in the field, at home, or on the train, but it can still be safety stored by uploading the research data when a more reliable connection is available.

A number of different research groups have attempted to develop digital systems to capture the notes made by researchers in the lab or field, in addition to those developed by commercial ELN providers or specialised electronic notebook applications such as Microsoft OneNote. Not all mobile notebooks rely on tablets or other smart devices for data entry, and many solutions are in fact hybrid forms that make use of digital pens and paper notebooks, whilst others use interactive screens located in laboratories rather than mobile devices. As mentioned in Chapter 4, a common solution to bridging between paper notebooks and electronic notebooks is to use a hybrid system of digital pen and paper. Many of these systems are ideal for use in the field where a digital pen is likely to be less expensive and more robust than mobile devices. ButterflyNet was an example of a digital pen based hybrid system mentioned in Chapter 4. ButterflyNet supports biologists in the field through the use of an Anoto pen and paper, together with a smart camera that can read barcodes and 2D barcode tagged envelopes to collect specimens.[14] These are used to link together associated notes, specimens, and photos, including facilitating the transfer of captured content into spreadsheets for easier collaboration. Other sensors can also be added to the system. Another example of a hybrid system, also using the Anoto pen, is Prism.[15] Prism enables the capture of handwritten text and drawings as time stamped images in an online journal and activity log that can include XML links to emails, web pages, and other online documents to bring relevant documents together in a central location. Another example, this time a field-based system for archaeologists, is RAMSES or Remote Archaeological Mobile Support Enhanced System. RAMSES uses a fixed station as an active object repository, whilst mobile units input archaeological evidence by means of an electromagnetic pen.[16] In these hybrid systems, paper is still the central medium for capturing notes. These systems are used primarily in the field and demonstrate the continuing importance of paper in these environments, but also the value of digital capture and linking of notes and other physical artefacts in field research. Not only are mobile devices such as tablets, tablet computers, and smartphones expensive and relatively fragile, there are other additional problems when compared with the digital pen in these environments. For example, in trials of a hybrid system for archaeologists called iTrench, which made use of Personal Digital Assistants (PDAs) and digital pen and paper, it was found that digital pens were successful, but the PDAs were not because of the extreme conditions.[17] Problems included the screen being too reflective to be seen in bright sunlight, the font size too small, and the stylus being uncomfortable to use. The PDAs have a sketch function, but this was found to be awkward and difficult to use, whilst the images generated were of inadequate quality. In contrast, the digital pens were seen as easy to use and actively useful. Although developments have been made in this area, technologies that deal with screen reflectivity and quality of screen drawing and writing tend to still be very expensive. We will talk more about the developments around technology for capturing drawing, digital pens, and writing on the screen in Chapter 12.

Hybrid systems can provide capabilities that go beyond the simple capture of handwritten notes, with the ability to perform system actions by interacting with special control elements on the paper notebook pages, or by making use of mark-up to assign meaning to the notes or sketches on the page. Examples of such systems designed for scientific researchers are the hybrid systems NiCEBook and A-book. NiCEBook is a prototype paper notebook that again uses an Anoto pen and paper to enable users to capture notes on paper as usual, while the notes are digitised in the background.[18] Special control elements are included on the notebook pages to enable users to assign notes to categories, define topics, and also provide the capability to 'dogear' pages. The physical notebook can be synchronised with a digital notebook that can then be used to search for content, view the content in different ways, and share with other programmes or users. The A-book project developed a series of prototypes making use of a graphics tablet (rather than a tablet computer or smart device), a paper notebook, and an inking pen connected to a computer.[19] The notebook included a command menu attached to the side and a pocket reader to detect the page number. The system recognises underlines and boxes; an underline indicates to the system that the information is text and can be used to open a URL or include details of a procedure; a box drawn around a sketch will save it as an image. The system enabled the user to browse the notebook on a PDA and included handwriting recognition, and the ability to create links to physical objects.

In the space between these hybrid solutions and the complexity of specialist ELN systems are the generic 'electronic notebook' applications for notetaking. There are numerous generic notetaking apps for mobile devices, and their functionality is similar. Microsoft OneNote represents one of the most advanced general applications for notetaking that works on both desktop and mobile devices.[20] The interface is intuitive to use, with a small number of menu items familiar to users of Word and other Microsoft software. The interface is primarily text based, with a fully featured text editor that enables formatting of text and font styles, justification, lists, and a spellchecking feature. The text editor also enables the user to highlight and mark-up content using predefined icons. Users can insert tables, equations, files, web-links, images, take photographs, and record audio. There is also a fairly sophisticated drawing tool that enables the selection of different pen widths and colours, and works much more effectively, with reduced lag and better accuracy, than similar drawing tools in other applications. Notebooks can be shared with other users enabling collaboration, and pages can be converted into PDFs and emailed to other users or sent to other applications on the device. Figure 9.3 shows a mock-up of the interface for OneNote on the iPad. The standard screen shown enables the user to add text content to their notebook, including 'tagged' content marked with an icon such as 'To Do', 'Important', 'Question', 'Idea', 'Source for article', and so on. A section of text can be marked up with more than one tag. The Draw menu allows the user to select from different kinds of pens and different colours to draw directly on the screen at any point in the notebook.

Evernote is another example of a general notetaking application, similar to OneNote, with a more basic text editor and the option to include

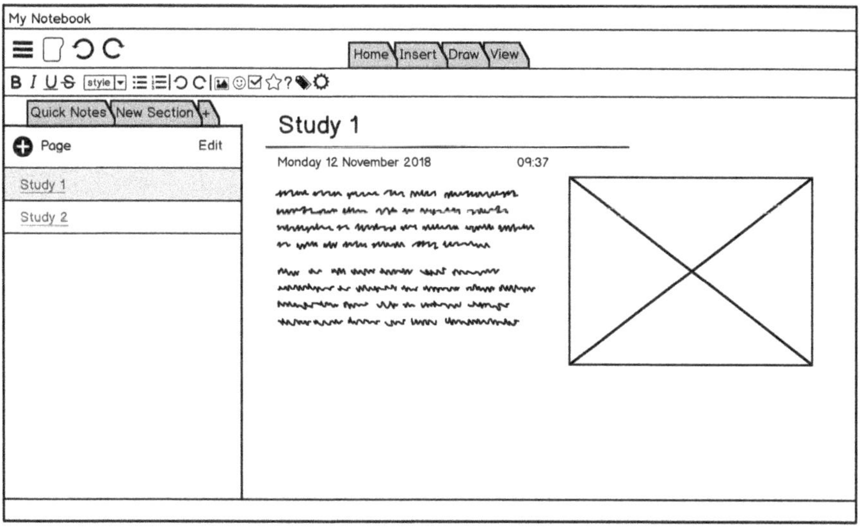

Figure 9.3 Mock-up of the Microsoft OneNote interface on iPad.

photographs, drawings, and audio clips in the notes.[21] The interface is a little different, making use of icons rather than menus, and creating drawings as an imported image, rather than as a layer over the top of the page content, changing the way that the images interact with the text. Notebooks can be shared with other users enabling collaboration and pages can be converted into PDFs and emailed to other users or sent to other applications on the device. Figure 9.4 shows a mock-up of the interface for Evernote on the iPad. Although the interface is similar to the OneNote interface, there are differences. In particular, the way that images are handled in Evernote is different, and selecting a drawing opens a new editor page to allow the image to be edited. Evernote also lets the user add audio to their notebook, providing a mechanism to capture notes 'hands free', which can be especially useful in the laboratory. Photographs can be added to a page, but can also be stored as separate 'snapshots' within the notebook.

An example of an older ELN that worked on an iPhone and iPad was the online eCat ELN. eCat made use of a simple and customisable template-based data entry system, enabling the creation and linking of constructs, protocols, and records.[22] eCat also provided the capability to perform searches, share records with others, and also manage samples.[23] The interface was styled to look like Google Docs to make it easier to use and more familiar for users. eCat has now been replaced with RSpace which provides a text editor based interface for capturing records and the ability to draw images. RSpace can also be used on mobile devices. Many other online ELNs will function on mobile devices, entirely or partially depending on how well the software has been designed with mobile in mind.

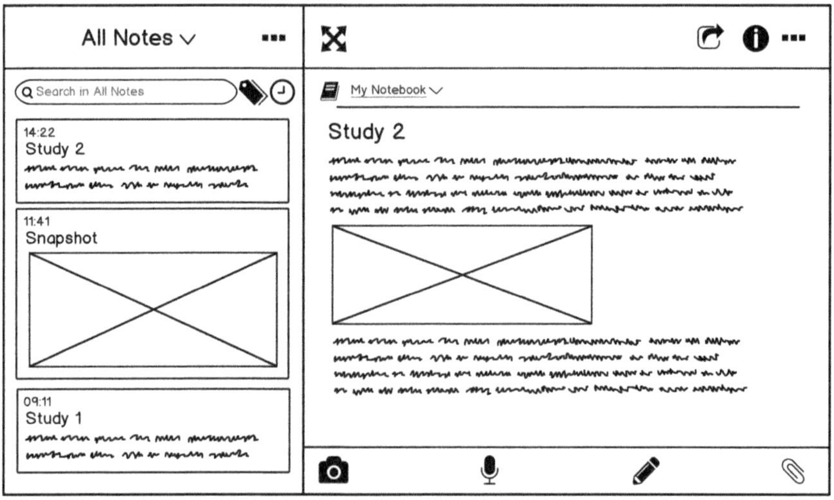

Figure 9.4 Mock-up of the EverNote interface on iPad.

A limitation of online versions of ELNs and other electronic notebooks is that they usually do not have permissions to access the local file system making it more difficult to include local files within the notebook or exporting files from the notebook.

An even earlier example of a mobile ELN was the Smart Tea project developed at the University of Southampton.[11] The Smart Tea ELN developed a Web-based planning tool and tablet software for capturing data during experiment execution in the lab. The system enabled the capture of the whole process of a chemistry experiment from planning to execution. The planning part of the system was a Web-form that extended the standard safety forms to include a definition of the steps of the experiment in addition to the materials to be used. The tablet part of the system was designed to be used within the lab as a replacement for a paper notebook to record the actual execution of the experiment. The notebook was pre-populated with the elements of the plan that were stored online in an Resource Description Framework (RDF) based back-end storage system. Users could capture their wet and dry measures through a numeric keypad, and additional context by providing annotations in the form of handwritten notes or drawings for each step of the experiment. The interface of both parts of the Smart Tea system were designed by computer scientists observing the experimental design and capture process by asking chemists to 'make tea' and record the process as though they were completing an experiment. This method of designing the interfaces using a familiar metaphor was so that the chemists could express the way they worked in a way that was understandable for the computer scientists. As a result, the design of the interfaces matches the normal routine of completing an experiment by planning using the health and safety forms and capturing measurements and information at each stage of the experiment. Both the planning and execution elements of the software made use of templates for the capture of the appropriate information, with the plan effectively operating as the template for the execution of the experiment. However, users did find that the rigidity of the templates meant it was difficult to make changes once the plan had been created, especially if something unexpected happened during the experiment and the protocol used needed to be changed. The Smart Tea ELN suited some disciplines and experiment types more than others, for example, suiting disciplines that typically follow a formal standard structure such as synthetic chemistry and analytical biology. As mentioned previously, such disciplines typically conduct the same experiments and analytical activities repeatedly using the same or similar protocols that do not vary greatly over time. The ability to duplicate protocols and reuse them can save time and effort for researchers in those disciplines. The rigidity of the template is less of a hindrance in those environments, as discussed previously.

CERF for iPad is an example of a native tablet client from a commercial ELN developer. It is intended for researchers to use in the lab but connects to an enterprise CERF ELN for storage.[24] A licence for the back-end ELN must be acquired before the mobile version of the ELN can be used, so this is not a stand-alone application. There is an offline function to overcome network issues, and the tool offers capabilities for Web browsing, text entry, audio recording, camera, specialised science functionality, and the ability to import files. The interface is relatively straightforward and predominantly text-based, but also enables handwritten text to be entered. Another mobile ELN example is Elsevier's Hivebench, which provides an ELN that can be run on desktop, tablet, smartphone, and even the Apple Watch wearable device.[25] Hivebench is free for online accounts and provides the ability to create and manage experiments, protocols, and results. The system is designed to be collaborative and enables users to share their notebooks. The interface is relatively straightforward to use, using a text editor to enter content, rather than allowing handwritten text. The user can also create entries for standard protocols and reagents that can be duplicated and imported into individual experiment records. Figure 9.5 shows a mock-up of the interface for Hivebench on the iPad and shows how different the interface for a specialist ELN application is in comparison with the generic electronic notebooks shown above. The ELN allows the user to create standard protocols, descriptions of reagents, and add data. The interface uses standard templates for capturing the information, although individual sections can be configured. There appears to be no way to add a drawing directly to the notebook, although data including images can be linked in the Results section of the template.

Several of the mobile ELN solutions described require that the researcher will have created a plan for their experiment in advance of entering the lab, enabling progress of the experiment to be displayed and often annotated to assist in the execution and recording of the experiment. For example, the

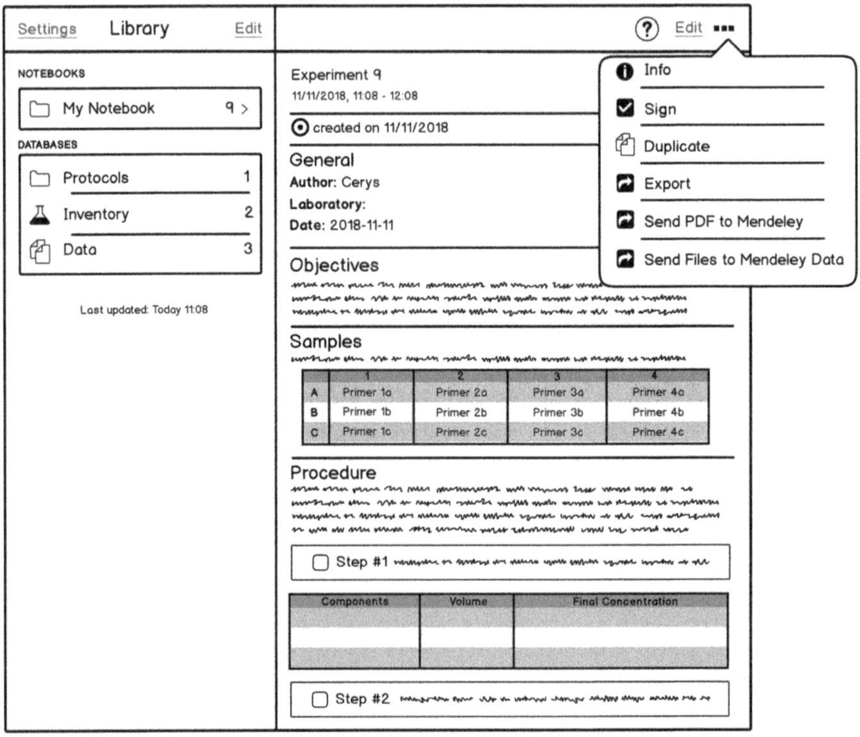

Figure 9.5 Mock-up of the Elsevier HiveBench ELN interface on iPad.

Smart Tea system makes use of the requirement for researchers to complete a Control of Substances Hazardous to Health (COSHH) form listing the materials they intend to use in advance of starting the experiment. COSHH is a set of regulations related to the control of exposure of hazardous materials by employers in the UK. As part of the regulations, scientists carrying out activities using hazardous materials must complete a risk assessment to identify potential hazards and the steps that will be used to mitigate them, including safety information and considerations about disposing of materials and managing spills. In the Smart Tea system, a Web-based form based on the elements of the COSHH assessment form asks the researchers to include extended descriptions of the steps for the experiment process, together with materials and quantities that will be used in order to create a plan.[11] Although completing the plan is extra effort, the benefits include helping the researchers to remember what to do next in the experiment and ensuring persistently legible results are available. Experiment plans can also be duplicated to use for repeating experiments or used as the basis for new plans. Through the use of experiment plans, ELNs can serve to augment the memory of researchers, both in the lab whilst executing experiments, and in the future by providing information about a previous experiment and its results.

The majority of both general notetaking applications and specialist notetaking applications, such as ELNs, focus on capturing text-based content, making use of familiar text editors, although there is often a way to add hand-drawn content. There are a number of likely reasons why input text *via* a keyboard interface tends to be used rather than through handwriting with a finger or stylus for example. Capturing drawings, and handwritten text in particular, is difficult to do with touchscreens. There can be problems with both lag between the touch and the mark appearing, and the accuracy of the mark. Tablet and smartphone interfaces are not yet at a point where they could be considered a genuine replacement for pen and paper. The drawing resolution through the touch interface is not as good as pen and paper, meaning that although you can make some basic sketches and rough notes, it is not yet possible to capture the detailed movements required for more natural handwriting or detailed drawings in the same way they are made on paper. This means that writing must be much larger than it would be if you were using a pen and paper, and therefore much less information can be fitted onto a page than the equivalent size pad of paper. Some applications use a 'zoom box' on the screen where the characters are written and then shrunk for display on the screen, but this is fiddly to use and takes away from the more natural experience of just writing where you want the characters to appear on the page. Touch gesture technology is improving all of the time, and in the future, we can expect to see increased touch gesture resolution that will help with such problems. We will talk more about related technologies and ways of interacting with computers that can help improve these issues in Chapter 12.

In the majority of mobile applications, handwritten text is captured as an image, and although it can be shared and reproduced, the meaning of the content is unavailable for other uses including search, data mining, and metadata creation. Some applications, however, do make use of handwriting recognition technology, which can be used to create more accurate character reproduction and also to access the meaning of the contents of the notebooks for other uses. There are two approaches to performing handwriting recognition, one is to include handwriting recognition libraries within the application to carry out the recognition tasks on the device itself, and the other is to use a Web service to do the translation. Some free and open source options for handwriting recognition exist, for example, the Tesseract Open Source OCR Engine, although these are currently less effective compared with the commercial options available.[26] Combining the use of a 'zoom box' together with commercially available handwriting recognition libraries enable mobile applications to incorporate handwriting recognition capabilities. Character recognition technologies are imperfect and, particularly for handwritten text, the process of correcting errors is likely to slow down the process of notetaking considerably compared to just writing on the screen. The use of a Web service to translate written characters on the fly is also possible, but this adds an additional requirement to have access to a data connection and may also slow down the process. A more manual process is to use online handwriting recognition services to translate the images of the characters captured in notebooks, which do not provide automatic handwriting recognition manually, by uploading exported image files by hand. However, this is a somewhat tedious process and will still contain the same kinds of errors as the automatic systems. Some of the commercially available electronic notebooks already include support for handwriting, for example, the iCerf ELN.[27]

Another 'natural' method of providing input into the notebook other than writing is to be able to dictate notes using audio input and a microphone. Being able to record notes 'hands free' using audio capture is efficient in the lab and can save time going backwards and forwards to the notebook, but does suffer from the same kinds of problems as text captured as an image. In fact, audio is even less convenient for retrieving specific content, because the whole audio file may need to be listened to in sequence in order to find the information that is being searched for. There are options available for voice recognition that can enable the conversion of speech to text, and in the same way as handwriting recognition, both offline and online options are available. One example for the iOS platform, for iPhones and iPads, is OpenEars, which can perform both offline and real-time speech recognition.[28] In many respects, including offline voice recognition to notetaking applications would be superior to the use of handwriting recognition because less work is required from the user to capture the information that they would like to record. This of course assumes that taking notes through speech is desirable and practical in any given environment. Big challenges for speech recognition are accuracy, dealing with different accents, and working in loud and busy environments. As anyone who has used one of the many voice-controlled digital assistants available, such as Apple's Siri, Microsoft's Cortana, and Amazon's Alexa will know, voice recognition is imperfect and can be frustrating, especially where there is any background noise interfering with the words being spoken. In the same way as handwriting, is it possible to perform the conversion between speech and text at a later time by capturing an audio file and uploading it to an online service. Although this may overcome some of the problems of real-time speech recognition difficulties, there are still likely to be mistakes and the quality of the audio recording will have an impact on the accuracy.

Additional challenges for both speech and handwritten notes come from the capabilities of the software to recognise scientific terminology, equations, chemical structures, species names, locality names, and so on that are meaningful to the researcher, but less likely to be included in the dictionaries used by translation technologies. Equations, chemical structures, and other scientific mark-up can be difficult, or at least less convenient, to enter using a text editor, and this is where specialised applications and interfaces can support scientists, where the general notetaking tools may be less usable. Specialised notetaking tools, such as ELNs, can also generate appropriate metadata and provenance information for the activity, such as information about a particular experiment or study that is harder to capture within a generic electronic notebook. For example, protocol information that is added to a notebook can have a unique identifier, author, and other information that can automatically be added to the experiment record that can help generate an appropriate provenance trail. A specialised tool can recognise scientific terminology and names and potentially highlight these in the interface and include them as significant information in the metadata about the record. We will talk more about capturing metadata for the scientific record in Chapter 10.

We have discussed above how mobile devices can potentially provide a bridge between paper notebooks and electronic notebooks by providing the portability and convenience of the paper notebook with the additional capabilities provided by digital devices. They can be used to provide not only a way to capture notes in the laboratory or field, but also provide a way to access materials such as experiment plans, maps, data, and relevant research articles. A variety of tools exist for capturing research notes, including both generic notetaking applications and specialist tools such as the extensions to ELN software, and even standalone ELN software. They can provide new and easy to use ways to capture information during research, for example, through the use of photographs, audio, and video. There are, however, challenges in developing digital applications for mobile devices. The market for ELNs is relatively small, and development of applications is expensive, and it is not simply a case of being able to take an existing solution and compiling it for use on mobile platforms. It is not appropriate to port a desktop application that is not designed for use in a 'touch-interface' environment to a mobile device, as mentioned in Chapter 8. Applications usually need to be developed from scratch and costs further increase with each different device you attempt to support. Each brand of device has its own programming language and applications need to be tested across the range of different devices and different operating systems. Other challenges for mobile devices include dealing with presenting information in a relatively small space and providing enough information to help researchers navigate the system and find the information or screen that they are seeking.[29]

A review of ELNs on tablet devices suggests at the current time that these are very basic, the vast majority do not provide many tools beyond those available in the generic notetaking tools, despite the great potential the capabilities of both ELNs and the devices for supporting scientists. There is an on-going need to bridge the gap between paper notebooks and digital tools, and the familiar interface and powerful capabilities of tablet and mobile interfaces have the potential to accomplish this challenging goal. In order to encourage the widespread adoption of ELNs and other digital tools for notetaking, we ideally need interfaces that allow natural input, and in particular handwriting and drawing, and we will talk more about the use of digital pens and other natural interfaces in Chapter 12. There are also many other ways we could be enhancing notetaking applications by utilising more of the capabilities of mobile devices beyond what is possible with a paper notebook.

9.5 Alternative Interfaces for Capturing the Experiment Record

Writing, text, and drawing are not the only way to capture the notes, measurements, observations, and links to physical objects that we record when capturing the scientific record. We have already talked about the way that mobile devices can easily add new methods of capturing information to the record in a digital form, such as digital photographs, sound recording, and video, as well as handwriting and speech recognition technologies. An alternative to mobile technologies in the laboratory is to have interactive workbench screens that can be used to view information, such as an experiment protocol or previous results, and also to capture the execution of the experiment. For example, eLabBench is a digital table top system supporting biologists in the organisation of personal information, capture of experimental work for later access and management in the office, and providing access to a variety of computational resources directly at the lab bench.[30] The system comprises an interactive table top with access to the kinds of tools that the biologists would have access to on their computers in the office as well as a Web-based notebook and an area for making rough notes. Users can interact with the system using mouse, keyboard, pen, tagged physical objects such as racks and instruments, and buttons, for example, to activate an overhead camera. The system is an alternate hybrid of computer and pen, but allows the user to interact with the system in a variety of ways and gives them access to both the notebook and office capabilities without needing to change location. LabScape is a similar system that makes use of touch interfaces positioned at different places in the lab.[13]

With digital technologies, we can also capture information through the use of sensors, barcodes, and Radio-frequency Identification (RFID) tags. Barcodes and RFID tags are commonly used for the management of physical objects such as biological specimens, geological samples, and archaeological artefacts in the field, and samples, specimens, and materials in the laboratory. Information associated with the IDs of these tags and barcodes can not only help track their progress and status, but also provide information such as expiry dates or reminders about actions that need to be taken. In addition to data we can collect digitally from instruments, there are hundreds of different kinds of sensors that can be connected by wire, Wi-Fi, mobile-data, and other communication technologies such as Ant+ and Bluetooth, to provide real-time data from the environment, both inside and outside the laboratory. These sensors can not only contribute by providing data specific for particular research or an experiment, but also help in explaining unexpected results or anomalies. For example, if an unexpected measurement is found during an experiment in the laboratory, looking at sensor data from thermometers and humidity sensors, and even sensors attached to doors, could suggest a reason for the unexpected result. An example ELN system that makes use of sensors is BioTISCH. The BioTISCH system replaces the traditional workbench with a glass covered table top system that presents information on the bench's surface and includes some augmentation of objects in the form of reagent sensors.[31] The system allows the biologists to access data from a database, record the steps in the experiment, and perform calculations, but is also prompted by the system to perform actions when the reaction sensors highlight the next reagent to be used.

Sensors that continually collect data can also provide information on trends and changes over time or related to conditions at particular times, days of the week, or seasons of the year that may influence research findings. Data from sensors can be automatically uploaded to or captured by some ELN systems, and then automatically incorporated in the appropriate activities, or be available for searching and modelling. Other interfaces that make use of sensor data could be real-time dashboards so that the environment can be checked before and after an experiment so that any problems can be addressed, and notification systems that send warnings if conditions change to no longer meet certain thresholds.[32] The sensor data can also be captured as part of the record so that it can be compared and examined over time or between different studies. Challenges for data from sensors comes in the potentially large volume of data and in designing effective interfaces for managing and interacting with the data in a meaningful way.

Other technologies for interfaces are in development, but are generally not commercially available for supporting scientists in their research. However, some groups are investigating potential uses for these technologies in a variety of areas. One example, the wearable digitisation system for life science experiments from Scholl and Van Laerhoven, is designed to augment the researcher in the lab.[33] Their system is designed for biologists to help in both the capture of experiments and in accessing information during the execution of an experiment. The system comprises a Google Glass head mounted display, a smart watch that records the wearers motion, and an RFID-reader which can support the scientist during the experiment by providing access to information and by recording elements of the experiment. For example, the watch can record wrist motions to use for activity detection, the RFID reader with tagged containers can identify materials used, and the Google Glass can take pictures and record comments on key observations. In Chapter 12, we will discuss more about potential interfaces to digital tools and uses of 'future technologies' such as wearables, artificial intelligence, virtual reality, augmented reality, and mixed reality. We will also talk more about sensors and connected devices in the Internet of Things (IoT), providing some further examples of how they can be used to support scientists in their research and also contribute to the creation of the scientific record.

9.6 Conclusion

In this chapter, we have discussed how different interfaces might impact on the information captured in the scientific record. The information captured in paper notebooks may be quite different to the information captured within digital tools, and in particular, template-based interfaces within ELNs. Different interfaces may be more suitable for different kinds of users and reflect our discussions in Chapter 8 about understanding the goals and behaviours of users. For those environments where experiments or other activities are repetitive, and where the information that needs to be captured is consistent and known, template-based interfaces are ideal. Templates can promote the capture of specifically requested information and improve quality and consistency of the record. For environments that are less formal and where there is significant variability in activities or the information that needs to be captured, then greater flexibility is likely to be more desirable. In such environments, a more free-form journal style of interface, such as those generally provided by the generic electronic notebooks may be more appropriate. Providing options for users to enable them to use a free-form interface or to customise their own templates provides the best of both worlds by enabling the user or their managers to select the most appropriate interfaces for recording activities.

Templates are not the only aspect of the interface and capabilities of tools that are influenced by the environment and goals of users. For some disciplines, the ability to capture information in the form of photographs or videos may be more useful than others. For some environments and disciplines, being able to hand-write notes and hand-draw sketches and images may be extremely important, and within such environments, it may be important to be able to utilise digital pens or a stylus with a tablet device in order to capture the information in the appropriate format. We also looked at how mobile technologies may help to facilitate the adoption of digital tools because of their relative ease of use and portability. Whether the interface is on a desktop computer, a laptop, a tablet, or a mobile phone, of vital importance is the usability of the application. A tool can have numerous useful functions, but unless the user is able to make use of them, they are unlikely to be adopted as essential tools for science. For mobile applications, complexity has to be limited by the nature of the platform. There is limited disk space, limited network connectivity to provide access to extra content, and limited screen size to present information, and well-designed applications will be simple in design and use, in order to focus upon the key tasks the user needs to complete. Cloud-based solutions often have the advantage that data captured on the mobile device is automatically uploaded, so that users can also make use of the 'full-featured' desktop versions.

Designing suitable interfaces for capturing the experiment record is difficult, there is no 'one size fits all' solution, and as we have seen, subtle differences in interface designs may have large consequences in terms of the information cued from users and what is captured in the record. The details of the experiment and the data produced are not the only elements of the research that needs to be captured in the record. As we discussed in Chapter 5, effective data management and curation of research is reliant upon the metadata that is produced to describe it. In Chapter 10, we will examine how metadata is used in science, investigate different methods of creating metadata, and the importance of capturing metadata 'at source'. We will also take a look at how interface designs within digital tools can facilitate the creation of appropriate metadata to describe scientific research. We will also examine other ways in which software can support the creations of software through automatic generation and smart assistance within digital tools.

References

1. A. Milsted, J. Hale, J. G. Frey and C. Neylon, *PLoS ONE*, 2013, **8**(7), e67460.
2. M. Elliott, *Scientific Computing*, 2004, Available at http://www.scientific-computing.com/articles/2004/08/its-not-about-paper? (accessed Apr 2016).
3. K. McGlade, C. Cargo, D. Fogarty, M. Boohan and M. McMullin, *Clinical Teacher*, 2012, **9**, 112.
4. D. Swinglehurst, T. Greenhalgh and C. Roberts, *BMJ Open.*, 2012, **2**, e001754.
5. C. Willoughby, T. A. Logothetis and J. G. Frey, *J. Cheminf.*, 2016, **8**(1), 9.
6. R. Tourangeau, F. G. Conrad, M. P. Couper and C. Ye, *Public Opin. Q.*, 2014, **78**(1), 100.
7. M. Elliot, *Sci. Comput.*, 2012, **August**, 9.
8. E. Dahlstrom and J. Bichsel, *Study of Undergraduate Students and Information Technology*, EDUCAUSE Center for Analysis and Research (ECAR), 2014.
9. D. C. Brooks and J. Pomerantz, *Study of Undergraduate Students and Information Technology*, EDUCAUSE Center for Analysis and Research (ECAR), 2017.
10. A. King, *Chemistry World*, 2013, Available at https://www.chemistryworld.com/review/notebooks-go-digital/6187.article (accessed Oct 2018).
11. H. Hughes, D. Mills, D. De Roure, J. G. Frey, L. Moreau, M. C. Schraefel, G. Smith and E. Zaluska, *Org. Biomol. Chem.*, 2004, **2**, 3284.
12. J. Giles, *Nature*, 2012, **481**(7382), 430.
13. L. Arnstein, R. Franza, G. Borriello and S. Consolvo, *IEEE Pervasive Computing*, 2002, **1**(3), 13.
14. R. Yeh, C. Liao, S. Klemmer, F. Guimbretière, B. Lee, B. Kakaradov, B. Stamberger and A. Paepcke, in *Proceedings of the SIGCHI Conference*

on Human Factors in Computing Systems, ACM Press, New York, 2006, p. 571.
15. A. Tabard, W. E. Mackay and E. Eastmond, in *Proceedings of the ACM 2008 Conference on Computer Supported Cooperative Work*, ACM Press, New York, 2008, p. 569.
16. M. Ancona, G. Dodero and V. Gianuzzi, in *Handheld and Ubiquitous Computing. (Lecture Notes in Computer Science)*, ed. H. W. Gellersen, Springer, Berlin, 1999, **1707**, p. 222.
17. C. Warwick, C. Fisher, M. Terras, M. Baker and A. Clarke, *Literary Linguist. Comput.*, 2009, **24**(2), 211.
18. P. Brandl, C. Richter and M. Haller in *Proceedings of the 28th International Conference on Human Factors in Computing Systems – CHI'10*, ACM Press, New York, 2010, p. 599.
19. W. E. Mackay and G. Pothier, *ERCIM News*, 2001, **46**, 52.
20. Available at https://products.office.com/en-us/onenote/digital-note-taking-app (accessed Nov 2018).
21. Available at https://evernote.com (accessed Nov 2018).
22. N. H. Goddard, R. Macneil and J. Ritchie, *Automated Exp.*, 2009, **1**, 4.
23. R. MacNeil, 2011, Available at http://elnblog.axiope.com/?p=956 (accessed Jun 2013).
24. Available at http://cerf-notebook.com/solutions/ipad-client/ (accessed Nov 2018).
25. Available at https://www.hivebench.com (accessed Nov 2018).
26. Available at https://github.com/tesseract-ocr/tesseract (accessed Nov 2018).
27. Available at http://cerf-notebook.com/blog/handwriting-recognition-notebook-entries/ (accessed Nov 2018).
28. Available at https://www.politepix.com/openears/ (accessed Nov 2018).
29. R. Cooke and M. C. Schraefel, *User Interface Software Technology (UIST), Mexico*, 2004, Available at http://eprints.soton.ac.uk/259251/ (accessed Nov 2018).
30. A. Tabard, J. Hincapié-Ramos and J. Bardram, in *Proceedings of the 2012 ACM Annual Conference on Human Factors in Computing Systems – CHI '12*, ACM Press, New York, 2012, p. 3051.
31. F. Echtler, M. Häussler and G. Klinker, in *Proceedings of the 28th of the international conference extended abstracts on Human factors in computing systems – CHI EA '10*, ACM Press, New York, 2010, p. 3439.
32. R. S. Gates, M. J. McLean and W. A. Osborn, *J. Res. Natl. Inst. Stand. Technol.*, 2015, **120**, 293.
33. P. M. Scholl and K. Van Laerhoven in *Proceedings of the 2014 ACM International Joint Conference on Pervasive and Ubiquitous Computing: Adjunct Publication*, ACM Press, New York, 2014, p. 1381.

CHAPTER 10

Creating Context for the Record Using Metadata

10.1 Introduction

In Chapter 5, we looked at the importance of data management and curation for ensuring that our research data is not just safely stored, but also that it is discoverable so that it can be reproduced, reused, and repurposed by ourselves and others. Data may be physical objects or electronic data files, but without providing appropriate context to describe them, they can be completely meaningless and impossible to use. For example, a biological sample collected from the field may be identifiable as a particular species, but without the additional context of how, when, where, and why it was collected, the sample alone is not particularly useful. A spreadsheet of instrument readings is also not useful if there is no additional context provided to describe what the spreadsheet contains, the meanings of row and column names, the details of the instrument itself, its settings, or the purpose, components, and conditions of the experiment that generated the data. Data is only useful when context and adequate descriptions are provided. As we have discussed, metadata is an essential aspect of these activities by providing context for the research data itself by describing the contents of digital files, providing provenance information that can aid trust in the data, and providing additional information about how the materials are organised and what changes have been made.

Digital tools such as electronic notebooks provide an opportunity to both capture and make use of metadata related to scientific research. As we discussed in Chapter 5, the majority of digital tools add some metadata to the files that they create. For example, at a minimum, metadata about when

Recording Science in the Digital Era: From Paper to Electronic Notebooks and Other Digital Tools
By Cerys Willoughby

Published by the Royal Society of Chemistry, www.rsc.org

files were created and last modified is automatically generated by operating systems, and software may also generate metadata such as information about the author, hardware and software versions, configuration properties such as the settings chosen on a device, and other descriptive, technical, and administrative information. The metadata added by software to describe files created by the software or to describe other content created within the software is known as machine-generated metadata. Metadata can also be generated by machines using automatic techniques such as metadata harvesting or data mining, which we will discuss in more detail later in this chapter. When metadata is automatically generated by software, it can be assumed to be relevant and useful in terms of managing an electronic file, but it does not necessarily provide very much information about the content of the file itself. For example, in a Microsoft Word file, metadata generated automatically includes only the location of the file, the size of the file, when it was created, and when it was last modified. This is useful information for locating the file, but not very useful for knowing what the content of the file might be about. The same is true of other kinds of files, for example, a data file or a photograph. The Exchangeable Image File (EXIF) data contained within a photograph provides a great deal of useful technical information about how the camera was configured when the picture was taken, but in itself it cannot tell us if the picture is of a cat, a balloon, or a person riding a bicycle! It can therefore be extremely useful to ask the creator of the file to provide additional metadata to describe the file or other content within a software system.

Ideally, the user should never need to add metadata that the software itself could provide; if the computer can provide the information, then it should be captured automatically. For example, if data is collected automatically by an instrument, then the information that the instrument knows about should be accessible to any software interacting with it, for example, the settings configured on the instrument, the manufacturer and version of the instrument, and location for devices containing a GPS sensor. Such metadata can be very useful for providing context to the data that is captured. However, there is information that can only be added by the user that would also be extremely useful to capture as metadata, such as metadata describing the content of the file or information about how the data in the file was collected. In order to enable the capture of this descriptive metadata content, many software tools allow users to add 'user-defined metadata' to describe their files and other digital objects. For example, in Microsoft Word, the user can choose to add a variety of user-defined metadata values including predefined types such as Client, Department, Editor, Keywords, Language, Project, Subject, and Status, but also custom properties to store values that are of relevance to the author and their organisation. These metadata properties are stored as name-value pairs within the file. Tags or keyword-like metadata are an example of another style of user-defined metadata. Tags are commonly used as a mechanism to enable users to associate descriptive metadata easily with files or other resources. For example, Flickr invites users to add tags to their photographs to describe the content of the image in addition to adding other metadata such as the location of the photograph and the people within it. Tags are also commonly used in blogs to describe the topic of posts, in email software to describe the content of emails, and music libraries where individual music files can be tagged to describe properties such as genre, time period, and mood. Metadata can also be assigned to collections of content, objects, or files as well as individual elements to describe the package as a whole. For example, in music libraries, an album, as well as individual songs, can have metadata to describe them including artist, year, genre, and mood. For a dataset, the individual files may have descriptive metadata associated with them, but also the collection of data files can have higher-level descriptive metadata and also structural metadata to describe the relationships between the different files.

Adding metadata to describe content within software can enable that software to make use of it and add capabilities that add value to users. For example, as mentioned in Chapter 5, Flickr makes use of both user-defined metadata and automatically generated metadata to aid discoverability when users perform searches, to filter content, collect related content together, and to provide links between related content. Within an electronic notebook, user-defined metadata can be used in the same kinds of ways to enable users to find research and experiments related to particular topics, specific samples, or localities, to group related content together, and to provide ways of navigating through the notes. Such capabilities can be powerful for supporting individual researchers with managing their research, but also the wider community in accessing the content and reusing it. However, there are a number of challenges to overcome. Interfaces for working with metadata need to be designed in such a way that they match the data management and curation practices of the individuals that are making use of the software. This is not an easy task given that practices differ between different disciplines, institutions, departments, and teams. Even teams working in very closely related fields may have very different methods of organising their data and use quite different terms to describe their research.[1] As we shall explore below, even if digital tools are developed with the capabilities to add user-defined metadata, there is also no guarantee that researchers will choose to make use of them. In this chapter, we will take a more in-depth look at how metadata is used in Electronic Laboratory Notebooks (ELNs) and other platforms to annotate scientific information. We will also take a look at some of the issues around metadata creation, the so-called 'burden of curation', and examine how we can use technology to mitigate those problems. We will also investigate different interfaces for facilitating the creation of metadata and managing research data. We will also briefly look at how collaborative metadata creation could enhance the creation of metadata for describing research data.

10.2 Metadata Use in Science

Metadata in science is typically used to organise information and to facilitate retrieval. As mentioned in Chapter 5, metadata has its origins in library catalogue cards to help users to find books and other resources within a collection. Outside of the library, metadata has long been used in science, although the use of the term 'metadata' is relatively recent. The information captured on cards to describe samples, specimens, and artefacts within science collections can certainly be considered to be metadata, in the same way as the details of items in the library collection captured on the catalogue cards are considered to be metadata. The various systems for classifying scientific objects and assigning nomenclature, including taxonomies, can also be considered to be metadata that define the meaning of terms and the relationships between different scientific elements. Figure 10.1 shows

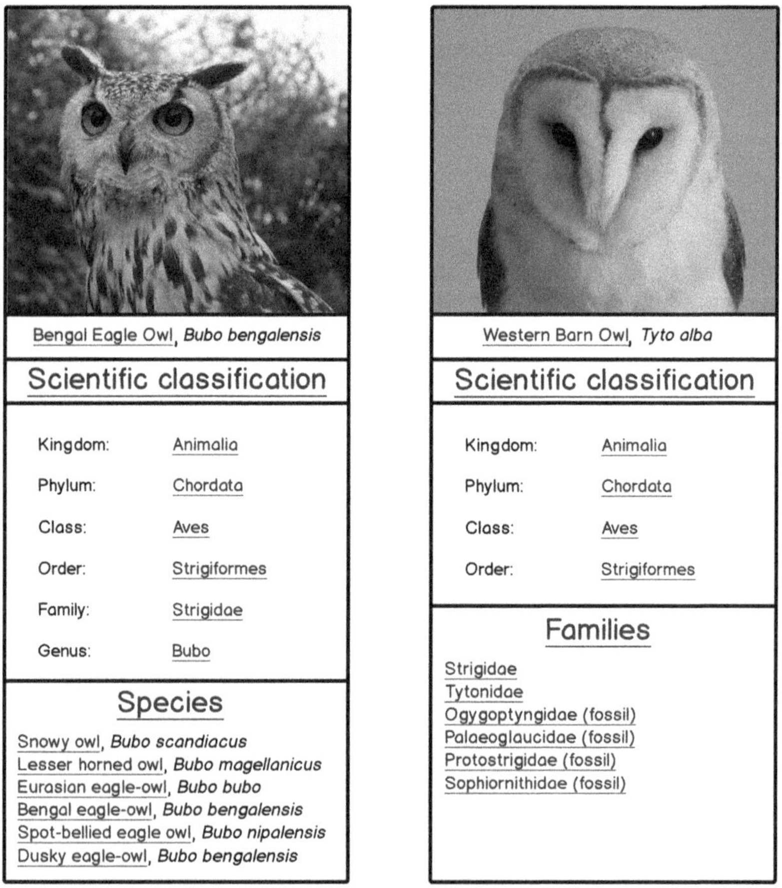

Figure 10.1 Example of taxonomy use in Wikipedia, showing how the taxonomy metadata allows users to follow links up and down the taxonomy tree.

an example of how taxonomy metadata can be used to facilitate the retrieval and organisation of information. Wikipedia makes use of taxonomy information about different species to enable users to follow links up and down the taxonomic tree. Wikipedia pages for species include information about their scientific classification. On the left in Figure 10.1 is an example of the scientific classification information for the Genus 'Bubo' or eagle owls. The user can click links to the different species of eagle owl in the Species section at the bottom, or alternatively click on one of the groups higher in the chain. For example, if the user clicks on the link to the Order 'Strigiformes', they are able to access information about owls as a taxonomic group as a whole and follow links to their living and fossil families, as shown in the example on the right of Figure 10.1. Wikipedia also includes similar classification and nomenclature metadata for other scientific artefacts, such as chemicals and astronomical objects.

Controlled vocabularies are often developed by different groups within the scientific community in order to organise information using standard terms that can later be used to facilitate retrieval. The terms are often specified with a standard definition and associated with thesauri to disambiguate between terms with the same name and different meanings, or where the meaning of a term differs between different fields. Ontologies extend classification schemes to include not just formal terms and definitions, but also properties and relationships between entities to describe a chosen domain in a machine-processable format. Providing vocabularies and ontologies in a machine-processable format means that they can be used within software to tailor the software for the benefit of specific disciplines, but they also enable interoperability facilitating the sharing of information across different systems. Relationships between entities are not just hierarchal, as in the taxonomies, but can be 'directed acyclic graphs', meaning that entities can have multiple parent entities. Also, the nature of the relationships can be more meaningful, for example, in the Gene Ontology, common relationship types between entities include 'part_of', 'derives_from', 'has_participant', and 'has_function'.[2] In the Chemical Entities of Biological Interest (ChEBI) ontology, relationships include 'is a', 'has part', 'is conjugate base of', 'is conjugate acid of', 'is tautomer of', 'has parent hydride', 'is substituent group from', and 'has role'.[3] Figure 10.2 shows an example of a ChEBI ontology representation of the mineral pyrite. This is a relatively simple representation, with only two types of relationships: 'is a' and 'has part', but also shows the directed nature of the relationships, with 'chemical entity' as the root entity that all the other entities ultimately derive from.

Ontologies can become very complex and become difficult to read and understand, but they are typically represented by structured formats that are machine-readable, and as such can be used by software to not only describe information and identify unique entities within a system, but also to facilitate advanced search capabilities, text mining and deriving meaning from data, combining data models and aggregating data, and finding

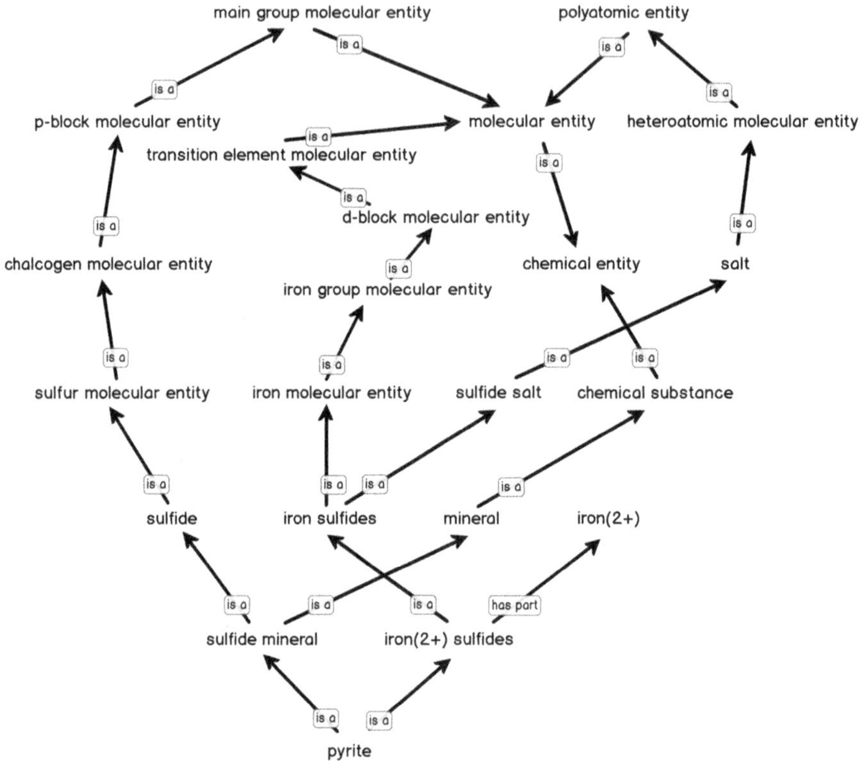

Figure 10.2 Example of an ontology, in this case, a ChEBI ontology representation of the mineral pyrite.

patterns in large-scale data.[2] Ontologies can also be used within digital tools to give a greater structure and meaning to a set of relationships. For example, Hughes *et al.* generated an ontology to describe the record of an experiment, which was used to design the storage system and interactions with an ELN.[4] The ontology records the relationships between materials and the processes carried out on them during the experiment.

Ontologies are also extremely useful for sharing scientific information. In order to be able to successfully share and reuse scientific information, there is a need for the information to be compatible between systems. Information needs to not only be human-readable, but also machine-readable so that data can be aggregated, searched, and displayed. Within organisations and institutions, information and data can be shared through internal networks, databases, and shared drives, but sharing information with a wider audience usually involves sharing data primarily *via* the Web. The idea of machines being able to access, understand, manipulate, and provide services based on a web of data is often known as the 'Semantic Web', a term coined by Tim Berners-Lee. There are a variety of tools and systems that enable the interoperability of information and data, with the generation of standard file formats, metadata, and protocols that are all part of the solution that enable machines to talk to machines. For the Semantic Web to be possible, data needs to be structured in a particular way to ensure it can be understood. Unique identifiers need to be generated to enable the resource to be unambiguously identified and locatable, using standard formats for the data and its metadata, and links to related data identifiers need to be included so that related information can be found. It is important that the structures that describe the data are also consistent and compatible. A common standard for describing information in an interoperable manner is Resource Description Framework (RDF).[5] RDF links together data objects and datasets using a graph structure, like that seen in Figure 10.2, using expressions with the format 'subject-predicate-object'. These expressions, also known as triples, describe the relationship between two items, with the predicate or property defining the nature of the relationship. For example, if we wanted to represent the relationship between a sensor and a measurement that it has taken, we might define a triple that uses 'Sensor' for the subject, 'takesMeasurement' for the predicate, and 'Value' for the object. The three parts of the triple typically have a unique identifier, so that relationships between specific individual subjects and objects, and what the predicates mean, can be captured and utilised. Each individual measurement can be linked to the sensor that generated it, and to other properties related to the taking of the measurement. The relationships between various objects and their properties are similar to the idea of concept networks that we discussed in Chapter 2. RDF itself does not describe the meaning of the data, to describe the meaning of the data, vocabularies and ontologies are required. The primary language for developing ontologies for the semantic Web is the Web Ontology Language (OWL), built upon RDF.[6] Additional meaning can be added to the triples in RDF using OWL by defining standard classes and sub-classes which the subjects and objects can belong to, and restrictions upon the relationships between them. For example, a specific sensor may belong to a 'Sensor' class, and a 'Thermometer' sub-class. A sensor may have many measurements associated with it, but each measurement can only be associated with a single sensor. Protégé is an example of an ontology editor that primarily creates ontologies using the OWL format and as an open-source editor it is the most widespread tool used for developing scientific ontologies to enable sharing on the Web.[7] Semantic technologies enable much more advanced search techniques to be used on the data, also enabling the links between documents and datasets to be exploited.

Much of the discussion around metadata in science focuses on the development of various different standards and schemas for describing data and terminology, and in particular the development of machine-readable formats. There are many advantages to developing metadata standards, not only to reduce ambiguity and inconsistency of metadata associated with research data, but also to enable interoperability and simplify the creation of software tools for managing and making use of the research data. Metadata structures and terms that support human readers of the content are likely to be quite different from those that support machine-readability. For example, geospatial metadata conforming to an appropriate metadata standard may not be easy for a human reader to understand if they look at the data or metadata file directly, but the standard structure enables the information to be more easily imported, interpreted, and displayed within appropriate software such as Geographic Information Systems (GIS).[8] The development of standards can also help to improve both the exchange of data and of the metadata standards themselves, particularly when they are adopted by software tools that are important or commonly used within any particular discipline. If such software adopts a particular standard, it is much more likely that the users will put the effort into adhering to the standard in order to get the best capability from the software tools that they want to use. As discussed in Chapter 5, standards are also beneficial for the preservation of data, because it makes it much more likely that files generated in an adopted and machine-readable standard format will continue to be supported by software over the long term.

It can be difficult to choose the best standard to use for any particular project. Often, the requirements of projects change during their lifetime, and these changes may affect choices about appropriate data formats and data management for a project. It is possible that the most appropriate formats for data may change even after the research is complete and the associated data has been published, because the data comes to be used by others whose needs could not have been anticipated at the beginning.[9] There are hundreds of different scientific metadata standards and schemas to choose from, not only representing the diversity of different scientific disciplines, but also the complexity and variety of scientific activities even within the same discipline. Much as no one digital tool or interface is appropriate to capture every piece of scientific research, there is no 'one size fits all' metadata schema that can describe every scientific data file or every research record. Different types of science have different needs and requirements for different types of metadata. Science that is primarily concerned with observation and measurement, such as geology, meteorology, astronomy, and ecology may focus on creating taxonomies and other classifications to distinguish between different types of objects.[10] Even for experimental science, the majority of schemas focus on describing the raw data created by the experiments. However, there are a relatively small number that have been designed to include descriptions of the processes and conditions of the experiment itself. For example, the oreCHEM ontology provides a structure for recording the planning an enactment of experiments;[11] the Scientific Experiment Markup Language (SEML) describes the experiment processes and interrelationships;[12] the CCLRC Scientific Metadata Model (CSMD) provides metadata for use across all sciences and for describing both the broader context of a study and the individual investigations, experiments, and simulations within it;[13] and the

Chemical Methods Ontology (CHMO) from the Royal Society of Chemistry that provides over 3000 classes that describe methods and equipment for handling materials and collecting data in chemical experiments.[14] Although the focus of metadata standards is typically on formats for representing the data itself, as we have mentioned before, without the context of the what, why, and how of the research, the resulting data is next to useless. Metadata schemas need to be able to describe both raw and processed data, and the various processes and methodologies that were used to capture them. In biological and biomedical sciences, there are community standards for metadata that describes both biological materials and the methods involved in extracting them, for example, the popular Minimum Information About a Microarray Experiment (MIAME) guidelines for microarray analysis of genomic data.[15] Standards such as these have often been adopted as required metadata for submission of data to both databases and journals, and have inspired the development of similar standards for other techniques and data types.

These standards and schemas are useful for annotating, organising, and manipulating certain information. However, even with such standards, the majority of scientists still need to manually create context for their electronic records in order to ensure that their research is discoverable and meaningful for themselves and others, as we discussed in Chapter 5. Outside of large-scale projects, the scientists who produce the data are usually solely responsible for their management and the creation of appropriate metadata.[16] The later in a scientific project that the metadata is created, the harder it will be to do so, and the more onerous the task. To capture adequate and appropriate metadata to describe scientific research and to enable the discovery and reuse of the scientific record, ELNs and other digital tools designed to help capture the details of research need to enable the creation of metadata as early in the project as possible, or as Bird *et al.* describe it, ensuring the capture of metadata 'at source'.[17]

10.3 User-defined Metadata in Science

Electronic notebooks, including ELNs, can potentially collect a variety of metadata to describe the contents of notebooks and related research data, including both machine-generated and user-defined metadata. Machine-generated metadata typically includes provenance type metadata such as timestamps, author details, and file versions, but can also include information derived from data files, instruments, and other data sources. As mentioned previously, although useful, machine-generated metadata is often limited in the information it can provide, and so the creation of user-defined metadata based on the expertise of the originator of the research data is often required to capture its full context and meaning.[18] There are also other benefits of enabling users to create their own meaningful metadata to describe their research data and other materials with which they work. Making use of formal taxonomies and ontologies is useful for validation and quality control where known terms are registered within an information system, but relies upon these schemas being available in the software and using the most relevant terminology to describe the content.[19] These mechanisms of formal classification are useful for organising information where the structure and terms are already known, but they are not so effective at matching the terms that scientists may use individually to describe their research. Schemas are also often limited to English-language terms and are by their nature restricted to individual terms with a single meaning. Enabling users to add their own descriptive metadata terms enables the capture of terms outside of the formal classification if these are more appropriate and to capture terms in the user's own language if this does not match the formal schemas. As we mentioned in Chapter 2, memory is more effective when users can generate their own cues and being able to describe content with their own metadata terms makes it more likely that the user will be able to successfully search and retrieve the content in the future.

As we discussed in Chapter 3, some of the mark-up that is added to paper notebooks can be considered to be highlighting information that is metadata, for example, common and scientific species names, localities and geographical information, along with topics and links to information from other sources. User-defined metadata can often be added in a limited form to even the most basic of electronic notebooks through the use of 'key-words' or 'topics'. ELNs certainly have the potential to support scientists by facilitating the creation of more sophisticated metadata, and perhaps even more importantly, they have the capabilities to be able to make use of metadata to provide additional functionality and value to the scientists who created it. Although ELNs undoubtedly have an important role in capturing metadata, it is less clear what metadata is being captured and whether it is fit for purpose. The kinds of metadata we might expect to be captured in an ELN include information such as the start and end times of experiments, the materials used in the experiments, techniques used, samples and specimen identifiers, instruments, environmental conditions, links to related resources, such as experiment plans, previously collected data, and the literature, and for some projects personnel, project identifiers, and locations. It is likely that the metadata that is captured varies between different ELNs, and not all ELNs enable users to add their own user-defined metadata.

10.4 Use of User-defined Metadata in LabTrove and Other Platforms

At the University of Southampton, we wanted to investigate how users were making use of metadata within our LabTrove ELN. We looked at both patterns of user behaviour and metadata use, and also attitudes towards metadata through interviews and a number of user activities to find out how users were using the facilities provided by the ELN.[20] In common with most other ELNs, LabTrove captures a range of machine-generated metadata elements for each entry within a notebook such as author, timestamps, and a URI, but also enables users to add their own metadata manually.[21] There are two user-defined metadata types that a user can add to their notebook entries. The first is a mandatory 'section' metadata field that must be added for each entry in the notebook, and that is meant to be used for categorising the entries. The second user-defined metadata option is the ability to add multiple key-value pairs, which provides a more powerful mechanism for adding metadata to notebook entries than simple tags. Figure 10.3 shows a mock-up of the LabTrove interface for creating a new entry in a notebook, with the interfaces for creating section and key-value pair metadata. Users can choose to add new options for section and for the key-value pairs, or they can choose metadata that has previously been used within the notebook from the options available in the drop-down lists. In

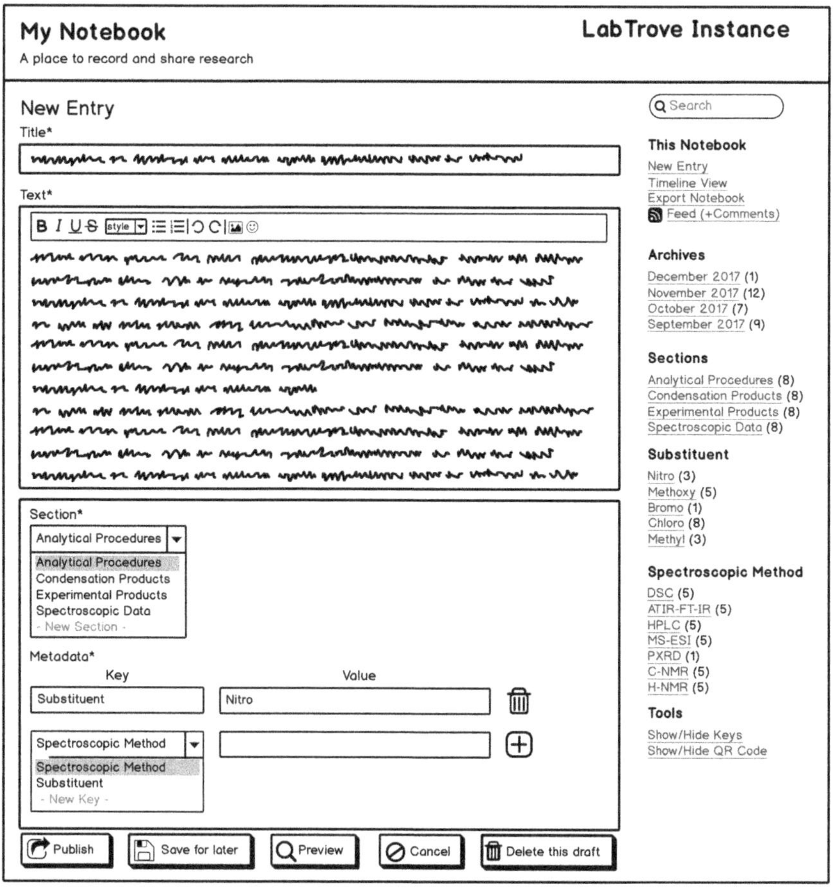

Figure 10.3 Mock-up of the LabTrove ELN showing a new notebook entry with section and key-value pair metadata.

this instance, the user has selected Analytical Procedures as the 'Section' metadata describing the type of notebook entry and added two key-value pairs to describe the content, the key 'Substituent' with a value of 'Nitro' and the key 'Spectroscopic method' with no value assigned as yet. The figure also shows how the metadata used within notebook entries is used to construct the navigation menu on the right.

The user-defined metadata elements in LabTrove are used in a number of ways to support the user in the interface, in addition to the visible navigation menus. For example, the provided metadata is used to enhance facilities for search and linking, and enable the grouping of related experiments. However, in our early discussions with researchers using LabTrove, it became apparent that many researchers found the interface for creating metadata difficult and were anxious about using it. Anxieties ranged from not knowing where to start with metadata to actively fearing the consequences of 'getting it wrong', and the amount of effort that would be involved in fixing any mistakes. As a consequence, many of the researchers we spoke to were struggling to add metadata for their experiments. The lack of metadata on their entries meant they were also finding it increasingly difficult to locate specific notebook entries and the details for particular experiments.

When we ran user testing sessions on LabTrove with researchers who did not use an ELN, we also found that researchers only added the Section metadata because the system required it, and they avoided adding any other metadata unless prompted. Even when they did add metadata, the terms used were inappropriate, even though examples of good metadata practice were provided as examples. When the researchers were subsequently interviewed, we found out that the researchers actually had very little understanding of what metadata was and why it might be useful. To investigate whether these issues of minimal metadata use and perhaps inappropriate metadata values were common in LabTrove, we conducted a survey of metadata use in over 100 LabTrove notebooks across the whole community. There were a number of interesting findings as a result of the notebook survey. The most significant finding was that the majority of users were not really using metadata at all, although we did find that there were a small number of individuals and groups who were using it successfully and appropriately. As discussed in Chapter 9, a big inhibitor to adding content in the ELN appears to be the 'blank canvas effect'. In this case, users were willing to add metadata but did not know where to start. The result of not knowing what to put, but the system requiring that something was entered, meant that a large percentage of users used a meaningless value for the Section metadata, such as a space or dash, or a generic term such as 'General'. Rather than creating different categories of entries to help organise the entries in the notebook, a generic term or blank value was used simply to satisfy the system. For the more sophisticated, but optional key-value pair metadata, less than a quarter of users were really making use of it, even though the function would have had benefits for both authors of the notebooks and their colleagues. We found that the amount of metadata used in a notebook increased with the number of authors using the notebook, suggesting that perhaps these users had a greater awareness of the usefulness of metadata for others, or that the greater number of entries made the use of metadata more beneficial in grouping and locating entries. In discussions with users, it appeared that after some time of use of the ELN, they realised that adding metadata values would be helpful, but they often had anxieties about going back and retrospectively adding metadata, as well as 'getting it wrong' for new notebook entries. This reflects the experiences that researchers also typically have with paper notebooks, as discussed in Chapter 3, where table of contents are added retrospectively to help with locating experiments, and where organisation is often learnt as a result of trial and error.

We had an expectation that metadata use may be associated with the privacy settings of the notebooks because users may consciously choose to add metadata for the benefit of other users. However, the addition of metadata in the notebooks did not appear to be correlated with the privacy of the notebook, suggesting that metadata was added for personal benefit as well as facilitating use by others. Adding metadata for both social reasons and personal organisation or future retrieval is in common with behaviour observed on other platforms for adding user-defined metadata.[22] Tools that are designed to enable the capture and display of metadata must support the use of metadata for both personal organisation and to facilitate more public discovery. There may also be a case for providing a mechanism to distinguish between metadata that is generated for personal organisation and for social reasons.

We also found that the metadata terms used in LabTrove were not necessarily particularly useful for the retrieval of information, although they may have helped with data organisation. The majority of terms used in the ELN represented 'labels' for the content of the entries such as 'samples', 'data', 'procedure', 'materials', and 'safety', rather than specific values such as the experiment or sample identifier, the name of the material used, or the specific technique that was used in the experiment, which would be much more useful for searching and navigation at a later stage. Metadata seemed to be being used most successfully where the notebooks were used by groups, for example, the X-Ray Group at the University of Southampton mentioned in Chapter 7, who use a shared LabTrove notebook as a hub for collaboration. Their shared notebook, at the time of the survey, had over 1600 entries with 25 authors, and made use of 58 sections to categorise the entries. Less use was made of the key-value pair metadata, but 16 different keys were included to help organise entries. Another example of successful use came from a bioscience group using LabTrove notebooks to record their experiments. Each member of the team had their own notebook and made appropriate use of both kinds of metadata. For the section metadata, the team appeared to use the same or very similar terms, and many of the keys were also similar; appropriate specific and individual values were used for the values associated with their own individual experiments. It appears as though the members of the team were either following the best practices of a 'champion' or they had collectively agreed to use the same metadata values for consistency.

The other notebooks where metadata was effectively used were individual notebooks where the author was well-organised and had an understanding of metadata before using the ELN. However, our findings from the survey of the ELN, and talking to the user community, indicated that the majority of researchers did not have a good understanding of the purpose and value of curation. In fact, many researchers did not actually know what metadata was. There was also a reluctance to provide additional information other than that necessary to overcome the requirements of the system. A similar problem of adopting a minimum effort approach has also been observed in other facilities of the ELN. For example, each time a change is made to an entry in the notebook, the system asks the user to provide a reason for the change. This is similar to the process used in a paper notebook of providing a signature and an explanation for any corrections made. However, users in the ELN typically provide the minimum amount of information required to satisfy the system, such as entering a single character as the response, even though an actual explanation would have been more useful for both the user and any reviewers or other users of the notebook in the future. It appears that even a small effort to create metadata at an appropriate time was too much of an interruption to the process of information capture or running the experiment. Researchers taking the 'minimum required' approach to metadata creation and data management is consistent with findings from other studies looking at data management and metadata practices in science as we shall discuss in more detail in the next section.

Following the survey of notebooks in LabTrove, we wanted to understand whether the behaviour and patterns that we saw in the ELN were unusual or consistent across other kinds of platforms. To find out similar surveys were conducted on a variety of other platforms where users are able to create user-defined metadata. The platforms included in the study were: Flickr, NASA blogs, chemistry-related blogs, and the myExperiment workflow repository. On these platforms, where arguably user-defined metadata is even more relevant for sharing and discoverability, we found some of the same kinds of problems and behaviours, in particular the use of minimum metadata. Additionally, the terminologies used were consistent to each individual platform, but each was recognisably different from each other, highlighting the importance of providing access to appropriate vocabularies within tools that are relevant to the community for which the tools are designed. From an interface design point of view, the two sets of surveys highlighted some key observations for encouraging users to add appropriate metadata. The various platforms that the chemistry-related blogs were held, such as Blogger and WordPress, and the NASA blogging platform, often provided a default generic metadata option to categorise the posts such as 'General' or 'Uncategorised'. The inclusion of the default option led to the majority of items being categorised under these default metadata options, so effectively everything was classified under the same meaningless term. Everything labelled with the same metadata value serves no useful purpose, and certainly does not help in the discoverability or organisation of information. The inclusion of a default metadata option also appears to deter users from creating their own more appropriate metadata values, possibly because of the 'minimum effort' approach already discussed. A possible way to mitigate such a problem may be to provide some examples of metadata values that the user can select from that are appropriate to the platform. This could be in the form of options the user can select from, or even as 'user assistance' in the form of text on the interface to prompt the users. The suggestions may act as cues for the users and they may select from the examples, rather than creating their own to begin with, but choosing from appropriate values is superior to labelling everything the same.

Based on the survey, it appears that actively providing invitations to add metadata and making use of the metadata in a meaningful and valuable way for the user encourages better metadata use. We will discuss more about using invitations to encourage the creation of metadata later in this chapter. The survey of other platforms also suggested that photographs may be easier to generate metadata for, based on the finding that the majority of users on the Flickr platform included tags on their photographs. The tags used were typically more diverse than the metadata used on the other platforms, and particularly useful for helping others understand the content and context of the photographs.

10.5 Metadata Friction and the Burden of Curation

The surveys described previously provide some insight into how interfaces might play a role in generating metadata within software. However, they also reinforce studies that have shown that even though metadata is essential for effective data management and curation, scientists do not see the creation of appropriate metadata for the benefit of others as a good use of their time.[23] There are a number of reasons why this may be the case. One of the major challenges associated with effective sharing of data in science is the 'burden of curation', so-called because of the significant amount of time and effort it takes to organise, manage, curate, and describe data.[18] The burden of curation increases as the audience of the data becomes less known to the originator of the data.[24] Curating and organising your own research data for future personal use is much easier than attempting to curate it for other audiences or uses that you have not considered. Even so, it can still be difficult to encourage researchers to create even basic metadata to describe their research data for their own use.[25] A related concept is 'metadata friction' defined by Edwards as the additional difficulty of attempting to create metadata by recovering the context of creation of historical data.[26] The further removed you are from the point of creation of the data, the harder it is to remember how the data was generated in the first place. It therefore becomes much harder to create adequate metadata to describe the research or data the later in the research process that the metadata is produced, and this increases the amount of time and effort involved in organising, managing, curating, and describing data. As we mentioned in Chapter 5, this contributes to the difficulty of sharing of research data because adequate metadata is needed to describe the data before it can be shared with others, and this is often created after the end of the research project when researchers are focused on other priorities. Capturing the metadata to describe the research data and other resources is much easier to do if it is created 'at source' or at the point where the data is created, as previously mentioned.[17]

The originators of scientific data do not necessarily see metadata creation and data management as part of their job, and therefore do only the minimum required, even when it is their responsibility to describe their datasets for deposit in repositories or for sharing with collaborators.[27] The primary concern for researchers is making use of the data. There is little interest in describing data for the benefit of an unknown audience with no apparently direct benefit to themselves, leading to various challenges associated with developing metadata to facilitate the interoperability and sharing of data.[25] In order to mitigate this reluctance to take on responsibility for creating metadata, some teams are supported by a data management expert to handle the processing and formal preservation of data, but for most, this support is unavailable.

These challenges of creating metadata to describe data can cause friction between researchers that can impede collaboration and the sharing of data as tools and services. Such challenges include the fact that many researchers are not really using formal metadata, individuals are responsible for their own data, and when metadata is created, it tends to be ad hoc for a specific purpose. There is also frequently disagreement and confusion about the meaning of terms. In the LabTrove survey, we found that some of these same themes were raised as concerns by the scientists. Even when the teams understood the benefits of curation and metadata, they did not know how to start because of disagreements over terms and standards. Not only was the time to create the metadata an issue, but also the potential impact and effort that may be involved in fixing problems or retrofitting a schema to previous research, a problem also identified by other studies. Confusion and conflict often occur over the use of terminology in the scientific community, even between experts in a particular field. As a result, it is also common for metadata schemes to be refined over time in order to correct mistakes and ensure a common understanding of terms and their relationships amongst users.

Another major problem impeding the creation of appropriate metadata to describe scientific results is that scientists are not as knowledgeable about metadata and curation as we might assume. We might expect that researchers from the so-called 'Google Generation' will have high levels of skills and information literacy because of their almost universal and regular use of technology, but the evidence shows that these expectations are unrealistic.[28,29] These observations are true even for physical scientists who are often perceived to be among the most technical.[30] Research suggests, in fact, that relatively few researchers actually know what metadata is or how to use it, and knowledge of data management and curation is uncommon among researchers across all disciplines.[31,32] From our discussions with students and researchers, it is clear that the understanding of metadata and how it is used ranges widely, ranging from at best familiarity with simple tagging systems, to understanding and using highly structured metadata schemas. Researchers may have experience of metadata from adding tags on social media, or more highly structured metadata schemas and standards if they have been required to use them for the deposition of data in formal repository systems. However, even those who have deposited data in a formal repository are not using this level of detail for organising and describing their own research, and many researchers have limited knowledge of data management practice and the use of repositories outside of their own institutions. This is true even for disciplines where there are established standards for data or metadata in the community, and, in fact, many researchers may not even be aware of widely adopted standards within their own disciplines.[27] Much of the research into improving metadata use suggests that more needs to be done to target scientists directly through data management education programmes, especially for researchers at a postgraduate level, but such programmes would also be advantageous for undergraduates.[31,33–36] Such programmes should aid researchers understanding of organising, classifying, and adding metadata to their own work for the benefit of themselves and others. There is some evidence that discipline specific and tool specific training may be of more value to researchers than generic data management education.[35] Providing the assistance of information experts to provide support and encourage best practices, together with active participation and assessment of compliance to best practices from management can also help.[34,37]

Although improving data management education is certainly important for improving the understanding of metadata in data management and encouraging effective curation by scientists, the effective design of digital tools is also critical to facilitate the process and to support users in order to take away some of the burden. However, metadata support is often lacking in tools developed for and used by scientists. The reasons for the lack of sufficient or appropriate support for metadata is often because it is not seen as a top priority, there are a lack of appropriate standards, or because there are incompatibilities between local practices and researcher needs.[25] As discussed in Chapter 7, tools must be designed to support how users are working. For supporting the organisation of research data and the creation of metadata, the tools need to reflect how both users and communities organise and search for their data.[1] Individuals may have their own styles of personal organisation and metadata schemes, or even no schemes at all, and this can present a challenge when sharing data with others. As we saw in LabTrove, when users collaborate as part of their research, they may adopt common schemes that provide consistency and insight into effective ways of organising and describing data that may be effective for a wider community, creating a kind of 'group information management' rather than 'personal information management'.

10.6 Automatic Creation of Metadata and Smart Assistance

There are a variety of ways that software can help to reduce the burden of metadata creation and curation for scientists. As mentioned before, users should never have to add metadata that a computer can automatically add for them, such as the basics of timestamps, author, and other provenance type information. For digital tools such as ELNs that can import various kinds of external files, if those files already contain metadata, that information should be included and available for use within the tool. Many tools, for example, are able to extract metadata from known file types such as Word and Excel. Another example discussed before is metadata for a photograph. The user should not need to enter manually the details about the camera, lens, film speed, aperture stop, image size, and so on for the photograph, because this information can be provided by the camera itself. If a system imports a photograph then it should be able to retrieve the EXIF information to not only make use of the metadata internally, but also provide useful ways of exposing the relevant metadata to the user. For example, using the EXIF metadata could enable users to search for images taken on a specific camera, with a specific lens, or with specific settings. The user may need to annotate the photograph with additional metadata that the software could not provide, such as the content of the photograph or the conditions under which it was taken. However, there are also tools that can support the user with such tasks. For example, Flickr, has image recognition tools that identify the content of the photo and apply suggested tags. The tools also enable photographs to be organised by content based on image recognition, such as if photographs contain people, vehicles, animals, buildings, or plants, and whether the image is taken at a particular time of day, includes sporting activities, and even the mood of the image. The photograph in Figure 10.4 was automatically tagged in Flickr with 'blackandwhite', 'monochrome', 'outdoor', 'serene', 'plant', 'surreal', 'tree', and 'sky', showing how sophisticated the classification through image-recognition can be. The classification is not perfect and sometimes inappropriate tags are added that need to be removed, for example, photographs of ski-jumpers in mid-flight have been misidentified and given the tags 'animal' and 'insect', although others have been identified successfully and tagged with 'sport' and 'snow'. Automatically generated tags based on image recognition are undoubtedly useful; however, the user may still want

Figure 10.4 Photograph of a landscape in England from Flickr with machine-generated tags.

to add additional tags to provide extra context, such as the location, the event, the reason for the visit, and who else was present.

For scientific purposes, image recognition tools can be trained to identify scientific subjects such as biological specimens, samples, analytical test results, geological structures, and other artefacts. If a camera was used in the lab to take photographs through a microscope, then it may be useful to know the conditions under which the photograph was taken such as the room temperature and lighting conditions, which could be automatically captured by sensors and the values included as metadata associated with the image. We will discuss in more detail scientific applications of image recognition with Artificial Intelligence and Augmented Reality in Chapter 12. Instruments that capture data are often carefully calibrated and have specific settings chosen; however, researchers often have to diligently record the settings manually because the instrument does not include the information in the digital data files it produces. The manual recording of settings is open to errors, as mistakes may be made in transcription or the user may record settings they thought they were using, rather than the actual settings on the instrument. Ideally, of course, the instrument itself would capture the correct information by incorporating metadata about the instrument itself and any adjustments made to settings within the digital data files that it produces. This information can then be incorporated into the research record as metadata that can be used as a permanent record of the reality of that experiment, a reference for troubleshooting, and for replicating the exact conditions of an experiment.

The content of the record itself can be mined for useful metadata. The most basic form of text mining is to extract the most commonly used words within a document or file to be used as potential metadata. More usefully, dictionaries or other schemas can be provided to match terms used within a document or notebook entry to those most likely to be useful in describing content in a particular context. For example, in the LabTrove ELN, chemical names used within entries in the notebooks are matched with records for different chemicals from the ChemSpider database.[38,39] The matched names are converted into links to the ChemSpider database for the appropriate chemical found, so that users can follow the links to get more information about the matched chemical compound. It would also be possible to extend the system to automatically generate metadata about the matched material and add it to the notebook entry. Another example of this kind of data mining of content comes from WordPress, where terms related to the content of a newly published post are 'suggested' to the user to use as tags for describing the post. The use of data mining and suggesting relevant tags to add to a post in WordPress solves a number of the metadata problems we have discussed. The user is prompted to add metadata, but is given relevant tags to add rather than just a generic tag; the effort of adding relevant metadata is reduced by the automatic generation of relevant tags, and the generation of tags also provides examples and a starting point for the creation of the user's own metadata. Given relevant examples, the user can choose to remove any that are irrelevant to their post and also to add their own if there are other more appropriate terms to include. Another example of a system for automatically extracting metadata from content comes from archaeology, where natural language processing techniques have been used to extract terms and phrases for metadata mapping from archaeological dig diaries.[40] Details and descriptions of objects, localities, people, dates, activities, and properties of the objects are extracted from the diaries and used to enrich the central register used to capture details about small finds from the digs. Using metadata mining in this way can enhance the research record, but in order for the metadata generated to be useful and meaningful, the appropriate information must have already been captured within the system and also captured in an appropriate format. For the ChemSpider example, if the researchers use appropriate chemical names within their notebook entries, then the data mining is able to identify the materials and potentially provide users with useful information about those materials. However, if the user uses their own abbreviations or spells the chemical names incorrectly within their notes, no match will be made, and the software's ability to make use of that information is lost.

The use of dictionaries and schemas within software also provide other ways to assist the user. For example, they can enable the digital tools to provide spell-checking of scientific terms and nomenclature and also to facilitate the 'autocomplete' of known terms that are entered to ensure consistency. As previously mentioned, selecting appropriate dictionaries and schemas for use in a tool can be difficult because of disagreements about terms and differing needs between individuals, let alone different scientific communities. Although different communities use different terms, there appear to be some high-level terms and topics that are consistent between most scientists such as materials, activities, locations, data types, people, events, and equipment, although each discipline uses more specific terms within these broader topics. For example, 'Materials' are common elements to most scientific research, but one or more of the following terms may be more appropriate depending on discipline: 'Inputs', 'Reagents', 'Strain', 'Products', 'Substance', 'Compound', 'Chemical', 'Molecule type', 'Batch', 'Sample', 'Specimen', and so on. Even basic schemas using common elements could be used to make interfaces a little smarter and help users develop their own extended schemas. Ideally, however, standard schemas and ontologies will be incorporated into tools to facilitate data mining, ensure consistent terminology, and interoperability with other systems. As mentioned previously, the challenge is in choosing the appropriate schema, but enabling flexibility and providing a way for users to incorporate their own schemas into tools could help to solve some problems, such as terminology conflicts and schemas changing over time.

10.7 Invitations to Encourage the Creation of User-defined Metadata

Inclusion of metadata for records helps with providing access, effective curation, improving search, and providing context for the record. As we found in LabTrove, providing researchers with the facility to add metadata to their experiment records does not mean that they will make use of it, or if they do that the metadata they add will be relevant and useful. Researchers need support from digital tools to create effective and meaningful metadata. Tools, such as ELNs, provide an opportunity to encourage researchers to include metadata in their records, but can also add extra value by making effective use of the metadata that is provided. In order to capture useful and appropriate user-defined metadata, we need to design interfaces that encourage the user to add descriptive information about their research records, including data and other resources within them. If we ask users to add metadata without any additional support, they may struggle or add inappropriate or unhelpful values. By providing additional context for adding the metadata, we can provide additional support to users of the tools to provide metadata of value.

In the survey of metadata use on other platforms, described previously, we found that many of the interfaces for these online tools made use of a variety of prompts and 'invitations' to encourage users to add tags, categories, or labels to their items. On these platforms, metadata is also displayed prominently in the interface, for example, the tags in Flickr, tag clouds in WordPress, and the use of tags for searching and filtering in MyExperiment. Invitations are a way to encourage users of software to provide information or to perform tasks. They often highlight a part of the interface and invite you to use it, for example, by providing a link or a button such as a plus-sign to encourage you to explore or by asking a question to elicit particular information. Invitations act in part to attract attention, but they also often act as prompts to record information. These kinds of invitation are used within many kinds of software, but they are particularly common on social networking platforms where the software needs to capture information from users to function, but also where users may not have a high level of technical expertise. For example, Figure 10.5 provides an invitation example from Facebook, where the user is prompted to record their thoughts or activities by being asked a question about what they are thinking. Users are also encouraged to add more detailed metadata about how they are feeling, what they are doing, where they are, and who they are with. 'What's on your mind?' could be considered a prompt for the free recall of information, whereas the other prompts are specific and can be considered as cues.

Figure 10.6 shows an example of an invitation from LinkedIn with a question used to elicit metadata about skills. The user is invited to add their

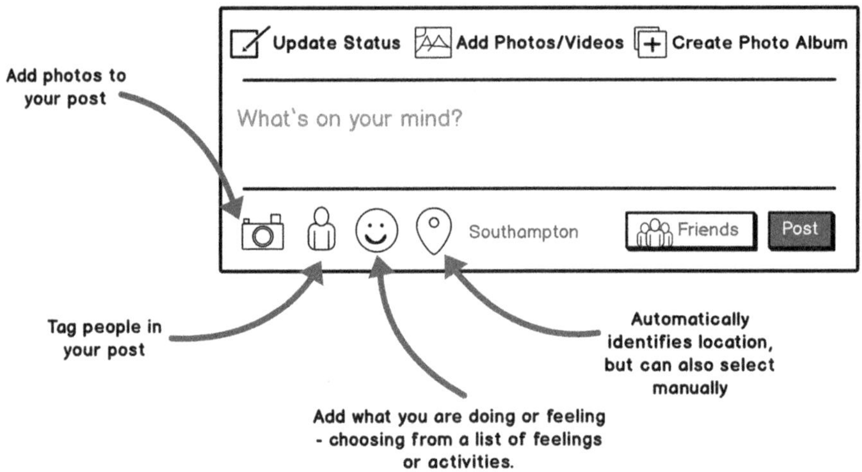

Figure 10.5 Example invitations: asking questions and adding metadata from Facebook.

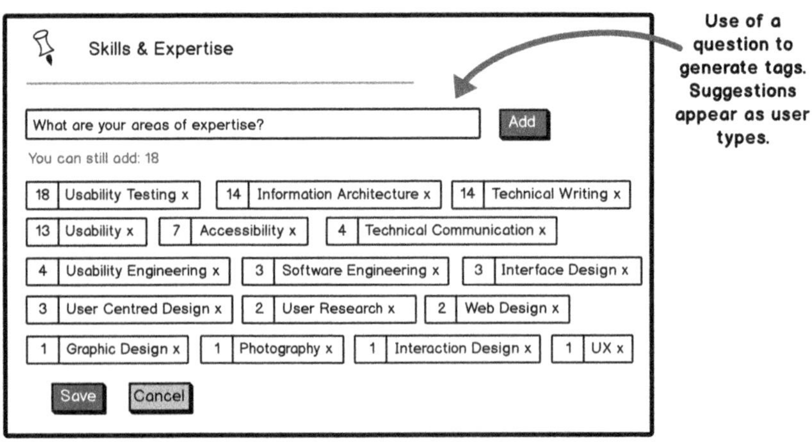

Figure 10.6 Example invitations: asking questions and tags from LinkedIn.

skills in the textbox and an autocomplete function matches them to skills that other users have chosen. The skills that are added are displayed prominently within the editing interface shown, indicating to the user the number of endorsements they have for each skill. The skills are also prominently displayed for the benefit of other users in their profile and included in search results, making it more likely that the user will be found and matched with appropriate contacts or career opportunities. Such prominent use of metadata provides an incentive for the user to add it, and for other users to interact with it.

Metadata is also collected by many of these social networking platforms in the form of 'profiles' where users are encouraged to complete a series of fields, usually about themselves or to describe an object. For example, Figure 10.7 shows a mock-up of an invitation from LinkedIn that encourages users to add information that has not yet been included in a personal profile. Invitations include the phrase 'Recommended for you', an arrow pointing at the interface, and the 'plus-sign' buttons. Clicking on the plus symbol reveals a form to capture the metadata for this particular content. Placeholders are left within the profile for sections without content as a reminder to complete them.

Figure 10.8 shows a mock-up of the Flickr photograph interface showing the 'Profile' for a photograph containing a number of invitations to perform actions and create new content, including metadata. The user is encouraged to add a title, comment, tags, add people in the photograph, and add the picture to an album or group. Instruction text is provided such as 'Add a person' and placeholders are provided as reminders to add missing information, such as the description, which has not been included yet.

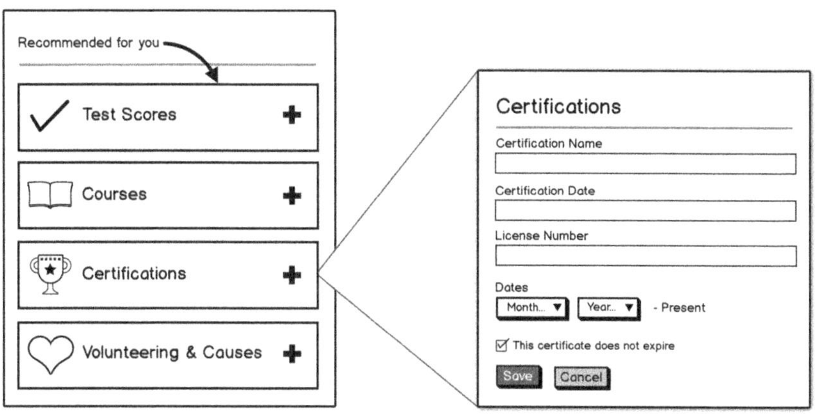

Figure 10.7 Example invitations: metadata elicitation example from LinkedIn. The user is encouraged to add content to sections with the 'Recommended for you text' and the '+' buttons. The window on the right shows the expanded Certifications section.

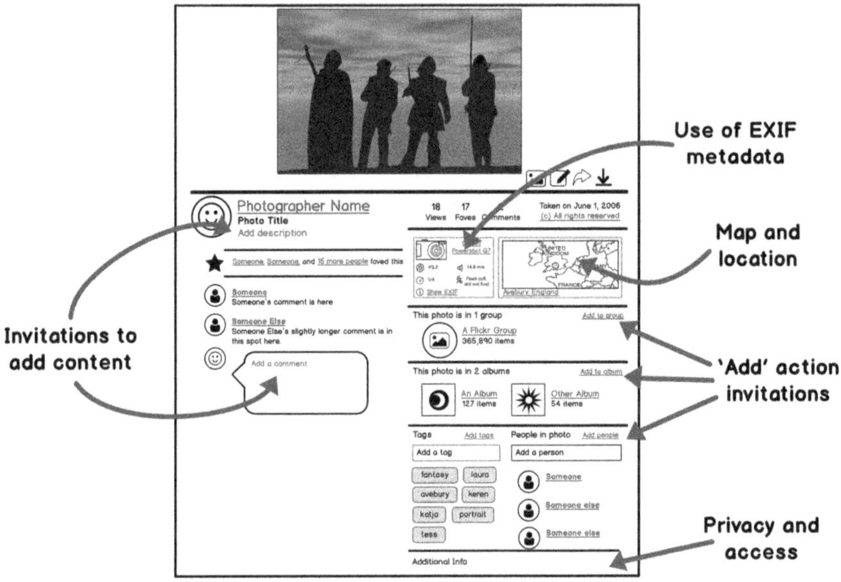

Figure 10.8 Photo 'Profile' from Flickr.

10.8 Encouraging Curation at Source by Using Cues and Changing Perspectives

An advantage with the invitation style of interface, as shown in the figures, is that most of them are acting as cues to generate information. As discussed in Chapter 9, using cues is an effective way to encourage users to record important information and particularly to capture specific information. As part of the studies looking at how the design of interfaces changed the information captured in the experiment record, we also took a look at how we might be able to influence the capture of metadata by making use of cues and the other methods of adding reminders to add metadata within interfaces.

Different techniques may be more successful at generating terms appropriate for different needs. Many social networking platforms collect metadata using a 'free recall' method, where invitations are used to encourage users to add metadata in the form of tags, categories, or keywords. These invitations act as a reminder to add the metadata, but are not directly acting as cues to prompt the capture of any information in particular. As we discussed in Chapter 9, cues can encourage researchers to remember and record specific information. For example, if we want a researcher to add metadata to describe an experiment, then asking about the materials, conditions, equipment, or techniques used, is likely to prompt them to provide relevant information. As we saw in the survey of LabTrove, when users are asked to add metadata without additional support, the values are not necessarily appropriate or useful, resulting in the production of meaningless or generic values that are useless for providing insight into the content of a record or to help locate a specific study. As we also discussed previously, it is also easier for the author of a record to create metadata of relevance for themselves than it is to create metadata that is likely to be of use to a wider audience. One of the ways of overcoming this difficulty may be to ask the creator of the metadata to 'change their perspective' or to consider what metadata would be of use to someone else. As discussed in Chapter 2, 'changing perspectives' is a useful technique to help recall additional information and to tailor that information to a particular audience. For example, making use of one of the examples from psychology research in Chapter 2, if a researcher was asked to add metadata tags to describe a photograph of a particular house, it is likely that the metadata terms chosen would be different if the user was asked to add tags when taking the perspective of a house-buyer compared with if they were asked to take the perspective of a burglar as shown in Figure 10.9.

At Southampton University, we were curious to find out what different metadata would be produced by using these different techniques and whether they could inform the design of interfaces for digital tools. In the studies described, three different techniques were used to generate metadata. The metadata terms generated were analysed to assess differences between the techniques and usefulness of the terms generated. The first study was a standalone investigation to find out whether using cues or changing perspectives would generate different and more useful metadata than free recall.[41] The study was also designed to investigate whether different metadata was generated for three different information types: photographs, structured textual information, and unstructured textual

Figure 10.9 Examples of different metadata produced to describe a house when taking the perspective of a house-buyer *versus* the perspective of a burglar.

information. The reason for examining whether metadata was different for text and photographs was because we suspected that it may be easier to generate metadata for photographs based on our Flickr study. The use of structured and unstructured text was to determine whether metadata generated for these two information types differed. The study asked participants to add ten words or phrases to the different information types under the three conditions: 'free recall', 'cued recall', and 'change perspective'. In free recall, the participant is asked to write up to ten words or phrases that come to mind when they look at the photograph or read the text. In the change perspective condition, the participant is asked to write up to ten words or phrases that come to mind for use in a search engine when they imagine helping someone else to find a similar picture or piece of text on the Internet. In the cued recall condition, the participant is asked to provide word or short phrase answers to a number of cues in the form of questions to elicit specific information about the photograph or text, including locations, people, equipment, activities or actions, other objects, and any other words or phrases of their choice to describe the material.

There were some interesting findings as a result of the analysis of the study. We anticipated that the different conditions and different information types would generate different descriptive terms, but the patterns we observed were more complex than initially imagined. Overall, more responses were produced for photographs than for structured and unstructured text, suggesting, as we already suspected, that it is easier to generate metadata for photographs and may explain why metadata is better used in Flickr compared with the other platforms we investigated as part of the LabTrove survey. The fact that images appear to be easier to describe, suggests it would be helpful to include functionality in interfaces for digital tools to enable users to add metadata, not only their textual research records, but also to other media generated during research such as photographs, sketches, graphs, plots, maps, and diagrams. Unstructured information appeared to produce more terms than the structured information, although many of these were personal opinions about the structure itself such as whether the content was well laid-out or the steps were in the right order, and the presence or absence of structure did not seem to make much difference to the creation of appropriate metadata as initially anticipated.

When participants generated metadata in the free-recall condition, they generated many appropriate metadata terms for the different items, but they also generated unexpected personal terms describing the nature of the information object itself, opinions about the content, or guesses about the context. For example, terms were generated that described the roles, tasks, or emotions of the people in the photographs; the smells, taste, and desirability of food in the recipes used for the structured information; and details of their own experiences or the imagined experience for the activities described by the unstructured text. These terms were very interesting and certainly provided additional context for the information items, but may or may not be appropriate for use by others. The terms generated in the study in the change perspective condition, where participants were asked to come up with terms that would help someone else find information in a search engine, were both appropriate and highly consistent between participants. Fewer terms were generated in this condition, but they matched terms that would make suitable keywords for use in a search engine. In the final condition, providing cues in the form of specific questions led to a significantly higher average number of responses compared with the other two conditions and provided the most consistent and balanced distribution of terms across the topics asked about. However, we observed a couple of problems arise from the use of cues. First, where a question about a relevant topic was not explicitly asked, then the metadata terms produced completely missed these topics. For example, in the terms used to describe the recipes in the structured text, there was no cue to prompt for ingredients. As a result, very few ingredients are recorded in the cued recall condition, even though ingredients were one of the most significant themes in the metadata generated by both the free recall and changing perspective conditions. The loss of these important terms reiterates the importance of asking the right questions, as we found with the use of templates for experiments in Chapter 9. If the information is not explicitly asked for, then it will most likely fail to be recorded. We also saw personal and 'guess' terms in this condition for those cues where the information object itself did not provide the information to answer the question. Users may be less likely to make guesses when generating metadata to describe their own research, but it is quite likely that ambiguous questions or questions that do not fit well with the data being describing are likely to lead to confusion and potentially incorrect or false information entered into such systems.

We conducted a number of other studies to see how these different methods of generating metadata terms would work for describing chemistry experiments, including a much larger study than the original with nearly 120 participants generating metadata terms for three different chemistry experiments.[42] We found that the results of these studies were similar. As before, the free recall condition often generated personal terms describing the experience of running the experiment, but mostly useful terms describing the conditions of the experiment, and materials, equipment, and techniques used. Also, as before, the change perspective condition generated the fewest terms, but also provided the most consistent and appropriate use of terminology, generating almost no personal or irrelevant terms at all. The cued recall condition continued to generate the highest number of terms and also highly relevant content for the specifically cued categories.

Perhaps one of the reasons why metadata use is so poor in most of the different platforms we looked at in the LabTrove study is because they make use of some form of free recall based methodology in the interfaces, for example, by asking the user to add tags, define categories, or add key-value pairs. Based on the above metadata user studies, it is clear that generating user-defined metadata through free recall may not produce the best metadata in terms of both quantity and quality. If the user is unsure what to add, they may add nothing at all, use a generic term to work around the system, or add inappropriate values. Adopting different strategies in the interfaces of digital tools to collect metadata, such as using cues and asking the user to change perspective by thinking about their audience, can help prompt users to add more relevant metadata to describe their research.

Each method of generating metadata has advantages and disadvantages. In general, free recall appears to be effective for generating a diverse range of terms that could provide context to the experiment, but may also produce a lot of irrelevant and personal information that is probably not valuable for use as metadata for sharing. These personal or opinion terms are probably of least value to others trying to make use of the experiment record, but they may well have value in terms of personal organisation, because it is likely that they have meaning for the individual who added them. Changing perspectives is effective for producing terms that would be relevant for search and discovery, reducing the amount of irrelevant information that is captured, and producing consistent values that are likely to be meaningful for different users. This condition produces the fewest responses, but specifically asking for search terms is much more effective for generating metadata that is beneficial for sharing compared with just asking for keywords. The terms used, as might be expected, are particularly relevant for generating metadata that would help with discoverability. The cued recall condition is very effective for producing high numbers of both relevant and diverse responses for the cued categories. However, two major downsides affect cued recall; if a cue is missing, important information may not be captured at all, and any ambiguity over the meaning of the cue can lead to confusion and a loss of information or the production of irrelevant terms for that cue.

It is likely that making use of a variety of these methods for prompting the capture of user-defined metadata within the interfaces of digital tools will generate a great deal of appropriate descriptive metadata to describe research and data. Making use of invitations or various types of prompts within the interface is also likely to help users overcome the 'blank canvas effect' by giving them some support in understanding what they need to provide.

10.9 Existing Interfaces for Generating Metadata

The majority of digital tools currently in use for creating metadata are tools for designing metadata schemas and for generating metadata to adhere to the schema, for example, to describe data to be added to a repository. In this section, we will take a brief look at some examples of the tools available for both of these tasks. Many of these tools generate separate files for describing the metadata associated with datasets, and in some tools, the data and the metadata may be stored in different places. In some cases, metadata is stored within the file itself, or as part of a package, but in many cases, the metadata is separate from the actual data files, and may be stored within the same system, or even in different systems. For data deposited in repositories, the metadata is used to describe the research data stored within them, and so is used in interfaces for search, discovery, and viewing information about the object, whilst the data itself is usually accessed and used in a different system. For example, only the data itself is likely to be downloaded for use on a local machine. The metadata for the dataset may also be used within a registry, whilst the data itself resides on a different system. In contrast, in some tools where the metadata is created together with the research data, both metadata and data are stored together and utilised within the same tool. These different styles of metadata creation for different purposes may lead to different requirements for the ways that the interfaces are designed, although there are some similarities that we will discuss at the end of the section.

The vast majority of metadata schemas are generated in XML format, and therefore at the most basic level, users can create or modify metadata in XML format using a simple text editor such as Notepad or TextEdit. There are also many native XML editors that can be used to create and edit metadata and schemas, such as XMLSpy, XMetaL, and Oxygen.[43–45] Many of these native XML editors provide functionality way beyond what the majority of scientists need for metadata creation. As a result, tools that have been developed or customised for metadata creation in an appropriate scientific discipline are more user friendly for the task. Many of these tools are designed to be used within particular disciplines and in conjunction with established metadata standards.[46] Such tools are often integrated with data repositories or software for managing information resources, but they may also be found as stand-alone tools or online services. The majority of these tools store the metadata as XML, but some can also export their content in other formats to facilitate interoperability with other systems.

A number of metadata tools have been developed to manage geographic metadata and related information resources. For example, ArcCatalog is a tool integrated into the ArcGIS suite of software for managing various geospatial data files and their associated metadata.[47] Machine-generated metadata that is viewable by the user includes Federal Geographic Data Committee (FGDC) metadata describing geospatial data including maps, images, and geographic information system files, and metadata appropriate to the type of file being described.[48] The editor also enables users to add user-defined metadata including titles, descriptions, credits, limitations of use, and keywords. The interface for editing the metadata is in a form style with labels for the fields and with limited user assistance provided.

The Metavist editor, developed by the US Forestry Service, is a popular editor for metadata creation, although an informal user guide authored by the US Geological Survey highlighted a number of usability issues with the original interface.[46,49,50] A very large number of metadata elements are required to be completed within the editor and it is very complex in places.

The interface format includes numerous pages that use a form style with simple labels for forms to collect the metadata. Although some sections provide user assistance, this does not appear to be provided for every field. The interface includes the Integrated Taxonomic Information System (ITIS) taxonomic classification standard as a default for describing biological data, but users can also choose to include other taxonomic classification standards.[51]

An example of a metadata editor that has been developed with the users firmly in mind is the CEDAR Workbench.[52] The product website expresses the philosophy of user centred design by sharing the strategies, requirements, user personas, and user scenarios generated and used during the design and development process.[53] The software is designed for managing and sharing resources for scientific experiments and comes with a metadata template designer and metadata editor. The tool supports interoperability by enabling users to export the metadata in a variety of Semantic Web friendly formats, and to be uploaded to external repositories using Application Programming Interfaces (APIs). The tool makes use of ontologies and controlled vocabularies, as well as using data validation and data analytics to speed up the process of metadata creation. The service links to BioPortal, a repository for biomedical ontologies, to enable template authors to choose appropriate ontologies for labelling the fields and to constraint the values that can be entered during metadata authoring.[54] The software also automatically extracts metadata from imported data files such as Excel.

An example of a metadata capture interface designed for use in field studies is Morpho, a tool for capturing field observations for ecological studies.[55] The interface is form based, but the forms are customised and pre-populated based on content created in Ecological Metadata Language (EML). A future development concept for the project was to provide templates that correspond to known ecological experiments or other activities that would enable users to choose the appropriate template and thereby see a customised interface for their study.

An example of a metadata editor for use outside of biological sciences is the Space Physics Archive Search and Extract (SPASE) XML editor.[56,57] There are two versions of the editor, a base version for metadata descriptions of observation data and the second for metadata descriptions of models and simulations. Each version of the editor provides a number of templates with appropriate fields for describing particular information or objects. For example, the base editor includes templates for annotations, catalogues and registries, data types, instruments, observatories, people, and services, whilst the version for simulations includes templates for models, simulation runs, and data outputs. The templates are customisable and the user can choose to complete them as a form or in text format. The text format is equivalent to editing XML in a text editor with a special mark-up, making the form version much more user friendly. User assistance is also provided in the interface to describe the meaning and appropriate values

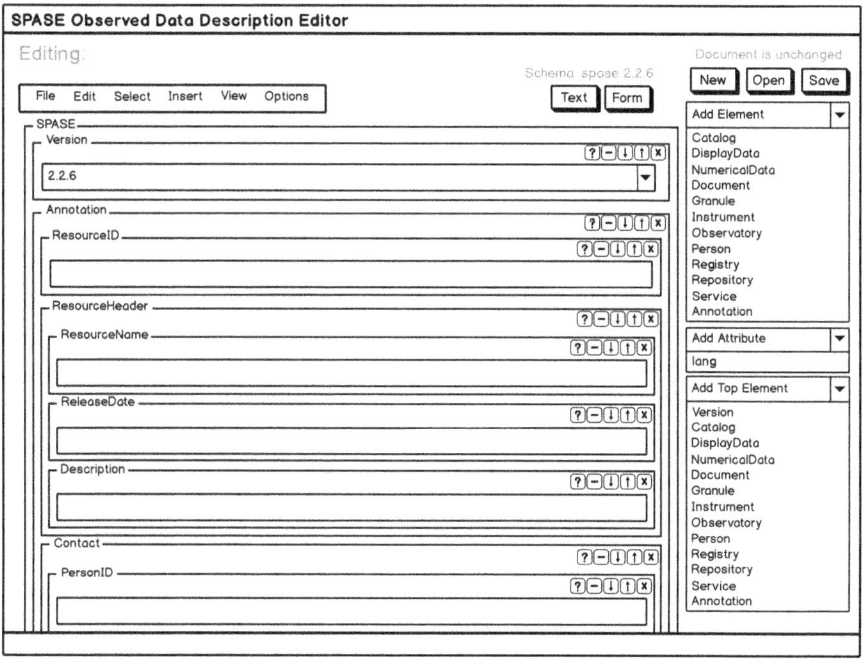

Figure 10.10 Mock-up of the SPASE Observed Data Editor.

for the element fields. Figure 10.10 shows a mock-up of the SPASE editor showing the form style of interface for completing the metadata schema. The user can choose from a variety of pre-defined templates for the type of information that they are describing. Additional elements can be added as needed, and the order of elements in the schema can also be changed.

The majority of metadata editors that have been designed for scientists make use of formal standards or schemas. Although the interface is therefore rigidly designed to fit the chosen standard, the advantage of such tools is that they enable users to quickly and accurately enter metadata conforming to the chosen standards.[58] As some of the examples above demonstrated, the creation of metadata templates enables metadata to be effectively created to appropriately describe different kinds of information objects, related resources, or activities. As mentioned in Chapter 9, providing a way for users to customise their own templates can make a tool more flexible for recording the information that the user really needs to record, whilst reducing the frustration of being rigidly constrained where unnecessary. The advantage of integration of formal standards, such as vocabularies, schemas, and ontologies, in digital tools for capturing metadata is that the constraints they place on selected values facilitates accuracy, consistency, and interoperability. Integration of standards within tools is particularly useful for disciplines and research where such standards have been adopted and are widely used. However, they may be too restrictive for more niche research areas or where the research is more diverse within the same discipline.

The metadata editors described previously are very similar in design, although they may be desktop or Web-based, they are basically form-based systems. They may have a single page of fields to complete, or multiple pages or tabs, or many levels to navigate. In some cases, drop-down lists or radio buttons may enable the selection of specific options, although most often the user has a blank text box to complete. The use of different kinds of control, such as drop-down lists and radio buttons, can assist the user by constraining the available options and making choices clear, whereas text boxes invite the user to add their own free form content with differing results. The constrained choices promote consistency and interoperability, whereas free-form content enables the user to choose the values most appropriate to their own research. The majority of the metadata editors available provide very simple labels for the fields, and in some cases, additional user assistance or guidance in the interface as to what the labels mean and the appropriate information to enter into the field. Often, user assistance is provided only for fields that are ambiguous or unclear in their meaning, but some of the editors provide a more usable and consistent experience by providing user assistance for every field. The labels act as cues for the user to prompt them to record information expected by the interface, but where the label is ambiguous, unclear, or lacks supporting user assistance, there is a chance of mistakes being made. Fields may be left blank, the wrong information may be entered, or inappropriate values entered. As mentioned previously, providing appropriate labels within an interface can help with overcoming the 'blank canvas' by prompting the user to record appropriate information, but the labels chosen need to be meaningful and supported by adequate user assistance. The appropriate interface controls need to be selected to allow the user to make appropriate selections for their input, such as drop-down lists or radio buttons as mentioned previously, or even more specific controls such as calendar controls for selecting dates or sliders for selecting numerical ranges. These are basic usability concepts as previously discussed in Chapter 8, but even the basics are still often lacking in interfaces for ELNs, metadata editors, and other digital tools used by scientists.

Poor usability in software leads to ineffective use, and in tools designed to capture metadata this means that metadata fails to be captured entirely or is of poor quality. In turn, the information resources described by the metadata are less likely to be discoverable and understandable, and therefore less likely to be used within the community, ultimately making it hard to replicate or make use of the original research that is described. These problems contribute to a dissatisfaction among scientists with the tools for creating metadata and also problems with finding data, including their own.[59] Usability is therefore just as important for the design of software where metadata is created and managed, as it is in the software in which the research and data is captured in the first place. How then can we improve the usability of software used to capture and manage metadata? Ideally, usability is considered from the very beginning of a software development project in order to avoid usability pitfalls, but it is possible to improve existing software. For existing software, performing a usability evaluation, as described in Chapter 8, enables improvements to be made to the interface based upon problems identified. Nasoz *et al.* provide an example of how a usability review was used to make improvements to a later version of the Metavist metadata editing tool described previously.[60] As well as identifying usability problems to be fixed, the authors were able to identify future feature improvements. A usability evaluation is a good first step towards improving usability, although it is unlikely to be able to solve really significant problems with the design.

10.10 Collaborative Metadata

Consistency of terminology is important within metadata, and as we have described previously, users often use the same or similar terms to describe the same information, and successful use of metadata often comes where teams are sharing the terms they use. But we also saw that users frequently generate metadata terms that are personal to themselves and that would not be useful or meaningful to others. Users creating metadata for their own personal use and for sharing is also observed within library catalogues and social media platforms, and suggests that designing interfaces that enable the creation of both public and private terms can facilitate the creation of metadata essential to the record, but which also supports the needs of both the author and other users of the content.[61] There may, however, be benefits of engaging in social metadata creation within collaborative digital tools such as social networks, electronic notebooks, and even data repositories. The presence and creation of 'public' metadata on shared materials within such tools provides opportunities to construct social vocabularies or 'folksonomies', to describe the data. Traditionally, materials are categorised by information experts, such as librarians or the author of the materials, with the terms generated representing a particular viewpoint on the data. For some projects, an information professional may be called in to help with metadata creation and curation of the research data to reduce the burden on the researchers. There are, however, potentially problems with the terms used by professionals to classify information. First, the language and terms used by information professionals may not represent the terms used by the authors themselves or other potential users of the data, but also the number of terms generated by professionals tends to be quite low and may not adequately describe the details that would add value to the record.[62]

Within environments where materials are shared, such as online social networking applications, users can add their own metadata to describe both their own content and also content uploaded by others. For example, many social networking applications allow users to add tags to describe their own and others content, for example, for bookmarking tools, photo-sharing, and blogs. For example, the Flickr Commons are publicly held photography collections which have been shared on the platform to provide a way for the general public to contribute information and knowledge about the content of the photographs.[63] Users are encouraged to add comments, but also metadata in the form of tags. Over time in social networks where tags can be added, a stable pattern of terms is generated on tagged resources, as consensus is reached on the terms that adequately describe the resource.[64] With increasing numbers of users and increasing numbers of tags, the most appropriate terms tend to surface as the most common terms. The same effect was seen in the metadata studies described earlier, where personal terms were only used once, compared with the most useful metadata terms that were used consistently and in high numbers. These folksonomies may provide more usable metadata that better facilitates the world view of the users that want to access and reuse such materials. Social collaboration in the production of metadata may also be able to address the problems of creating metadata that is appropriate to different audiences. For example, metadata generated by different disciplines interested in the same biological material are likely to be quite different, in the same way that the search terms used by the different disciplines to find research related to such material are also likely to be different.

Folksonomies also have the advantage that they can capture important terms in the languages of the users, rather than just the usual English of formal vocabularies. Differences in terms used can also reflect cultural as well as language differences. Folksonomies can, however, be effectively combined with formal vocabularies and ontologies in digital tools so that users searching for particular keywords still see objects that have been tagged with different words with the same meanings. Folksonomies have often been used as a basis for more formal vocabularies and taxonomies by providing a way to find out the terms commonly used within a particular community.

A major challenge with any form of metadata creation and curation is actually encouraging users to put in the time to create content. Motivations are required to encourage users to do this, but given the difficulties of encouraging users to create metadata to describe their own work, it seems like an even bigger challenge to encourage users to add metadata to the work of others as well. Beyond personal benefit, collaborative environments typically provide incentives to interact with the system through trophies, rankings, and other techniques for gamification, as well as the desire to contribute and the visibility associated with activity on such systems. An important aspect of encouraging interaction on collaborative platforms is around the development of reputation, and other ideas for encouraging the creation of social metadata include awards and rankings for curation and metadata creation, such as 'best curated' or 'most effective curator' awards and metadata league tables.[65]

10.11 Conclusion

We have identified a number of pitfalls and strategies for capturing user-defined metadata by examining the various platforms that are effective or ineffective for metadata capture as mentioned in the preceding sections. Providing the users with an option to add simple default metadata is counterproductive. If we want to capture good quality metadata to describe an information object, whether that is a description of an experiment, a data file, a model, a workflow, a biological sample, or a photograph, we need to provide support to the user to prompt them to record metadata that is useful. We must also ensure that we are not missing important prompts. If we fail to ask the right questions, then important information will not be recorded. Various strategies such as using invitations, cues, and changing perspectives can be used to encourage users to add their own user-defined metadata to capture metadata to describe their research in a user-friendly manner.

Digital tools can provide additional support to users by using mechanisms such as data mining and metadata extraction to make suggestions to users about appropriate metadata to include. Where digital tools can generate appropriate metadata automatically to describe the research record, ideally, the metadata should be captured and used within the tool to reduce the burden on researchers and to provide additional value for users. Digital tools for both designing and capturing metadata need to be usable and to encourage users to add meaningful content that can then be used to provide benefits for the user to help them organise and find their own research.

In Chapter 11, we will look in more detail at how we can facilitate collaboration and the sharing of scientific information when we discuss Open Science. We will also take a look at public engagement with science, how open access to data can provide opportunities for collaboration, and how we might overcome some of the barriers to sharing. Finally, we will examine digital tools in Open Science, including those for sharing data, replicating research, and communicating scientific knowledge.

References

1. H. C. White, *J. Libr. Metadata*, 2010, **10**, 156.
2. J. Hastings, in *The Gene Ontology Handbook*, ed. C. Dessimoz and N. Škunca, Humana Press, New York, 2017, p. 3.
3. ChEBI, Available at https://www.ebi.ac.uk/chebi/ (accessed Nov 2018).
4. G. Hughes, H. Mills, D. De Roure, J. G. Frey, L. Moreau, M. C. Schraefel, G. Smith and E. Zaluska, *Org. Biomol. Chem.*, 2004, **2**(22), 3284.
5. Available at https://www.w3.org/RDF/ (accessed Oct 2018).
6. Available at https://www.w3.org/OWL/ (accessed Oct 2018).

7. M. A. Musen, *AI Matters*, 2015, **1**(4), 4.
8. United Nations Economic Commission for Europe, *Making Data Meaningful Part 4: A Guide to Improving Statistical Literacy*, United Nations, Geneva, 2012.
9. J. D. Tenenbaum, S. A. Sansone and M. Haendel, *J. Am. Med. Inform. Assoc.*, 2014, **21**(2), 200.
10. C. Davenhall in *Curation Reference Manual*, ed. J. Davidson, S. Ross and M. Day, Digital Curation Center, Available at http://www.dcc.ac.uk/resources/curation-reference-manual/scientific-metadata (accessed Aug 2018).
11. M. Borkum, C. Lagoze, J. G. Frey and S. Coles, in *IEEE Sixth International Conference on e-Science*, IEEE Computer Society Press, Los Alamitos, 2010, p. 316.
12. G. Kaestle, E. C. Shek and S. K. Dao, in *Proceedings of the 11th International Conference on Scientific on Scientific and Statistical Database Management*, IEEE Computer Society Press, Los Alamitos, 1999, p. 168.
13. S. Sulfi and B. Matthews, in *Knowledge and Data Management in GRIDS*, ed. D. Talia, A. Bilas and M. D. Dikaiakos, Springer, Boston, 2007, p. 135.
14. Available at https://github.com/rsc-ontologies/rsc-cmo (accessed Nov 2018).
15. A. Brazma, *Sci. World J.*, 2009, **9**, 420.
16. C. L. Borgman, M. S. Golshan, A. E. Sands, J. C. Wallis, R. L. Cummings, P. T. Darch and B. M. Randles, *Int. J. Digit. Curation*, 2016, **11**(1), 128.
17. C. L. Bird, C. Willoughby and J. G. Frey, *Chem. Soc. Rev.*, 2013, **42**(20), 8157.
18. C. L. Bird, C. Willoughby, S. J. Coles and J. G. Frey, *Inf. Stand. Q.*, 2013, **25**(3), 4.
19. T. Hammond, T. Hannay, B. Lund, J. Scott, *D-Lib Magazine*, 2005, **11**(4), Available at http://www.dlib.org/dlib/april05/hammond/04hammond.html (accessed Nov 2018).
20. C. Willoughby, C. L. Bird, S. J. Coles and J. G. Frey, *J. Chem. Inf.*, 2014, **54**(12), 3268.
21. A. Milsted, J. Hale and C. Neylon, *PLoS ONE*, 2013, **8**(7), e67460.
22. M. Ames and M. Naaman, in *Proceedings of the SIGCHI Conference on Human Factors in Computing Systems*, ACM Press, New York, 2007, p. 971.
23. G. C. Bowker, *Memory Practices in the Sciences*, The MIT Press, Cambridge, 2008, p. 116.
24. C. L. Borgman, *Learn. Publ.*, 2008, **21**(1), 29.
25. P. N. Edwards, M. S. Mayernik, A. L. Batcheller, G. C. Bowker and C. L. Borgman, *Soc. Studi. Sci.*, 2011, **41**(5), 667.
26. P. N. Edwards, *A Vast Machine: Computer Models, Climate Data, and the Politics of Global Warming*, The MIT Press, Cambridge, 2010, p. 318.
27. M. S. Mayernik, *J. Assoc. Inf. Sci. Technol.*, 2010, **67**(4), 973.

28. S. D. Smith, G. Salway and J. B. Caruso, *The ECAR Study of Undergraduate Students and Information Technology*, EDUCAUSE Center for Applied Research (ECAR), 2009.
29. I. Rowlands, D. Nicholas, P. Williams, P. Huntington, M. Fieldhouse, B. Gunter, R. Withey, H. R. Jamali, T. Dobrowolski and C. Tenopir, *Aslib Proc.*, 2008, **60**(4), 290.
30. J. Goodger and W. Worthington, *Research Data Management Training for the whole project lifecycle in Physics & Astronomy research Final Report*, University of Hertfordshire, 2014.
31. E. T. Mitchell, PhD Thesis, University of North Carolina, 2010.
32. M. Whitton and K. Takeda, Available at https://eprints.soton.ac.uk/196243/ (accessed Nov 2018).
33. S. Sharma and J. Qin, *J. Assoc. Inf. Sci. Technol.*, 2014, **51**(1), 1.
34. A. Swan and S. Brown, Available at https://eprints.soton.ac.uk/266675/ (accessed Nov 2018).
35. G. Earl, W. White and P. Wake, Available at https://eprints.soton.ac.uk/196237/ (accessed Nov 2018).
36. S. Hodson, Available at http://www.rin.ac.uk/system/files/attachments/Research_Data_Challenge_briefing_paper.pdf (accessed Nov 2018).
37. T. R. Bruce and D. Hillmann, Available at https://blog.law.cornell.edu/voxpop/2013/01/24/metadata-quality-in-a-linked-data-context/ (accessed Nov 2018).
38. Available at http://www.chemspider.com (accessed Nov 2018).
39. A. Day, Available at http://www.chemspider.com/blog/search-and-add-from-chemspiderdirectly-from-labtrove-and-other-blog-based-websites-which-use-tinymce.html (accessed Jul 2017).
40. M. Henninger, *J. Inf. Sci.*, 2017, **44**(5), 658.
41. C. Willoughby, C. L. Bird and J. G. Frey, *Int. J. Digit. Curation*, 2015, **10**(1), 18.
42. C. Willoughby and J. G. Frey, *Int. J. Digit. Curation*, 2017, **12**(2), 1.
43. Available at https://www.altova.com/xmlspy-xml-editor (accessed Nov 2018).
44. Available at https://xmetal.com/content-xmetal-author/ (accessed Nov 2018).
45. Available at https://www.oxygenxml.com (accessed Nov 2018).
46. I. San Gil, V. Hutchison, M. Frame and G. Palanisamy, *J. Libr. Metadata*, 2010, **10**(2-3), 99.
47. Available at https://www.esri.com/en-us/arcgis/ (accessed Oct 2018).
48. Available at https://www.fgdc.gov/metadata (accessed Oct 2018).
49. D. Rugge, 2005, Available at https://doi.org/10.2737/NC-GTR-255 (accessed Nov 2018).
50. D. Walworth, *Metavist User Guide*, Alaska Science Center, 2017.
51. Available at www.itis.gov (accessed Nov 2018).
52. R. S. Gonçalves, M. J. O'Connor, M. Martínez-Romero, A. L. Egyedi, D. Willrett, J. Graybeal and M. A. Musen in *The Semantic Web – ISWC*, ed. C. d'Amato, M. Fernandez, V. Tamma, F. Lecue, P. Cudré-Mauroux, J. Sequeda, C. Lange and J. Heflin, Springer, Cham, 2017, p. 103.

53. Available at https://metadatacenter.org/purpose (accessed Nov 2018).
54. Available at https://bioportal.bioontology.org (accessed Nov 2018).
55. C. Jones, C. Blanchette, M. Brooke, J. Harris, M. Jones and M. Schildhauer, *Ecol. Inf.*, 2007, **2**(3), 270.
56. Available at http://www.spase-group.org/tools/xmleditor/base.html (accessed Nov 2018).
57. Available at http://www.spase-group.org/tools/xmleditor/simulation.html (accessed Nov 2018).
58. M. Martínez-romero, M. J O'Connor, R. D. Shankar, M. Panahiazar, D. Willrett, A. L. Egyedi, O. Gevaert, J. Graybeal and M. A. Musen, *AMIA Annu. Symp. Proc.*, 2017, **2017**, 1272.
59. C. Tenopir, E. D. Dalton, S. Allard, M. Frame, I. Pjesivica, B. Birch, D. Pollock and K. Dorsett, *PLoS One*, 2015, **10**(8), e0134826.
60. F. Nasoz1, R. C. Bryce, C. J. Palmer and D. J. Rugg, in *Human Interface and the Management of Information. Interacting with Information*, ed. M. J. Smith and G. Salvendy, Springer, Berlin, 2011, p. 122.
61. Õ. Mets and J. Kippar in *Research and Advanced Technology for Digital Libraries*, ed. J. Kamps, G. Tsakonas, Y. Manolopoulos, L. Iliadis and I. Karydis, Springer, Cham, 2017, 421.
62. G. Alemu, B. Stevens, P. Ross and J. Chandler, *Journal of Library Metadata*, 2012, **12**(4), 311.
63. Available at https://www.flickr.com/commons (accessed Jan 2019).
64. S. A. Golder, *J. Inf. Sci.*, 2006, **32**(2), 198.
65. D. De Roure, C. Goble, S. Aleksejevs, S. Bechhofer, J. Bhagat, D. Cruickshank, P. Fisher, D. Hull, D. Michaelides, D. Newman, R. Procter, L. Yuwei and M. Poschen, *Concurrency Comput. Prac. Exp.*, 2010, **22**(17), 2335.

CHAPTER 11

Open Science and Sharing the Research Record

11.1 Introduction

Just as the digital revolution has transformed the way we live our lives, so too has it transformed science. The way that scientists are working is shifting, in part as a result of the proliferation of computers and the digitisation of science, but also because of the increasing globalisation of the scientific community, and the need for greater collaboration to solve the major challenges facing society. Facilitated by technology, including the Internet, greater processing power, and cheaper storage, these changes impact across the whole research lifecycle. There are a variety of emerging trends away from science conducted in private and published behind paywalls, to greater collaboration, increasing participation in science, sharing of ideas, methods and data before publication, and the dissemination of results online. Technology facilitates a more open way of working with global colleagues, with neither distance nor time-zones a limitation to access, contribution, or discussion of results. The Internet, in particular, provides new ways to share information, not just the raw data from which published results were generated, but also ideas and unpolished preliminary results generated during the research process. Social media allows us to create and publish our own content providing opportunities to engage in discussion with the wider scientific community and even the public. The term 'Open Science' means different things to different people, and encompasses the ways that technology is facilitating these changes in society, enabling more open dissemination of scientific knowledge, the development of tools to support collaboration, and bridging the divide between science and society.[1]

Recording Science in the Digital Era: From Paper to Electronic Notebooks and Other Digital Tools
By Cerys Willoughby

Published by the Royal Society of Chemistry, www.rsc.org

Open Science helps to make science reproducible, to improve scientific processes, and to accelerate the progress of science. As discussed in Chapter 5, reproducibility is an essential aspect of science. In order to reproduce and validate the results of scientific research, it is necessary to have access to the data, methods, and materials, including software, that were involved in deriving the results. In traditional publications, access to both in-depth data and methods may be limited. Often, journal articles contain only summaries of results data without providing access to the raw data and any intermediate transformations generated by an analysis. The methods describing how the data was obtained and analysed may also be summarised, and key steps are often missing, preventing accurate reproduction and scrutiny. Open Science provides more opportunities for validation, peer review, and reproduction of scientific results by increasing access to both scientific data and associated resources, including in-depth methods and the source code for analytical tools that enable direct replication of experiments and analysis. Efficiency of scientific research can also be improved by reducing duplication and learning from the successes and failures of other teams. Although rarely published, failed experiments are just as important as successful experiments to provide context to the research as a whole and so that others avoid duplicating effort unnecessarily.[2] Best practices in science can be promoted by the open availability of research tools, methods, and data. Earlier access by the scientific community to methods and data can also help to accelerate the progress of science and improve processes by enabling other research groups to make use of the data for novel applications or to adopt best practices, and to build upon and improve methods more rapidly. Sharing can also promote the adoption of standards in science, because standard formats for data enable it to be more easily shared and reused with different software tools. Standardisation of data formats mean that groups using different tools are still able to use the data and associated resources without needing to re-collect the data or transform the data to a different format before it can be used.

Much research, and data collection in particular, is very expensive to perform. It would be wasteful in terms of time and money to duplicate the work, and it is unlikely that other research groups would do so. For many research projects, it may even be impossible to recollect the data at a later time. If the data is unavailable because it is stored on an individual's machine or in a paper notebook in a filing cabinet, it is less likely to be found and is much more difficult to use. The only option for accessing such data is for researchers to contact the authors of a publication directly to ask for it, and by that point, the data may be lost or not in a fit state to share. In such a scenario, money and time spent on the data collection and associated research is wasted in comparison with data which is easily accessible because it has been shared through deposition in a repository or within a Web-based ELN. Ensuring that data is not lost, and that money is not wasted is particularly important for publicly funded research, and numerous funders now require that research data, materials, and code are shared as soon as possible to facilitate innovation, social impact, and wider engagement with the research. In addition, open access to research publications from publicly funded projects ensures that research is not in effect being paid for twice, once through funding and then again through subscription costs to journal publishers. Some funders, the Wellcome Trust as an example, cover the cost of article processing charges to ensure that research can be published in peer-reviewed open access journals to ensure that the research is freely available to all.[3] Requiring that data is available and shared, at least at the point of publication, makes it more likely that consideration will be given to data quality and ensuring that the data is in a fit state for others to use. Requirements from funders and other organisations for Data Management Plans (DMPs), as discussed in Chapter 5, ensure that data quality and sharing needs are considered at the very earliest stages of research.

The idea of science as 'public knowledge' is a relatively new concept. Before the sixteenth and seventeenth centuries, there was an element of secrecy in researching the secrets of nature.[4] This secrecy was to prevent knowledge getting into the hands of those considered unworthy and to protect the commercial potential for private exploitation of discoveries and technical developments. The scientific revolutions of the sixteenth and seventeenth century have ensured that the scientific record, or at least the part that is published, is available to the wider scientific community and theoretically available to the interested public. The journal publication system also became the way that researchers gained the appropriate credit for their work, reducing the need for secrecy, and encouraging the idea that scientific knowledge should be built upon the research of others. There are, however, challenges with this traditional model of scientific communication relying on the publication of results in journals and books. The scientific community builds upon the results presented in journals, but there are a number of problems with traditional publication that slows down this process. Results may take many years to be published, and once published, the underlying data was historically only available on direct request to the author, making replication and validation of results difficult. The 'closed' nature of scientific research and publication can also lead to a lack of transparency in the processes used to obtain the results. One of the most significant issues with traditional publications are the costs. Access to scientific journals is extremely expensive and the costs of subscription have increased significantly over time, often leading to a relative reduction in access to the overall scientific output, even for researchers.[5] Also, not all research is published, for example, negative results and failed experiments, leading to unnecessary duplication of research, as mentioned previously. Traditional journals are also an inappropriate format for communication to the general public, and the lack of wide access to articles may mean there are delays before important scientific information is communicated by traditional news media.

In a variety of ways, both technology and shifting attitudes towards public engagement with science are changing the way that science is both recorded and communicated. Open Science looks to move from traditional publication in journals to sharing scientific knowledge as early as possible and seeking greater engagement with wider society.[6] In order to do this, it is necessary to address a number of different issues to reduce the barriers to sharing scientific knowledge. Such barriers include restrictions on access to published research and data through paywalls and the use of technical innovations controlled by patents. Other barriers to sharing come from researcher and organisational behaviour, including poor data management practices and the use of proprietary software, leading to the loss of scientific data. Perhaps most importantly of all, there is a need to overcome organisational resistance and the anxieties of researchers about the sharing of data, which we will discuss in more detail later in the chapter. Various related 'open' initiatives are providing potential ways of overcoming these various barriers, including 'open data', 'open source software', 'open access', and 'open notebooks', as well as developing mechanisms that are beginning to give researchers credit for sharing their ideas, code, methods, and data outside of the traditional publication cycle, and also new ways of engaging the public with science. We will take a deeper look at these initiatives in the next section.

11.2 Open Initiatives

A key enabler of Open Science is the availability of scientific publications and related research resources. Although the majority of scientific publications are now available online, there is usually a significant financial cost associated with accessing them. The 'open access' movement advocates for free access to all publications and other research outputs provided in electronic form, such as scholarly journals, historical scientific publications, e-prints, e-books, theses, and conference papers. Access to such materials may be through an individual's website, open access repositories such as institutional, funder, and disciplinary repositories, or *via* an open access or commercial publisher's website. Although truly open access refers to access without cost, scholarly journals and e-books are still most commonly available *via* a paid for service, whether a charge for individual access or through an institutional subscription. Whether the publications are truly open access or paid for electronic publications, the ability to be able to easily access published materials on the Web is of enormous value to researchers across the globe. In theory, the time saved from needing to visit the library to locate and access materials can be more effectively spent conducting science in the laboratory or field.

There are a number of aggregator services that collect details of open source electronic resources and provide indexes, directories, and search facilities for their contents. For example, the Directory of Open Access Journals (DOAJ) provides a subject directory and search facilities for over

12 000 open access journals from across the globe;[7] the Open Access Library provides access to free scientific publications;[8] the Registry of Open Access Repositories provides directory and search facilities for e-prints;[9] arXiv provides access to e-prints in Physics, Mathematics, Computer Science, Electrical Engineering, and a variety of other numerical sciences;[10] and OAPEN, which provides access to freely accessible academic books.[11]

For transparency and reproducibility of scientific research, we need to share and have access to both the raw data used in a project and the methodology that was used to create it. Open data advocates the free access and use of data within the scientific community, so that the data can both be validated and reused. Within the scientific community, the term 'Open Research Data' is often used to distinguish scientific data from data provided by open government initiatives. Although some data is provided in publications and as supplementary materials, it is less commonly available for reuse because restrictions are imposed by the publishers, the data is difficult to extract, the format is not suitable for reuse, and the data is missing necessary context. As previously mentioned, it is possible to ask authors for access to their data, but studies have shown that even when researchers are theoretically willing to share, many authors do not actually make their data available when asked, even when their research is published in open access journals.[12–14] As a result, much of the discussion about open data revolves around ways to encourage researchers to share their data, from funders mandating it to mechanisms for providing credit to researchers. We will talk more about overcoming the barriers to sharing later in this chapter.

In some disciplines, sharing data has become the norm, thanks to agreements across the community to do so. For example, the Bermuda Principles, agreed at the First International Strategy Meeting on Human Genome Sequencing in Bermuda, state that all human genomic sequence information should be made freely available and in the public domain in order to benefit society.[15] As a result, if researchers do not share this information, for example, through the Genbank repository, they will not receive future funding.[16] As mentioned, many funders and publishers are now requiring that data be deposited in a publicly accessible repository as a condition of funding or publication. There are also a number of journals now available specifically for publishing data, for example, *GigaScience*,[17] *Geoscience Data Journal*,[18] *Earth System Sciences Data Journal*,[19] *Scientific Data*,[20] *Biodiversity Data Journal*,[21] and *BioMedical Data Journal*.[22] As mentioned in Chapter 5, there are also many scientific repositories for depositing and sharing data. These repositories may be discipline-based, institutional, country-based, and even commercial solutions for scientists without an appropriate official option. Examples include NCAR Research Data Archive,[23] Open Science Framework,[24] Dryad Digital Repository,[25] EMBL-EBI Protein Data Bank,[26] Zenodo,[27] Figshare,[28] UK Data Service,[29] Australian National Data Service,[30] and the EU Open Data Portal.[31] OpenDOAR, provides directories of repositories by geographic location and search facilities that provide details of the name, type, country, and software for the repository, together with various policy descriptions and metadata.[32]

In addition to open data, there are some researchers that go further and share the methods used in their research. Sharing the methods as well as the data obtained is important for both transparency and reproducibility. As discussed previously, data is generally of very little use without information that explains how it was generated and how any conclusions drawn from it were derived. There are typically two ways that methods are shared, the first is through a website specifically for sharing methods or protocols, and the second is through the use of so-called open notebooks, or 'open notebook science', to share the whole of the research process as it is carried out. Websites for sharing methods and protocols include OpenWetWare,[33] protocols.io,[34] Protocol Online,[35] and myExperiment for sharing scientific workflows.[36] There are relatively few sites for sharing protocols compared with sharing the data itself. There are also relatively few platforms that enable the sharing of notebooks online, but these typically have the advantage that both methods and data are presented together, and research is available very quickly, so that it can be understood and validated rapidly. Some Web-based Electronic Laboratory Notebooks (ELNs) can facilitate open notebook science by enabling open access to notebooks. Examples of ELNs that are commonly used for open notebook science include LabTrove and OpenWetWare, discussed in Chapter 6.[33,37] It also appears that many scientists engaged in open notebook science have created their own websites for the purpose, often based on a blogging platform such as WordPress. There are also some open notebook networks to facilitate the creation and discovery of open notebooks, for example, openlabnotebooks.org and the Open Notebook Science Network.[38,39] The fact that many scientists are choosing to make use of their own websites for creating open notebooks suggest there is the opportunity to develop new tools to help facilitate open notebook science in particular, but also that the blogging format successfully supports the needs of many scientific content producers. We will discuss scientific blogging in more detail later in the chapter, but there may be other reasons why some scientists are choosing blogs as a platform for their open notebooks.

Access to the data alone is not enough to be able to replicate the data or to be able to make use of it in scientific studies. For scientific studies that rely on computation, even access to data and methodologies is insufficient, and we also need to make available the software and code that was used to generate or to manipulate the data.[40] It can be extremely difficult, if not impossible, to replicate results without access to the original software on which the computation was performed. Simply providing a description of what the software does is not sufficient to be sure that the programme actually does that, or to be unambiguous enough to be directly replicated without the actual code. In the 1960s and 1970s, it was usual practice for software to be freely shared across academic and corporate laboratories.[41]

This sharing included the source code, making it possible for others to modify and build upon the original software for their own needs. Although the majority of software has since been commercialised, and its use restricted through licences, there has been a long history of open software, free or paid where the source code can be reviewed and modified and improvements shared. Such software, often called 'open software' or 'open source' is frequently developed collaboratively using volunteers. It is now becoming more commonplace for scripts, codes, and tools developed by scientists for manipulating and analysing data to be included in computing and software orientated repositories, such as GitHub,[42] for public use, sharing, and feedback. There are also various directories of tools for particular disciplines available. For example, the NeuroImaging Tools and Resources Collaboratory (NITRC),[43] Free Software Directory,[44] nanoHUB for simulation tools,[45] and the Research Software Directory.[46] There are also examples of other websites that share specific scientific resources, for example, WikiPathways enables users to share and access biological pathways.[47]

Some authors advocate the packaging up of research data and related resources for deposition in repositories or sharing websites. For example, creating Research Objects (ROs) is suggested to enrich workflows by aggregating them with other related resources and metadata so that all the appropriate context and documentation is available for others wanting to reuse the workflow.[48,49] The RO should also contain all the appropriate scripts and software required to enable the workflow to function. The idea is that the object includes all of the information required to make the workflows understandable and reusable, but also contains appropriate metadata and description of the relationships between the contents to make the RO machine-readable and interoperable. Making the RO machine-readable and interoperable enables software programmes to make use of the contents and to make them discoverable on the Web.

Open Science can also go beyond the sharing of data, methods, and computational tools into the sharing of actual materials. For example, the Open Access Malaria Box, is a box of 400 anti-malarial compounds designed to accelerate research into anti-malarial drugs and other neglected diseases.[50] The box of materials is available for free for researchers working in the area on condition that users publish and share the results of their research in the public domain as soon as possible. Assistance is also provided to help researchers publish their results in the ChEMBL Malaria data repository specifically.[51]

11.3 Open Science and Collaboration

Open Science should not just be about working on research projects and putting out results without thinking about how others may use them. Science is becoming increasingly collaborative and multidisciplinary. Results and data from one field can help inform another, both reducing the costs of research and raising new research questions never imagined during the original research design. Working together can speed up the process of science and lead to greater impacts for society, such as addressing environmental problems or developing more effective drug treatments. Conducting Open Science can help to build communities of interest around particular topics or problems, encouraging the sharing of expertise across institutional, geographical, and disciplinary boundaries. An example of such a project, where the entire primary research record is made available online, is the global Open Source Malaria project.[52,53] The project is attempting to find new medicines to cure malaria by improving the synthesis of analogues of compounds, identified by major pharmaceutical companies, to make them more potent and more effective at tackling the disease. All data and ideas in the project are open and the project actively encourages any and all participation. The project makes use of an instance of the LabTrove ELN to share data and results.[54] Within the LabTrove ELN instance are individual notebooks used by the different participating laboratories that share both the methods and results of experiments carried out in the laboratory. The LabTrove ELN was chosen as an alternative to paper notebooks because the ELN facilitates collaboration with external colleagues and also provides mechanisms for linking and classification that enable the team to demonstrate reproducibility between different samples.[55] Resources and project updates are shared on the Web through sites such as GitHub and the OpenWetWare wiki.[56] For the longer term, materials are preserved in the ChEMBL chemical information database.[51] Engaging the community is vital to Open Science projects so that feedback can be given, project status can be discussed, and new collaborators can be recruited. The Open Source Malaria project makes use of a number of different tools for community engagement. Teleconferencing tools are used to run regular online meetings and create recordings to help discuss strategy and guide the research. Social media sites such as Twitter, Facebook, and LinkedIn are used to push drives for help with specific problems and to provide project updates. The social media sites also provide opportunities to connect and interact with interested parties.

For an Open Science project such as the Open Source Malaria project, collaborators are typically institutions that are able to conduct wet lab experiments to help progress the knowledge of the wider team and the development of materials and processes needed for the project. For other Open Science projects, different levels of collaboration may be useful with opportunities for engaging collaborators with differing levels of expertise and access to equipment. Sometimes collaboration can help to solve a problem by dividing it into smaller pieces and assigning the tasks between a group of collaborators, as in the Open Source Malaria project, or a large problem can be tackled by a group of people sharing ideas or resources. An example of how open collaboration can speed up the process of solving problems comes from the Polymath Project, originally conceived by Timothy Gowers as a way to solve a specific problem in mathematics by inviting contributions on a blog and engage in 'massive online collaboration'.[57] The idea behind the post was that increasing the number of contributors would bring different knowledge, experiences, and even different personalities to the discussion, and enable unanticipated connections to be made.[58] With just a description of the problem, some background materials, and ground rules for the collaboration, it only took a few weeks for the initial problem to be solved by dozens of contributors from a variety of backgrounds. Not only did the idea provide the desired solution, but it also provided a record of the whole process of how the mathematical result was discovered, showing how ideas grew, improved, and got discarded along the way. The success of the original project has led to a number of new mathematical problems to be set and solved. Each of the formal papers that have been published so far link to the original discussion trails of how the problems were solved.

Open Science also provides new opportunities for public engagement. Although public interest in science is high, there is a lack of trust, especially associated with the way science is conducted by government and industry.[59] There are a number of reasons for the lack of trust, and many of them result from the apparent opacity of scientific processes and problems in the way that scientific research has been communicated to the public.[60] The public has also been excluded from discussions in science, and yet often there are important debates about the ethics, morals, and values underlying research that has public concern. There are also circumstances in which an informed public may have a greater knowledge to share than a professionally trained expert, for example, local knowledge may be beneficial to research in ecology and environmental sciences, and patient groups on social media are likely to have much more knowledge with their combined experience about a medical condition and its treatment than a non-specialist doctor.[61]

Open Science also forms part of the drive towards making the processes and activities of science more transparent to address these issues of trust. Communication of science to the public can be complex, and miscommunication or deliberate manipulation of results can have significant impacts on society and increase mistrust. Although Open Science practices do have risks for public communication of science, for example, data may be released before results are fully understood or the data may be misused, there are also advantages. In addition to transparency of methods and practices, open access to research provides opportunities for peer review to validate the accuracy and credibility of research. Researchers can also have control over how their research is explained to add appropriate context for the public interest, including both potential benefits and limitations of the research and how it fits with previous research and conventional wisdom. In terms of engaging with the public, there is a fine balance between complete openness and making research openly available before it is 'ready' for public consumption. The appropriateness of full openness is likely to be different for different kinds of projects, but tools for Open Science should help with controlling and facilitating access, and also ensuring that errors are effectively communicated when they are discovered.

11.4 Citizen Science

Engagement with the public through the medium of Open Science goes beyond communication of science into actual participation in science. Citizen science is a relatively new term, but there has been a long history of amateur scientists contributing to scientific knowledge. Ornithologists and other wildlife enthusiasts have long been taking part in field surveys; amateurs have been building weather stations and recording the weather; and many important discoveries of new celestial objects have been made by amateur astronomers. During the last 10–15 years, there have been an increasing number of scientific projects that have made use of amateur scientists and interested members of the public to help perform their research. The Web has provided an easy way to upload results for surveys, such as those run by the RSPB,[62] Woodland Trust,[63] and OPAL.[64] The Web also provides a way for scientists to access volunteer resources to perform tasks that are impossible or error prone using computers. One of the first citizen science projects was the NASA Stardust@home project, which made use of volunteers to search for particulates of interstellar dust collected in space by the Stardust spacecraft.[65] Shortly after this project began, the Galaxy Zoo project was launched to invite volunteers to classify galaxies from images derived from the Sloan Digital Sky Survey (SDSS).[66,67] The project has made a number of significant discoveries through the work of hundreds of thousands of volunteers and published over 60 papers so far. Following on from the success of the Galaxy Zoo project, the Zooniverse portal was developed to host a variety of different citizen science projects.[68] The site now contains projects across a range of disciplines including arts, history, and literature, as well as biology, climate, medicine, nature, physics, and astronomy, with datasets from both local and global level projects. Example projects include transcription of library and museum records, data collection, counting and classifying objects, and image segmentation and annotation. Many of the projects are utilising the volunteers to make comparisons with computer-based analysis and training in order to improve techniques for computer automation and Artificial Intelligence (AI).

Crowdsourcing is sometimes used as a term to describe these kinds of activities of recruiting large numbers of people to collect information or to provide input into a task, usually using the Internet as the medium of communication. For citizen science, the activities are usually voluntary, but they may also be completed for a fee. For example, Amazon Mechanical Turk (MTurk) is a crowdsourcing marketplace where organisations can connect to individuals to perform tasks that are currently beyond digital automation.[69] Workers on the platform typically complete tasks in exchange for monetary payment. The platform has been used particularly for social sciences research to access large and diverse numbers of participants for surveys. An example of a crowdsourcing platform targeted directly at the scientific community is InnoCentive, set up originally by the pharmaceutical company Eli Lilly to tackle a problem in organic syntheses.[70] The platform enables users to submit solutions to challenges set by organisations, including governments, industry, and charities. The challenges come with a financial prize for solving the problems in the challenges, many of which have significant value, for example, a $100 000 prize from the US government for the best suggestion for halting the progress of invasive mussels.[71] The challenges cover a diverse range of topics, including electronics, technology, chemistry, materials, engineering, physics, environmental science, health, drugs, food, and agriculture.

Before the development of interactive websites for citizen science, volunteers could download software that makes use of the computing resources of volunteers. For example, SETI@home uses these computing resources to analyse observational data from telescopes to detect radio signals that might be produced by extra-terrestrial life.[72] Similar projects, Rosetta@home and Folding@home, made use of volunteered computer resources to model the structure of proteins and mechanisms of protein folding, with potential implications for medical research.[73,74] An interactive game version of the Folding@home software, called Foldit, enables volunteers to not only use their computer resources, but also to actively engage with designing new protein structures by directly manipulating the proteins structure within the software to increase their score using various intuitive tools.[75,76]

Smartphone use has also provided new ways for researchers to engage with citizen scientists. The built-in functionality of GPS, the digital camera, and other sensors within smartphones has provided many ways to easily capture data that can then be rapidly shared over the Internet. An example citizen science application for smartphones, Project Noah, is used to capture photographs or videos of wildlife or plants together with classification information, a GPS location, and descriptive metadata.[77] The results uploaded by users can be viewed by and utilised by other citizen scientists, but researchers can also recruit users for both local and global missions to capture data in specific locations or for specific types of animals or plants. There are a number of similar smartphone applications for tracking wildlife or plants that combine the idea of providing a field guide with the ability to capture sightings to enable the data to be reused by others and for scientific research, for example, Leafsnap and BirdsEye.[78,79] There are also a number of environmental monitoring applications for citizen science. For example, the NoiseTube app uses the microphone in smartphones to measure sound exposure in different locations to identify noise pollution and examine differences in perceptions of noise.[80] Another example is the Secchi Disk project that encourages seafarers across the globe to measure levels of phytoplankton in the oceans using a simple instrument and storing measurements in the application until they can be shared.[81]

For citizen scientists with more skills or a desire for greater interaction, there are also groups such as the do-it-yourself biology community (DIYbio) who are not only designing biological materials with computer-based tools, but also conducting experiments in synthetic biology. For example, participants are engineering bacteria using ready-made biological components in community labs such as Genspace, Biocurious, La Paillasse, and Madlab.[82] These community labs are associated with the growing hacker and maker communities, providing new ways for amateurs to engage with science and technology, and potentially solving problems or developing solutions that are too niche for government or industry to fund. Although there are issues with such activities around safety, skills, and access to appropriate materials, this is likely to be an area that will continue to develop in the future. Encouraging Open Science, sharing of data, and capture of the record from such activities will have value to both the scientific community and the wider public. Capturing and sharing the record from such activities will also be important for providing transparency of the activities and providing opportunities for peer review and recognition for the amateur scientists taking part.

11.5 Barriers to Open Science

There are a number of barriers to Open Science. There is clearly a tension between science as public knowledge and science as commercial enterprise. There is an anxiety amongst organisations that Open Science takes away opportunities to capitalise on research that may have been expensive in terms of money, time, and effort, and enables others to unfairly benefit without contributing to the cost. Although many Open Science projects choose to refrain from the patent process, it is recognised by funders and organisations that there are situations in which it is appropriate for researchers to protect their intellectual property to facilitate commercialisation of ideas, products, and technologies with a positive impact for society. Even in industry, where commercialisation of products and technologies may be of primary importance, Open Science can have a role to play. For example, the pharmaceutical company Eli Lily has created an online crowdsourcing platform to make use of Open Science principles for the development of anti-malarial drugs.[83]

There are situations, within industrial and governmental research organisations in particular, where privacy, safety, and security, can limit the extent to which materials can be openly shared. Materials that cannot be shared for privacy reasons are usually sensitive documents or materials that contain information that may enable the identification of individuals. Privacy is a concern particularly for medical research and clinical studies, but also for studies in genetics, psychology, and the social sciences, because it can be very difficult to completely anonymise data. Even though names and other personal information such as locations and dates of birth can be removed, it may still be possible to identify individuals from the information that remains, particularly when information from different datasets are combined. Some scientific research can also be sensitive in terms of safety and security, for example, research that has potential military applications or that could be misused for harmful purposes.

An example of research with safety concerns and potentially harmful uses comes from a recent controversy about the publication of information about the creation and transmission of the H5N1 strain of avian influenza viruses. In order to research the potential transmissibility and pathogenesis of the disease, two NIH-funded studies made the virus transmissible to ferrets (often used as a model for human transmission of airborne viruses) by engineering the virus using already well-documented and published protocols.[84] There are two safety concerns that caused the controversy with these experiments, first, that the engineered virus could escape and cause a pandemic, and second that the research could be used by bioterrorists to cause significant harm. The papers were eventually allowed to be published, but the controversy highlights the importance of evaluating the potential benefits and security and safety risks of projects up front. This is particularly important for Open Science to avoid sensitive or potentially dangerous information being released inadvertently.

Even if researchers are not concerned about the potential financial rewards of their research, there are concerns about how their data may be used by others. Although studies suggest that researchers are willing to share data, one of the largest barriers to sharing is that researchers frequently do not believe that anyone else would want to use their research data.[14] This partly stems from their own experiences where most researchers were neither asking others for their data, nor being asked to share their own data. Even those that were asked to share their data were not necessarily depositing it within repositories and instead sharing it on websites for the purpose of personal exchange rather than public availability. In the past, collaboration was usually carried out with known colleagues with a mutual respect and common interest, whereas open data means that the data is shared with an unknown audience.[85] Concerns may relate to misuse of the information, for example, will the data be reused in an inappropriate way, or will it be misrepresented in such a way that the reputation of the original researcher may be damaged? There is also the potential for competitors, or even collaborators, to take the open research and 'scoop' the original researchers who have performed the work, or to pass off data as their own. There is, however, the counter argument that anyone trying to pass off data as their own or to misuse the data in some way are more likely to be caught if the research is already open. Access and negotiation of use of data has gone from being personal to potentially being open to anyone and completely anonymous, although restrictions can be added through the use of licence conditions or requiring sign-in and approval.

Concerns about being scooped or others taking credit for work also closely relates to the issues of how researchers can gain appropriate recognition for their work. A great deal of time and effort is involved in the collection of research data and preparing it for sharing. Recognition for research is important for the development of careers for individual researchers, but also for the acquisition of further funding for both the researchers and their organisations. Traditionally, recognition for research comes from formal publishing, but gaining recognition for data and other materials shared in Open Science is more difficult, especially if there is the risk that someone else takes your data and reaches publication before you do. More effort and time is required to prepare materials and data to be shared compared with traditional publishing, and therefore researchers need to be provided with suitable incentives to share. It is also important to ensure that the researcher and their organisations get the appropriate recognition for sharing.

Traditionally, recognition in scientific research comes from the number of articles published and citations of those articles, with often the 'Journal Impact Factor' considered more important than the impact of the article itself. Bibliometrics, metrics retrieved from bibliometric databases such as Web of Science, Scopus, and even Google Scholar, are used to measure research outputs and assess the impact of journal articles in particular.[6] Usage metrics, such as the number of views or downloads of an article, are also valuable as a way of measuring the broader interest in a publication. Usage metrics have the advantage that they can also be applied to non-traditional publications, such as reuse of open data, workflows, and software, and other publication platforms such as blog posts or open notebooks. For social media platforms that enable research and ideas to be shared, such as Twitter, ResearchGate, Mendeley, and the various blogging platforms, alternative metrics or 'altmetrics' can be used to measure impact and the influence of traditional and non-traditional publications, including eprints, blog posts, news articles, data, protocols, and code. Different platforms provide different measures, but many track the numbers of followers, likes, shares, views, downloads, and comments. Information may be shared across many platforms and potentially engage a much wider audience than traditional publishing.

Some sites provide these metrics within pages, but there are also aggregator services that collect these metrics and enable researchers to publicise their work. For example, Altmetric.com provides monitoring of published research with a Digital Object Identifier (DOI) including the number of times the research is blogged about, mentioned by news outlets, Wikipedia, Twitter, Facebook, and reference management sites such as Mendeley and CiteULike.[86–88] The service provides badges to encourage the promotion of 'the buzz' around an individual article or an entire institution's research, and to create engagement with potential collaborators. ImpactStory provides some similar features, but the ImpactStory profiles provide altmetrics for the individual researcher, in addition to metrics for individual articles.[89] The altmetrics are in the form of achievements, such as rankings for software reuse, most popular publication, global reach, sharing, followers, and online mentions over time.[89] There are also more spurious achievement categories. For example, according to ImpactStory, my publications have good ratings for open access, but I am also presented with the somewhat dubious 'Rickroll' achievement of having co-authored a publication by someone with the name Richard! ImpactStory provides access to links to the various blog posts, news articles, and social media sites in a timeline format that is useful for reviewing the attention the research received over the last few years. Kudos is another tool for altmetrics, but has a slightly different focus by encouraging the researcher to claim their publications and add additional resources such as web links, presentations, links to expanded datasets, and related publications.[90] In Kudos, the author is also able to add explanatory content to their materials in the form of responses to questions that enable the author to not only describe the article in a user-friendly way, but also to provide information about why the research was done and why it is important.[91] For each article, a web page is published containing the publication title, appropriate metadata, descriptive content, resource links, and a link to the publication, which can then easily be shared across social media as a very professional and user-friendly 'landing page' for promoting the research. The platform also provides a dashboard with information about the various metrics for articles, including views and downloads, as well as the usual social media and online interactions.

At the present time, it is unclear whether altmetrics are useful or have any relation to future citations of articles, but there is bound to be an effect where an article that receives a lot of attention and publicity is more likely to be read, more likely to be shared, and ultimately therefore more likely to be cited.[92] Although the popularity and relevance of articles and other materials related to scientific research are likely to wax and wane over time, the rapid open publication of materials, associated metadata, and altmetrics provides the opportunity for the development of more sophisticated tools. For example, tools that can pull content from aggregator services and give researchers up to date recommendations on the most significant research in their area of interest, as well as providing new avenues and incentives to promote their own research.[93]

In traditional publishing, individual contributions are recognised through co-authorship, and in some cases, the specific contribution is called out. In Open Science, contributions may come from anywhere, collaborators may have a range of different skills and backgrounds; and the amount and value of each person's contribution may differ considerably. Credit can be given for specific activities, such as data collection, data analysis, or review, whilst co-author status can be negotiated. An alternative strategy is to make use of the open discussions between collaborators on projects to document the 'who did what' of the activity directly, without singling out any particular individual for credit or removing anyone who contributed less. This is the approach used by the Polymath project, discussed earlier, where the problems of attributing credit are solved by signing all papers 'DHJ Polymath' and providing a link to the full working record.[57] The contribution of any individual can be determined by examining the full working record of the project. In many of the citizen science programmes, credit is given to participants in the form of recognition for significant discoveries, but also for the amount of work that they have done, for example, by ranking 'top' contributors and providing metrics on the activities. For example, participants may be ranked by transcriptions performed, number of images annotated, types of objects classified, photographs captured, videos tagged, and so on. A balance of course needs to be struck in encouraging participation and encouraging quality input into a scientific problem or research project. The use of metrics in this way can be thought of as 'gamification' or making use of game-like elements within tools as an approach to encouraging engagement with an activity or desirable behaviours. Such gamification elements could be used within software tools such as ELNs and repositories to encourage positive sharing and collaborative behaviour. For example, the workflow sharing website myExperiment includes metrics and rankings for both individual users and groups, these include 'Most Credited' and 'Highest Rated' users, and 'Most Members', 'Most Shared Items', and 'Most Credited' for groups.[36] The use of rankings, trophies, badges, and reputation points are also commonly used on specialist social networking sites, for example, those that encourage the interaction of participants or through collaborative assistance to help answer questions or solve problems that other users may have. Higher rankings or other gamification elements can encourage participation through increased visibility and prestige within a particular software environment or social networking platform. Gamification may be most useful in specialist environments where users can make use of their expertise to assist other users, promote their research, and to share materials of value to the wider community.

11.6 Digital Tools for Open Science

Digital tools are essential to Open Science, for not only generating and processing the data, but also in facilitating and encouraging the sharing of data with the wider community. Digital tools are needed to access and find the data, but they can also provide ways to engage and interact with the research directly. Interaction with the research is not just limited to accessing and reusing data, but also access to other resources, for example, plans, methods and processes, workflows, and code. Collaborative tools enable the scientific community to peer review research as it happens, share ideas, share results, and potentially even influence the direction of research. Tools can also be developed to create greater inclusion in science by not just 'informing' the public about scientific research, but also by engaging interested individuals in citizen science.

Facilitating Open Science involves the support of many different kinds of activities. Tools required for storing, sharing, and finding scientific information need to work at large scales, handling both large quantities of data and large amounts of interaction. Repositories have a primary role in Open Science enabling both the dissemination of scientific materials and facilitating their discovery. Repositories may contain materials from a particular institution, funded research, or disciplinary field. For the potential viewer of the research, a major challenge comes in knowing where to look for the data. However, there are some services that provide ways to search for research data repositories to help locate relevant data. For example, OpenDOAR, previously mentioned, and the Registry of Research Data Repositories.[94] In addition, many libraries and institutions compile lists of different data repositories, for example, the Open Access Directory list of repositories is maintained by the School of Library and Information Science (SLIS) at Simmons College.[95]

The way that data in a repository is used, analytical requirements, and the scale of individual datasets determines the nature of infrastructure needed for any given repository. If analysis is to be performed on very large datasets within a repository, then it makes sense to co-locate the processing resources and analytical software on the same servers as the data to avoid the need to transport large quantities of data across networks before computations are attempted. The use of metadata within repositories not only enables search and discovery, but also provides ways to bring together related materials so that they are more easily found and manipulated. The use of persistent identifiers is important for both being able to locate information within repositories, but also to create links between data stored in any particular repository and that located elsewhere. For example, to maintain links between a published paper or report and a dataset within a repository. Interfaces for repositories need to enable users to be able to find and make sense of the data within them, and where appropriate to conduct appropriate tasks, such as performing analyses, asking questions, or seeking permission to download and use the data or code. Repositories also need to have interfaces designed to effectively capture appropriate metadata from the researchers depositing their research materials and data within them.

An alternative to the publish and deposit in a repository model for sharing research is the 'open notebook' model, as used by the Open Malaria project discussed previously. In 'open notebook science' tools such as Electronic Laboratory Notebooks, blogs, wikis, and Google Drive are used in conjunction with open file and code sharing tools, such as GitHub and SourceForge, to share all of the material from scientific research. Open notebook tools have many of the same challenges as repositories, but there are also additional challenges involved in providing interfaces that are usable and functional for many different roles, including research authors, collaborators, managers, and other information seekers as described in Chapter 8. Such tools may need to provide very specialised interfaces for capturing and displaying data, but may also need to handle a variety of different data formats in order to be useful to researchers from varied disciplines. Collaboration is an important part of Open Science, and for open notebook tools in particular. Facilities need to be provided to enable collaborators, and potentially a much wider audience, to be able to contribute and discuss the research materials in a meaningful way. This is particularly important so that open notebook tools can provide mechanisms to validate the research. Such mechanisms for validation may include both informal 'buddy checking' of work within a project team and more formal open peer review by the wider scientific community.

Open notebook science provides an opportunity for researchers to access all the methods and other materials associated with the research so that it can be replicated, but more may be needed to enable full understanding and reproducibility of scientific research. Being able to replicate the research in the same environment as the original research was carried out would enable the precise conditions of the study to be matched. Although this is often impractical for the data collection elements of field and laboratory work, it is entirely possible for computational elements of the research, such as *in silico* experiments, data processing and analysis, and work involving modelling and simulations. These computational elements should in theory be easy to replicate using technology. In reality though, manually replicating a research environment from a published paper is extremely difficult. To fully replicate an environment, the same hardware, operating system, software stack, code, data, and configurations are needed, and all software and code needs to be at the correct version. The exact set-up steps for all of these elements need to be thoroughly documented in order for them to be repeated, but published papers cannot go into enough detail to accomplish that task. Even where perfect documentation is provided as supplementary material or as part of an open notebook, there are likely to be discrepancies in the set-up between different machines or instances of software. A potentially better solution is to enable the user to explore the results of the research without needing to directly duplicate the environment itself.

A relatively simple approach to the replication of computational elements of science is to use 'interactive computing', where the user can view and run code or computations using an interactive notebook. For example, Juptyer notebooks are interactive online notebooks that can be used to create and share documents that contain text, visualisations, equations, and code in a variety of scientific programming languages including Python, R, and Julia.[96] The content of the notebook is both executable and interactive, so that calculations and computations can be re-run, plots and graphs can be recreated, and data can be changed to model effects and generate new results. The notebooks are both human and machine-readable, enabling them to be shared and stored as part of the scientific record, and come with options to enable the contents to be exported both statically and interactively in a variety of formats. They can therefore be used as support material for journals or books, but in theory can also be embedded on web pages for sharing. In addition to providing a tool for sharing results at publication time, they can also be used as working tools for exploration and refining processes iteratively throughout the research.

These kinds of tools are suitable for executing small amounts of code or equations, but do not scale up for very large datasets and more complex software. An alternative to this approach is the 'executable research paper' where readers can replicate the results inside a paper published online.[97] Although these sound very similar to interactive notebooks, the executable papers instead provide a link to a virtual machine containing the research environment and data, so that users can explore the data and results using the original software used. Virtual machines have the advantage that they provide complete transparency, enabling users to view the raw data and transformations at each stage of the research, instead of just the end results that are usually provided by supplementary material or open data. They also enable large datasets to be examined without the need to download them to a local machine. There are a number of different potential solutions for creating and hosting such executable papers and their associated virtual machines, but there are some challenges. Authors of the papers are likely to be responsible for creating the virtual machines, and this often involves installing operating systems and software, implementing security measures to prevent unauthorised access to other materials, and there may be licensing issues to consider in creating virtual machines for users outside of the research organisation.[98] Another challenge is that users of such environments need to have the appropriate technical knowledge to understand and use the software that is provided in the virtual research environment.

An alternative to requiring a complex research environment for replicating results are frameworks for conducting repeatable experiments. Such frameworks include workflow systems that enable scientists to perform complex data transformations, computations, and visualisations using built-in services and without requiring the development of complex code and bespoke software. Built-in services for workflow systems typically provide commonly used functions, such as remote data collection, pre-processing and data cleaning, modelling, analytics, statistics, visualisation, machine learning, data mining, and data deposition. Many of the workflow systems available are open source and built-in services are often supplemented by third-party extensions and custom programming functions. Workflow systems such as E-Science Central,[99] Taverna,[100] Pipeline Pilot,[101] KNIME,[102] Galaxy,[103] Vistrails,[104] and Kepler,[105] typically provide a 'flowchart' style interface where users can select built-in functions in the form of 'nodes' that can be connected together to create flows of data. Data may be transformed by the nodes or decisions may be made about which processing path the data may take based on the contents of the data. As shown in Figure 11.1, workflow tools usually provide drag and drop interfaces that are relatively easy to use and understand, and much easier for non-programmers to use than scripting or other executable software environments. Many workflow systems can also capture a provenance trail for each run of data through the workflow. The provenance trail can capture a record of each action in the workflow, dependencies, and intermediate changes to the data that can be used to inform how a particular end dataset was created by the workflow. Workflows enable the automation of complex

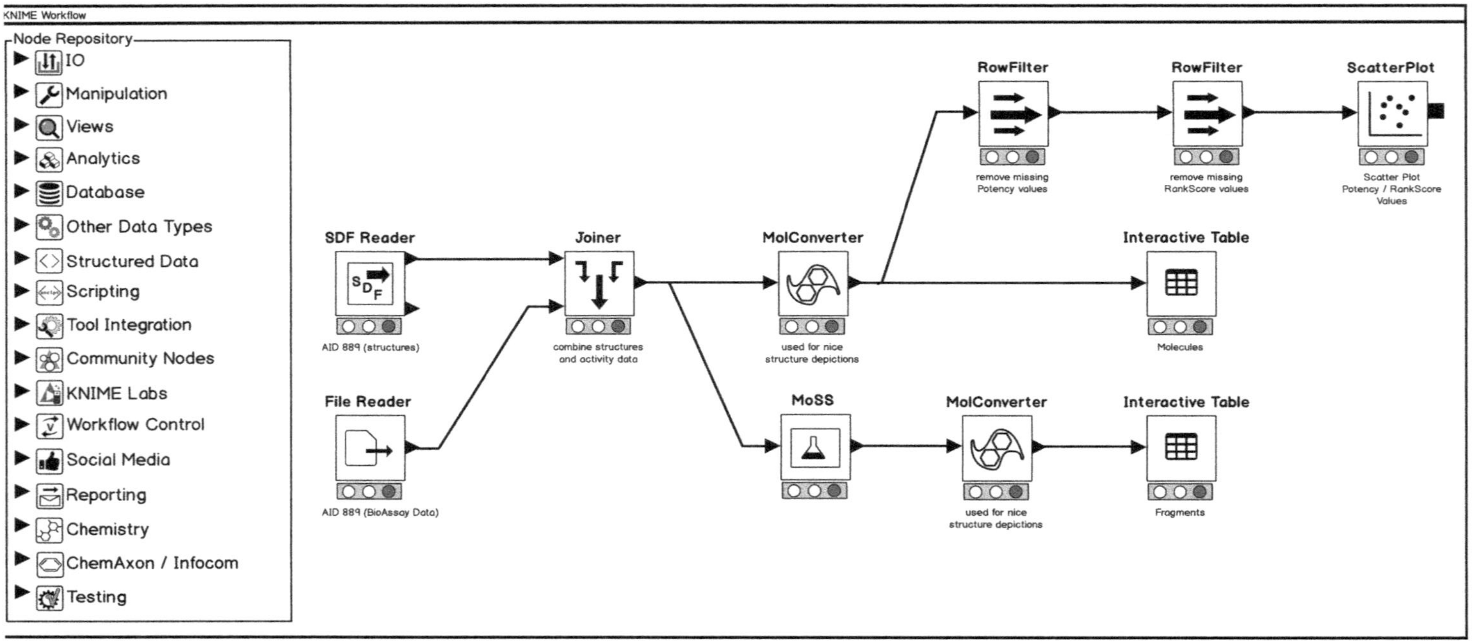

Figure 11.1 Mock-up of the KNIME interface showing the selection of node categories available and an example workflow. In the workflow, input nodes extract data from external files and the data flows through the connections between nodes and undergoes processing at each node. The results are made available *via* output nodes, in this case, in the form of interactive tables and a scatter plot.

repetitive tasks that are relatively simple to develop but are also easily shared and understood by others. They can also include embedded data so that others can directly reproduce results, but they can also be reused by adding new data. Workflows can be developed and optimised to address particular processing tasks and be further built upon by other teams for their needs. Many workflow tools have sharing capabilities built-in to the software, for example, through online and cloud-based versions of the software, organisational servers, and through file-sharing software. For wider access there are also workflow sharing repositories and websites, such as myExperiment and PBase that enable users to upload and describe their workflows or search for existing workflows that may be useful for their research.[36,106]

There are a number of potential Web-based tools for collaboration, beyond day-to-day communication tools such as e-mail and messaging. Social networks provide opportunities for communication, discovery, and making connections. Digital tools with social capabilities enable users to not only interact and collaborate, but also to be aware of the comings and goings of colleagues.[107] For example, messaging type programmes such as Messenger, Slack, and Skype enable users to be aware of the presence of other users through displays showing whether users are logged in and active. Social media sites with 'newsfeed' facilities also provide a way of seeing the activities and activeness of other users on the platform. These platforms also typically enable the user to get involved with the conversation if they want to, by adding comments, 'liking' content, or otherwise responding to the activities or questions of others. Although social network systems are typically considered to be focused on creating sophisticated profiles, social relationships, and social interactions, other types of Web-based services are also often discussed under the same umbrella including research portals, networking tools, researcher identification, citation management tools, blogs, and microblogging sites.[108] In fact, many services developed for scientists contain elements of more than one of these capabilities.

In 2014, a survey of over 3500 scientists across the world investigated social network use in science. The results of the survey by Nature suggested that the majority of scientists are using one or more social networking sites professionally.[109] The survey contained a mixture of social network platforms, including both scientific and non-scientific. The most regularly visited sites were Google Scholar, ResearchGate, LinkedIn, and Facebook with smaller numbers of regular visitors for Google+, Twitter, ResearcherID, Mendeley, ORCID, Academia.edu, BiomedExperts, and Microsoft Academic Search. Levels of engagement differ across the different platforms, and regular visiting does not imply professional use. For example, the majority of Facebook users were not using the platform for activities related to their research, and many users of scholarly networks ResearchGate and Academia.edu were largely maintaining a profile on the platform in case someone wanted to contact them. Twitter appears to be a platform that is used by regular visitors to follow discussions, comment on and discuss research, and to post information about the scientists' own research, as well as discover peers, jobs, and literature recommendations. A follow-up survey focused on researchers from the humanities and social sciences added in additional platforms such as YouTube and blogs, which although less popular than the academic focused platforms, suggests that professional use of social networks is not just about discovery and collaboration, but also about making use of opportunities to communicate and promote the research to a wider audience.[110] Although most researchers in these surveys were not using Facebook as a tool for discussing research, there are some active science communities on Facebook, often involving younger researchers or students. For example, there are a number of community groups on Facebook with an interest in drug discovery and development that share job openings, technical questions, resources of interest, and general musings about the topic.[111] One of the advantages of Facebook for collaboration may be that the informal nature of interaction removes some of the inhibitions that users might have on more formal platforms. The informal nature of social networking sites can also be one of the reasons that users are reluctant to use them professionally, because they do not want to mix elements of their personal and professional lives. The microblogging site Twitter has the advantage that it is easy to set up multiple accounts, and use one account to promote your own research, to share citations, and engage in discussion, and use a different account, with an entirely different audience, for personal content.

There have been a number of attempts to create social networks specifically for scientists, but without significant success. For example, Nature Network, a site with blogs, groups, forums for questions, and profiles for making connections, closed in 2013 after moving the blogging portion of the site to a new location and recognising that the site was no longer hosting the discussions and collaborations that it once had. Other examples have also disappeared, SciLink, a social network that mined relationships from scientific literature to build connections disappeared by 2012; Epernicus Network, a social network for researchers to make connections and post questions, appears to have disappeared in 2016 when the owner was acquired by another company; and 2collab from Elsevier, a tool for sharing references and bookmarks with other scientists was closed in 2011 so that the company could focus on integrating with commercial bibliographic management and bookmarking software. Another example, AtmosPeer was a social network platform for connecting atmospheric scientists and providing document sharing, discussion forums, newsfeeds and access to conferences and literature, but it is unclear if the service is still available. Scientist Solutions, a discussion forum website for questions and issues in life sciences and related topics, started in 2004, is still running and appears to be active.[112]

Blogging is potentially interesting as a tool for both communication and collaboration in science. As discussed previously, many of those scientists engaged in open notebook science make use of blogging platforms to do so. Blogging about science is done by journalists and enthusiasts as well as scientists, although the target audience and purpose of posts is likely to be different.[113] Scientists are most likely to be blogging about the process of research, although those engaged in open science may also be sharing their research activities and findings with peers, collaborators, funders, and even the public. Blogging can be used for promotion of research, networking activities, and also as a way of keeping up-to-date with related activities.

A majority of the public think that scientists themselves should be the ones telling the stories of their scientific research, rather than journalists, consumer organisations, or government representatives.[114] Traditional journalistic and media coverage of science is often distorted into 'extreme' science, for example, blowing things up (literally) in the name of science, explanations of everyday phenomena, or reporting of alleged success stories. None of these convey the issues and complexity involved in the scientific process. Blogs have the potential to provide the public with the story of the research from the perspective of the scientist and to expose the process of science. Blogs encourage a less formal style than traditional publications and the more formal structured records that some use in their day-to-day notebooks. A more narrative style can be used that is better suited to storytelling and a more natural way of expressing ideas and experiences. Storytelling is an effective way of communicating 'tacit' knowledge, and the social and interactive elements of blogs can also contribute to the sharing of tacit knowledge which is more difficult in formal publications.[115] Pictures and videos can be embedded on blogs to demonstrate and explain concepts, and these elements can make the content more accessible to a less technical audience. The success of any endeavour to communicate with the public is likely to hinge on the authors ability to communicate with their intended audience. As Cribb and Sari point out, scientific writing more often resembles an impenetrable train wreck than an act of communion with the reader.[116] With the Web becoming the dominant way that we all seek information and news, there is more and more demand in society for instant communication and for scientists themselves to become the communicators of their own research; however, there may be risks involved in expecting researchers to be up to this challenge.[1]

Blogging is unlikely to replace the traditional publication process, but it can supplement it by providing a forum to publish content that does not fit into formal publications such as discussion of literature, early results, unproven hypotheses, and new findings after publication. Although it is possible to have a certain amount of discussion in traditional publications with 'comments on' and reviews of journal articles, published articles are rarely updated and so blogging potentially provides a platform for researchers to put out updates to the published literature. That there is a need for a forum for publishing information outside of the formal journal article does appear to be recognised by some of the academic publishers. For example, the journal *BMC Research Notes* publishes short research communications and scientifically valid research outputs that could not be published as full research or methodology articles including updates to previous work, additions to established methods, short publications, null results, research proposals, and data management plans.[117]

Attitudes towards blogging are mixed, both from the point of view of potential authors of scientific blogs and from commentators upon them. One the one hand, blogs are seen as a potential way to effectively communicate science to the public, being less technical and more accessible than traditional publications, but able to provide more in-depth information and explanation than traditional news media. Journal articles also often are missing 'tacit' information that the target scientific audience may already know, but that journalists or the public accessing the material may not, and blog posts provide the opportunity to convey this knowledge to a lay audience.[118] However, not all authors agree with this point of view, and suggest that science blogs can add to the potential confusion and mistrust of science, especially when they are critical of formal publications, arguing that instead criticism of journal articles should take place in the peer-reviewed literature.[119] There are certainly potential risks with scientific blogs for misinformation and confusion. Anyone can write a blog without any scientific credentials, and even with credentials, blogs are not peer reviewed, and the content can easily be taken out of context or misused. Some bloggers choose to deliberately spread misinformation in the guise of proven scientific knowledge. Such information can be accessed by the public or propagated by the media, or other social media, generating 'fake news' and other conspiracy theories. In reality, most science blogs authored by scientists are read by other scientists. Opportunities for discussion and review provided by comment threads on the blogs are also typically utilised by members of the same communities.

The question of providing 'tacit' knowledge to a lay audience is being addressed in part through Wikipedia. Many scientists contribute to Wikipedia updating pages on scientific topics to ensure their correctness, but there are still many topics that are completely missing. In 2012, the journal *PLoS Computational Biology* launched a project called 'Topic Pages' to encourage scientists to improve Wikipedia's coverage of computational biology by offering to credit authors for their contributions.[120,121] Contributed pages are written in the style of a Wikipedia article and are openly and publicly peer reviewed on the PLoS Wiki before being added to Wikipedia and then published in a PLoS journal. Although the creation of such articles benefits both the scientific community and the public, they take a lot of effort to create, and providing an incentive in the form of a published article makes it worth the effort compared with independently creating such content.

11.7 Conclusion

Technology facilitates more openness in science. It provides easier ways to access the scientific record, to share it, and to discuss it. The importance of collaboration and connectedness comes across as a strong theme in discussions with many researchers. Teams actively engaged in Open Science desire digital tools that can help them to attract, inform, update, and interact with collaborators, but even outside of Open Science, tools that facilitate these activities are important for enabling more effective sharing and collaboration within an organisation.

Even where the protection of intellectual property, opportunities for commercialization, and sensitive data are not significant issues, there is a reluctance from many scientists to embrace technologies for openness and to share research outside of the traditional publication format for the reasons that we have discussed. The changes around encouraging the deposition of data are having significant effects in making data available together with publications, but there are still problems of missing methods and a disconnect between the explanation of the research, the data, and the code that was used to create it. Currently, there are limited potential solutions to this particular problem. Open notebook science, where tools such as ELNs can be used to capture the research record, can also act as a repository for the associated data, code, and other digital artefacts related to the research. All the information is then provided together in one system so that it can be discovered and reused. Another option to overcome the disconnect is to bundle all the resources together for deposition in a repository at the end of the research process as supplementary material for a formal publication. The advantage of the research bundle is that the release of the information is controlled by the author, and that all the information is together for a potential user when they want to use it. The downside is that the information cannot be used until the project is complete. Also, it is likely that much of the effort involved in preparing the resources for sharing, such as putting the data into appropriate formats and adding descriptive and structural metadata, is likely to be done at the end of the research. For the open notebook, all the information is available, but it is also available during the research, so that data and methods are open to be used and validated as early as possible. Preparation for sharing is unnecessary because metadata can be generated and collected at source, and the tool itself should be able to export in appropriate human and machine-readable formats. Digital tools, such as ELNs, also provide opportunities to encourage sharing of data by reinforcing the value of research through effective and responsible online communities where participants are given the opportunity to present personal research stories, rather than the comparative anonymity of archiving of data in a repository.

Developing positive online communities requires effort and responsibility. As previously discussed, researchers are unlikely to engage in activities that are perceived as a waste of time or lacking personal value. One way to encourage engagement and effort by users is to provide mechanisms for recognition and reward. Gamification elements, such as trophies, badges, ratings, rankings, and reputation points, can be used to encourage sharing and interaction on social media platforms. Although altmetrics can potentially capture some of these gamification elements, they are unlikely to be sufficient incentives for scientists to spend effort on digital platforms without other benefits. Ideally more initiatives like the PLoS Topic Pages and open data publications can be implemented to encourage activities by providing real publications as a result of creating scientific content for non-traditional publication.

Science is changing, in part as a result of the impacts of technology and the capabilities they provide. More data is being produced, and more of that data is becoming available as more data and more research is shared. In Chapter 12, we will take a look at how new technologies might impact on how the scientific record is captured and shared in the future, including how technology might support us with overcoming the data deluge and discovering and managing knowledge that might be relevant to our research.

References

1. B. Fecher and S. Friesike, in *Opening Science*, ed. S. Bartling and S. Friesike, Springer, Cham, 2014, p. 17.
2. B. A. Nosek *et al.*, *Science*, 2015, **384**(6242), 1422.
3. Available at https://wellcome.ac.uk/funding/guidance/charity-open-access-fund (accessed Nov 2018).
4. P. A. David, *Capitalism Soc.*, 2008, **3**(2), 1.
5. S. M. Bachrach, *J. Cheminf.*, 2009, **1**(2), 1.
6. J. Wilsdon, J. Barilan, R. Frodeman, E. Lex, I. Peters and P. Wouters, *Next-Generation Metrics: Reponsible Metrics and Evaluation for Open Science*, European Commission, 2017.
7. Available at https://doaj.org (accessed Nov 2018).
8. Available at https://oalibrary.org (accessed Nov 2018).
9. Available at http://roar.eprints.org (accessed Nov 2018).
10. Available at https://arxiv.org (accessed Nov 2018).
11. Available at http://www.oapen.org/home (accessed Nov 2018).
12. B. Nelson, *Nature*, 2009, **461**, 160.
13. C. Tenopir, S. Allard, K. Douglass, A. U. Aydinoglu, L. Wu, E. Read, M. Manoff and M. Frame, *PLoS One*, 2011, **6**(6), e21101.
14. J. C. Wallis, E. Rolando and C. L. Borgman, *PLoS*, 2013, **8**(7), e67332.
15. The Royal Society, *Science as an Open Enterprise*, 2012, Available at https://royalsociety.org/~/media/Royal_Society_Content/policy/projects//sape/2012-06-20-SAOE.pdf (accessed Nov 2018).
16. Available at https://www.ncbi.nlm.nih.gov/genbank/ (accessed Nov 2018).
17. Available at https://academic.oup.com/gigascience (accessed Nov 2018).
18. Available at https://rmets.onlinelibrary.wiley.com/journal/20496060 (accessed Nov 2018).
19. Available at https://www.earth-system-science-data.net (accessed Nov 2018).
20. Available at https://www.nature.com/sdata/ (accessed Nov 2018).

21. Available at https://bdj.pensoft.net (accessed Nov 2018).
22. Available at https://www.biomed-data.eu/ (accessed Mar 2018).
23. Available at https://rda.ucar.edu (accessed Oct 2018).
24. Available at https://osf.io (accessed Oct 2018).
25. Available at https://datadryad.org (accessed Oct 2018).
26. Available at http://www.ebi.ac.uk/pdbe/ (accessed Oct 2018).
27. Available at https://zenodo.org (accessed Oct 2018).
28. Available at https://figshare.com (accessed Oct 2018).
29. Available at https://www.ukdataservice.ac.uk (accessed Oct 2018).
30. Available at https://www.ands.org.au (accessed Oct 2018).
31. Available at https://data.europa.eu/euodp/data/ (accessed Oct 2018).
32. Available at http://v2.sherpa.ac.uk/opendoar/ (accessed Nov 2018).
33. Available at https://openwetware.org/wiki/Protocols (accessed Nov 2018).
34. Available at https://www.protocols.io/protocols (accessed Nov 2018).
35. Available at http://www.protocol-online.org (accessed Nov 2018).
36. D. De Roure, C. Goble and R. Stevens, *Future Gener. Comp. Syst.*, 2009, **25**, 561.
37. A. Milsted, J. Hale and C. Neylon, *PLoS One*, 2013, **8**(7), e67460.
38. Available at https://openlabnotebooks.org/ (accessed Nov 2018).
39. Available at http://onsnetwork.org (accessed Nov 2018).
40. D. C. Ince, L. Hatton and J. Graham-Cumming, *Nature*, 2012, **482**, 485.
41. J. Willinsky, *First Monday*, 2005, **10**(8).
42. Available at https://github.com (accessed Nov 2018).
43. Available at https://www.nitrc.org (accessed Nov 2018).
44. Available at https://directory.fsf.org/wiki/Category/Science (accessed Nov 2018).
45. Available at http://www.nanohub.org (accessed Nov 2018).
46. Available at https://www.research-software.nl (accessed Nov 2018).
47. Available at https://www.wikipathways.org/index.php/WikiPathways (accessed Nov 2018).
48. S. Bechhofer, D. De Roure, M. Gamble, C. Goble and I. Buchan, *Nature Precedings*, 2010, DOI: 10.1038/npre.2010.4626.1 (accessed Nov 2018).
49. D. De Roure, 2009, Available at http://wiki.myexperiment.org/index.php/Research_Objects (accessed Nov 2018).
50. T. Spangenberg, J. N. Burrows, P. Kowalczyk, S. McDonald, T. N. C. Wells and P. Willis, *PLoS One*, 2013, **8**(6), e62906.
51. Available at https://www.ebi.ac.uk/chembl/malaria/ (accessed Nov 2018).
52. M. N. Robertson, P. M. Ylioja, A. E. Williamson, M. Woelfle, M. Robins, K. A. Badiola, P. Willis, P. Olliaro, T. N. Wells and M. H. Todd, *Parasitology*, 2014, **141**, 148.
53. Available at http://opensourcemalaria.org (accessed Oct 2018).
54. Available at http://malaria.ourexperiment.org (accessed Oct 2018).
55. K. A. Badiola *et al.*, *Chem. Sci.*, 2015, **6**, 1614.
56. Available at https://openwetware.org/wiki/OpenSourceMalaria (accessed Oct 2018).

57. T. Gowers and M. Nielsen, *Nature*, 2009, **461**, 879.
58. Available at http://gowers.wordpress.com/2009/01/27/is-massively-collaborative-mathematics-possible/ (accessed Oct 2018).
59. House of Lords Select Committee on Science and Technology, *Science and Society – Third Report*, 2000, Available at https://publications.parliament.uk/pa/ld199900/ldselect/ldsctech/38/3801.htm (accessed Nov 2018).
60. The Royal Society, *Science and the Public Interest: Communicating the Results of New Scientific Research to the Public*, 2006, Available at https://royalsociety.org/topics-policy/publications/2006/science-public-interest/ (accessed Nov 2018).
61. G. Bowker, *Memory Practices in the Sciences*, MIT, Cambridge, 2008, p. 117.
62. Available at https://www.rspb.org.uk/get-involved/activities/birdwatch/ (accessed Oct 2018).
63. Available at https://naturescalendar.woodlandtrust.org.uk (accessed Oct 2018).
64. Available at https://www.opalexplorenature.org (accessed Oct 2018).
65. Available at http://stardustathome.ssl.berkeley.edu/ (accessed Oct 2018).
66. K. Land, A. Slosar, C. Lintott, D. Andreescu, S. Bamford, P. Murray, R. Nichol, M. J. Raddick, K. Schawinski, A. Szalay, D. Thomas and Jan Vandenberg, *Mon. Not. R. Astron. Soc.*, 2008, **388**(4), 1686.
67. C. J. Lintott, K. Schawinski, A. Slosar, K. Land, S. Bamford, D. Thomas, M. J. Raddick, R. C. Nichol, A. Szalay, D. Andreescu, P. Murray and J. Vandenberg, *Mon. Not. R. Astron. Soc.*, 2008, **389**(3), 1179.
68. Available at https://www.zooniverse.org (accessed Oct 2018).
69. Available at https://www.mturk.com (accessed Nov 2018).
70. Available at http://www.innocentive.com (accessed Oct 2018).
71. K. Ridler, 2018, Available at http://www.kmvt.com/content/news/US-scientists-try-crowdsourcing-to-stop-invasive-mussels-474512133.html (accessed Nov 2018).
72. Available at https://setiathome.berkeley.edu (accessed Oct 2018).
73. Available at http://boinc.bakerlab.org/rosetta/ (accessed Oct 2018).
74. Available at https://foldingathome.org (accessed Oct 2018).
75. S. Cooper, F. Khatib, A. Treuille, J. Barbero, J. Lee, M. Beenen, A. Leaver-Fay, D. Baker, Z. Popović and F. Players, *Nature*, 2010, **466**, 756.
76. Available at https://fold.it/portal/ (accessed Oct 2018).
77. Available at http://www.projectnoah.org (accessed Oct 2018).
78. Available at http://leafsnap.com (accessed Oct 2018).
79. Available at http://www.birdseyebirding.com (accessed Oct 2018).
80. Available at http://www.noisetube.net (accessed Oct 2018).
81. Available at http://www.secchidisk.org (accessed Oct 2018).
82. T. Landrain, M. Meyer, A. M. Perez and R. Sussan, *Syst. Synth. Biol.*, 2013, 7(3), 115.

83. D. L. Shaw, *Yale J. Biol. Med.*, 2017, **90**(1), 147.
84. A. S. Fauci and F. S. Collins, *Science*, 2012, **336**(6088), 1522.
85. J. Kaye, C. Heeney, N. Hawkins, J. de Vries and P. Boddington, *Nat. Rev. Genet.*, 2009, **10**(5), 331.
86. Available at https://www.altmetric.com (accessed Oct 2018).
87. Available at https://www.mendeley.com/ (accessed Oct 2018).
88. Available at http://www.citeulike.org (accessed Oct 2018).
89. Available at https://profiles.impactstory.org (accessed Oct 2018).
90. Available at https://www.growkudos.com (accessed Oct 2018).
91. Available at https://blog.growkudos.com/2016/11/21/experiences-with-kudos-why-mobilizeresearch-matters/ (accessed Oct 2018).
92. M. Thelwall, S. Haustein, V. Larivière and C.R. Sugimoto, *PLoS One*, 2013, **8**(5), e64841.
93. J. Priem, D. Taraborelli, P. Groth and C. Neylon, *Altmetrics: A Manifesto*, 2010, Available at http://altmetrics.org/manifesto (accessed Oct 2018).
94. Available at https://www.re3data.org (accessed Oct 2018).
95. Available at http://oad.simmons.edu/oadwiki/Data_repositories (accessed Oct 2018).
96. Available at https://jupyter.org/ (accessed Oct 2018).
97. P. Van Gorp and S. Mazanek, *Procedia Comput. Sci.*, 2011, **4**, 589.
98. G. R. Brammer, R. W. Crosby, S. J. Matthews and T. L. Williams, *Procedia Comput. Sci.*, 2011, **4**, 658.
99. Available at http://www.esciencecentral.co.uk (accessed Nov 2018).
100. Available at https://taverna.incubator.apache.org (accessed Nov 2018).
101. Available at http://www.3dsbiovia.com/products/collaborative-science/biovia-pipeline-pilot/ (accessed Nov 2018).
102. M. R. Berthold, N. Cebron, F. Dill, T. R. Gabriel, T. Kötter, T. Meinl, P. Ohl, K. Thiel and B. Wiswedel, in *Data Analysis, Machine Learning and Applications*, ed. C. Preisach, H. Burkhardt, L. Schmidt-Thieme and R. Decker, Springer, Berlin, 2008, p. 319.
103. Available at https://usegalaxy.org (accessed Nov 2018).
104. Available at https://www.vistrails.org/index.php/Main_Page (accessed Nov 2018).
105. I. Altintas, C. Berkley, E. Jaeger, M. Jones, B. Ludascher and S. Mock, in *Proceedings of the 16th International Conference on Scientific and Statistical Database Management*, IEEE Computer Society Press, Los Alamitos, 2004, p. 423.
106. V. Cuevas-Vicenttín, P. Kianmajd, B. Ludäscher, P. Missier, F. Chirigati, Y. Wei, D. Koop and S. Dey, *Int. J. Digit. Curation*, 2014, **9**(2), 28.
107. J. Donath, *The Social Machine: Designs for Living Online*, MIT Press, Cambridge, 2011, p. 12.
108. M. Entwich and R. König, in *Opening Science*, ed. S. Bartling and S. Friesike, Springer, Cham, 2014, p. 107.
109. R. Van Noorden, *Nature*, 2014, **512**, 126.

110. M. Staniland, 2017, Available at http://blogs.nature.com/ofschemesandmemes/2017/06/15/how-do-researchers-use-social-media-and-scholarly-collaboration-networks-scns (accessed Nov 2018).
111. D. S. Bailey and E. D. Zanders, *Drug Discovery Today*, 2008, **13**(19-20), 863.
112. Available at https://www.scientistsolutions.com (accessed Nov 2018).
113. C. Puschmann, in *Opening Science*, ed. S. Bartling and S. Friesike, Springer, Cham, 2014, p. 89.
114. A. Gerber, in *Opening Science*, ed. S. Bartling and S. Friesike, Springer, Cham, 2014, p. 73.
115. S. Panahi, J. Watson and H. Partridge, *J. Knowl. Manag.*, 2016, **20**(2), 344.
116. J. Cribb and T. Sari, *Open Science: Sharing Knowledge in the Global Century*, CSIRO Publishing, Collingwood, 2010, p. 15.
117. Available at https://bmcresnotes.biomedcentral.com (accessed Nov 2018).
118. G. Schmidt, *Nat. Geosci.*, 2008, **1**, 208.
119. M. Allen, *Nat. Geosci.*, 2008, **1**, 209.
120. D. Mietchen, S. Wodak, S. Wasik, N. Szostak and C. Dessimoz, *PLoS Comput. Biol.*, 2018, **14**(5), e1006137.
121. Available at https://collections.plos.org/topic-pages (accessed Nov 2018).

CHAPTER 12

Futures for Technology and Scientific Research

12.1 Introduction

Memories fade and change. If we fail to capture the information inside of scientists' heads, it will be lost forever. An individual researcher may be able to rack their brains to try and remember the information, but this option is not available a few months after the event once the researcher has moved on, and certainly not in a hundred years. The more information we can capture in the research record, the less reliance there is on fallible memory and the more opportunities there are for discovering, making-sense of, and reusing the information so that we can build upon the work and gain new perspectives.[1] Although people often think of the published record as being equivalent to the scientific record, in the majority of cases, the published record is missing significant details that enable research to be repeated and reused. There is a drive to encourage scientists to publish their data as well as their results, but the data must also be furnished with appropriate metadata to describe its provenance and provide context, together with the details of how it was captured in order to be reusable. Providing access to the record is also a major challenge. If there is no access to the record for the scientific community, the research within cannot be reused, and there is the risk that other scientists will waste time, effort, and funds repeating research, collecting data that has already been collected, and going down research paths that are already known to have failed.

In this book, we have explored how researchers record their research and how they can use electronic notebooks and other systems to capture and organise their research. We have investigated how different interface designs might impact the information that is recorded in digital systems

Recording Science in the Digital Era: From Paper to Electronic Notebooks and Other Digital Tools
By Cerys Willoughby

Published by the Royal Society of Chemistry, www.rsc.org

and how different designs might encourage the capture of metadata to describe the data to enable discoverability and reuse. The various experimental studies described in the book have been focused primarily on the capture of information relating to experimental records, but along the way scientists have also provided an insight into their desires for additional capabilities for digital tools that go beyond just replicating the 'traditional' way of working more effectively, but to actually change and improve the way they work. These findings provide a valuable glimpse of 'the bigger picture' that extends beyond merely the need for data archives or replacing paper with electronic notebooks. The desires users have for additions to technology are usually limited by their own experience of what technology can do, and how they perform tasks without technology. It is not a large stretch to ask for a way to automate a task that is currently being done manually, for example, to ask for an electronic notebook to automatically post a link to a new entry on Twitter. The limitations of technology can also provoke ideas for better ways of interacting with the technology. For example, in the laboratory it is difficult to work with computers, often requiring that the user leave the lab to look up information or to record the details of their experiment. Many researchers recognise that one solution to this problem is to enable 'hands-free' operation of the computer by using voice control or dictation using a microphone. In this chapter, we will investigate how developing technologies might influence the ways that scientists conduct their research and how digital tools might support the scientific process and scientific record-keeping in the future.

Science is changing in a number of ways. There are increasing numbers of researchers, growing at 3% per year, to between 7 and 9 million in 2015, publishing some 2.5 million journal articles per year in 30 000 peer-reviewed journals.[2] Not only is there a deluge in data, but also a deluge in scholarly publications. Although online and open access may make it easier to access more of the published literature, it also becomes more difficult to be able to find research that is relevant, and to have the time to be able to read and absorb it. Where once it might have been possible for a scientist in their field to have a good knowledge of both their subject and the latest research, this is now impossible. At the same time, the scale of research has grown significantly, with more funding for larger scale enterprises or 'Big Science'.[3] For example, funding for large scale research in high energy physics and astronomy have enabled the development of enormous and expensive instruments such as the Large Hadron Collider and the European Spallation Source (ESS), and sky imaging telescopes such as Panoramic Survey Telescope and Rapid Response System (PAN-STARRS) and Large Synoptic Sky Telescope (LSST), and even more expensive projects such as space exploration.[4–7] The scale of research in other fields such as life sciences, geology, and environmental science is also increasing, for example, the Human Genome Project, EarthScope, and the Ocean Observatories Initiative.[8–10] Such projects change scientific working practices, and in particular, increase collaboration, not only with greater numbers of researchers working on a project, but also with researchers working in different locations, different time zones, and across many different disciplines.

It is difficult to predict what the next piece of disruptive technology is likely to be within any domain, and particularly within science. In theory, science is on the leading edge of technological developments, with many of the innovations in technology emerging from scientific organisations, both academic and industrial. However, as our continued reliance on the trusty paper notebook demonstrates, not all technological advances are adopted with equal enthusiasm. The computer has made an enormous difference to how science is carried out, in particular through the ability to capture data in digital format and to be able to manipulate, analyse and visualise the data to produce results. The Internet has had an important role in not only fostering and facilitating collaboration, but also enabling the concept of Open Science, where the published scientific record and the data and materials that support it can be easily discovered and reused. Data can be stored and shared in repositories, and thousands of publications on any given topic can be found and accessed at the click of a button. More recently, the proliferation of sensors and smart devices has provided new ways to work, to stay in contact, and to gather information. Technological advances contribute to the challenges of 'big data' and the 'data deluge' by providing new ways to generate more data more quickly, but without technology it would be impossible to solve the challenges of capturing, handling, analysing, managing, and sharing the data. Technology provides the opportunity to ask and answer new questions in science, and to make use of existing data to get smarter. In this chapter, we will take a look at a variety of different technologies, how they are being used in science, and how they may affect how we interact with and create the scientific record in the future. In Section 12.2, we will discuss how futuristic sounding technologies such as Virtual Reality (VR), Augmented Reality (AR), and Artificial Intelligence (AI) have provided new ways to make use of and interact with our scientific data. In Section 12.3, we will examine the role of the Internet of Things in not only capturing data through the use of sensors and connected systems, but also how we can use such devices to interact with our data and capture the record of our activities. We will also look at the potential uses of wearables and smartphones in science in Section 12.4. Technology can make our lives simpler or more complicated, and will we also examine how different kinds of interfaces, some familiar and some less familiar, might make it easier to interact with our computers or provide novel ways to engage with them, when we discuss natural interfaces in Section 12.5. All of the technologies discussed in this chapter work in conjunction with each other and are interlinked in a variety of ways. In Section 12.6, we will take a look at how the technologies might be brought together to make the environments in which we work smarter and to create 'intelligent scientific assistants' that might help us in our research and in capturing the record.

12.2 Artificial Intelligence, Augmented Reality, and Virtual Reality

Artificial Intelligence describes computer systems that are able to perform tasks that are usually thought of as requiring 'intelligence', for example, speech recognition, visual perception, learning, and problem solving. The main requirement of AI systems is that they are able to perform tasks that they have not been explicitly programmed to perform by making use of data provided to the system. For example, an AI system may be taught the rules of a game and left to choose which moves to play, or another AI system may make use of data from sensory inputs to determine information about their environment and use that data to make decisions. Although AI is considered a relatively young field, research in AI for scientific purposes has been carried out since at least the 1960s, and AI in the form of 'expert systems' have been used in a number of disciplines. An early example of AI research comes from the area of mass spectrometry, with the 'Heuristic Dendral' programme designed to take data from a mass spectrometer and generate candidate molecular structures to match the data.[11] The programme was written in the LISP programming language, commonly used for AI. The programme was designed to remove any molecules inconsistent with the data and to present users with a list of potential candidates in decreasing order of likelihood of a match. An example of an early AI application in the clinical environment was 'MYCIN', which was used for providing physicians with guidance on microbial therapy for patients with bacterial blood infections.[12] During that period, it would take a long time to get microbiology results from the testing laboratory, so it would be necessary to try and start an appropriate treatment before the results with the specific bacterial cause would be available. Physicians often made poor treatment choices, for example, treating those who did not need treating or prescribing broad-spectrum antibiotics to cover a lot of potential causes, even though clinical signs would suggest a more specific and less toxic treatment regime. In order to be trusted by the physician, the system needed to be able to provide the same advice as an infectious disease expert, and therefore be able to provide an explanation for the advice given. The system also needed to be able to acquire new knowledge over time. The system asked the physician for details of the condition of the patient and then provided advice together with the explanation. The physician was also able to query the system to find out why particular questions were asked or particular responses given.

MYCIN is a typical example of an expert system, where information in the 'knowledge base' is easy to change and is not 'hard-coded' within the programme. Rules for knowledge-based systems are often described in more natural language than programming syntax and describe the attributes and logical relationships between facts in the knowledge base. There are a number of different methods of generating information from a knowledge base, using 'inference engines', depending on the desired outcome of the system. For example, data-driven approaches can attempt to derive as many facts as possible from the data in the knowledge base, whilst goal-driven approaches look for solutions to a specific problem. Tracking back over the rules used to make an inference allows the expert system to provide an explanation for the 'fact' or conclusion reached. If the system finds that it is missing information that it needs to continue, it can ask the user to provide the missing information.

One of the biggest problems with expert systems is the time it takes to acquire the knowledge and create the rules for the knowledge-base, and the need to anticipate all potential inputs. An expert would be needed to provide input into the system or the developer of the system themselves would have to be an expert. An alternative to manually generating the rules is to use machine learning techniques. Rather than having to manually define the rules for the system, machine learning techniques allow the system to be presented with example data and to learn to distinguish data that is related or matches a particular label. Data mining and machine learning are closely related concepts, with data mining using unsupervised machine learning methods and statistics to locate patterns in large datasets, whilst machine learning uses algorithms and mathematical models to improve the system performance on a given task. There are some tasks that are impossible for a programmer to write code to perform, for example, speech recognition. It would be impossible to write code that could not only translate every different word, but also every possible voice and accent. Machine learning enables systems to be developed that can complete these tasks by using vast quantities of example data, predictions, and probabilities to match new data, and in the case of speech recognition, the likelihood of what is being said can be determined. For machine learning, both supervised and unsupervised methods of learning can be used. In supervised learning, the system is provided with example data that is pre-classified, for example, to create an AI system for detecting animals in photographs taken by a trail camera, the system could be presented with thousands of example photographs with and without animals that are labelled appropriately. The system could learn to distinguish photographs containing a wild animal of interest from photographs that are empty or show domestic animals. When unclassified photographs are added to the system, the system should consequently be able to label them correctly. In unsupervised learning, the data is not pre-classified, and relies upon mathematical and other techniques, such as cluster analysis and topic modelling, to identify commonalities, content, and patterns within the data. Machine learning methods have been developed to explore data from across scientific disciplines and to analyse high-throughput experimental data in novel ways. In particular, so-called 'deep learning systems' that have led to major improvements in certain areas, such as speech recognition and image processing.[13] Deep learning is a learning method known as 'representational learning'. In representational learning, the system uses raw data and automatically discovers the representations needed for detection or classification, using multiple layers of processing, and applying representations learnt from one layer as input into the next layer.[14] Such methods have been used successfully for particle identification and tracking in physics, generating structure maps and identifying disease causing gene variations in biology, and for applications in drug discovery, such as discovering protein structures and functions, medicine development, genomics modelling for drug repurposing, and drug interactions.[15–19]

AI and machine learning are important tools for analysing scientific data and generating scientific results, but they are also used in everyday applications, including search engines and voice assistants. Using similar algorithms, machine learning can utilise the scientific record itself to generate 'intelligent scientific assistants'.[20] As previously mentioned, it is impossible for any individual researcher to process and keep up to date with all the published research within the scientific record, but computer systems have the ability to take the information stored in the record, use inference to generate knowledge from it, and the generated knowledge can then be exploited by researchers. Machine learning systems can automatically extract information from external sources for processing and in order to keep their knowledge bases up to date. External sources could include databases from within an organisation or public repositories and the published record such as online journals, eBooks, and scholarly databases. As shown in Figure 12.1, the external data can be 'read' from these external sources to produce a knowledge base and to track knowledge provenance. Once extracted, techniques can be used to generate a knowledge network to represent relationships between the elements in the network and to calculate probabilities. Based on these findings, knowledge of interest can be reported to the user of the system using visualisations or other tools that enable knowledge-based analysis of the data.[21] These assistant applications could identify and summarise research from across the whole record, not

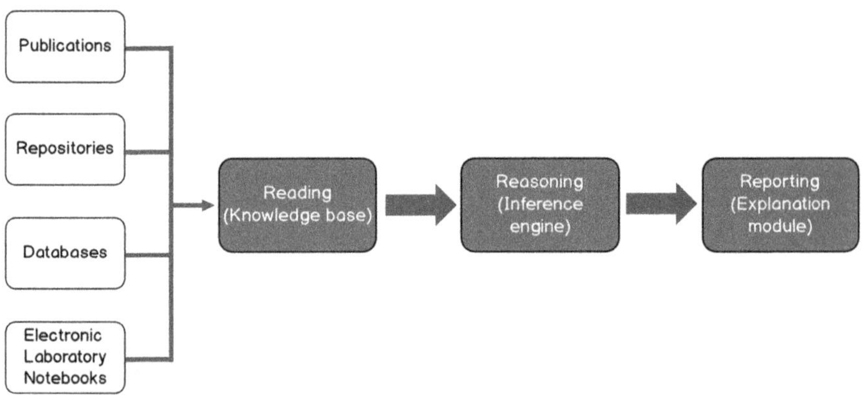

Figure 12.1 Example of an AI system for extracting 'knowledge' from part of the scientific record.

just published research, but also blogs, preprints, forums, repositories, and open notebooks. The suggestion is that these tools could use the vast wealth of scientific information to find relevant research and rerun analyses automatically, leaving researchers to spend more time on new research activities. Such systems can also function in a similar way to the expert systems described above, by providing ways to help researchers and clinicians to keep up to date with the latest research in science and healthcare.

An example of an intelligent scientific assistant is the Hanalyzer tool developed to accelerate investigations into the relationships between gene expression and craniofacial development.[21] The Hanalyzer system reads data from peer-reviewed databases related to molecular biology and related literature on biomedical research, targeted by genes of interest based on experimentation. To extract meaning from the imported information, structured data is parsed, whilst unstructured information is processed using biomedical language processing. Reasoning methods are used to create a knowledge network by finding relationships and common properties between different genes and assigning reliability scores that assess the likelihood of a relationship between two genes. A simple visualisation tool shows meaningful clusters of related genes and enables users of the system to explore these relationships and their shared attributes. The Hanalyzer system also combines the knowledge network with a data network generated with results from gene array expression experiments to enable users to discover whether an apparent relationship seen in an experiment matches known relationships in the literature or whether a potential novel relationship has been found.

Another example of an intelligent scientific assistant comes from the Sunfall project which is also designed to take large quantities of complex information from a source, and monitor and refine it into a form that improves researcher efficiency, although in this case, the source data is astronomical.[22] The Sunfall project is a collaborative visual analytics system for an international astrophysics experiment searching for supernovas and investigating dark matter in the universe. Sunfall uses a variety of AI techniques in order to analyse a variety of complex astronomical data sources to generate an interactive dashboard interface that enables researchers to easily see all the information in a very simple and usable format. The AI takes away the complexity of the data by automatically generating the dashboard, enabling the researchers to get the information they need, but also enabling them to focus on other research tasks rather than interpreting the data themselves.

Artificial intelligence is also often used in combination with software involved in AR, with the AI often providing interpretation of the reality onto which computer-generated elements are superimposed. For example, a project by Google coupled real-time AI with a standard optical microscope to detect cancer cells within tissue samples and improve diagnostic accuracy.[23] Images from the microscope are continually captured and fed through the AI system, which has been trained with digitised pathology slides classified with a diagnosis. The system generates inference data such as a heatmap of the likelihood of cancer for each pixel in the captured images. An augmented display, added to the optics of the microscope, displays an overlay of the outline of the heatmap area. For the researcher looking at the slide, they use the same processes that they previously used to diagnose cancer, but now additional information is available to help them identify areas of metastasis and tumours within tissue samples. Such a system can be enhanced or used in a variety of ways, for example, colour changes could be used to indicate increased or decreased confidence in the diagnosis. The technology supports the researcher with the diagnosis, but does not replace the need for their specialist skills and experience, and can benefit the scientific record by providing additional information and visual representations that can be used to justify the conclusions of the diagnosis for other users of the record or at a later time. Such a system could also easily be applied to other areas of research and diagnosis, for example, microorganism identification, measuring tumours, and in other medical imaging such as radiology and Magnetic Resonance Imaging (MRI).

Augmented reality and virtual reality can be thought of as two ends of a spectrum, with various stages of 'mixed reality' in between, depending how much of the real world is involved.[24] Augmented reality is closest to the reality end, and as described previously, in augmented reality, the real world is typically overlain with virtual elements. Augmented Virtuality (AV) is closer to the virtual end, where data from the real world, such as tracking of an object or data from a sensor, is typically used to change something in the virtual world. For example, if the user's hand is tracked using sensors, then it can be used to point, grab, or manipulate objects within the virtual world. Between these ends of the continuum is mixed reality, where some of the world is modelled digitally. In VR, the whole world is modelled, and the user is fully immersed in an imagined or replicated world. Although we might tend to think of these technologies as being relatively recent, in fact, many of the earliest forms of these technologies were developed during the 1960s. Virtual reality technology typically uses VR headsets or, less often, a multi-projection environment, known as a computer-assisted virtual environment or cave automatic virtual environment (CAVE). In a CAVE, images are projected on walls and sometimes the floor and ceiling, so that multiple people can share the same immersive environment. VR is most commonly known for games and cinematic experiences, but it is also used for learning (for example, in surgical, sporting, and dental environments), and also for the control of remote vehicles. VR is not well-used within science, and has generally been better used in fields such as archaeology, where models of architecture and archaeological sites can be produced that enable researchers and others to explore what these environments may have looked like in the past, including identifying elements of structures that may not be visible from the archaeological record itself.[25] There are, however, a number of areas where VR is used for scientific data visualisation, for example, fluid flow visualisation using a virtual wind tunnel; modelling in cosmology, astronomy, and planetary exploration; magnetohydrodynamics; molecular modelling; 3D models of Geographic Information Systems (GIS) data; and the manipulation of materials under a Scanning Electron Microscope (SEM).[26–28]

Both AR and VR have been used extensively in education and training. For example, a VR model of the solar system can be used for teaching astronomy, whilst VR systems using imagery generated from patients and coupled with haptic feedback mechanisms can be used to teach or practice surgery before an operation.[29,30] Haptic interfaces within VR systems allow users to interact with a computer system using physical movements and to receive tactile feedback. For example, in surgical training, the user may manipulate surgical instruments as they would in the operating room and receive feedback from the system on the position of the instruments and the forces needed to deform tissues that are being operated on. Another example comes from the VR and SEM example described above, where a haptic interface allows the user to 'feel' the surface they are viewing under the microscope.[28] Another example of VR use with a haptic interface comes from projects such as Project GROPE, where users can manipulate physical objects in the VR environment using force-feedback manipulation.[31] One of the developments of Project GROPE was a haptic arm device and molecular force-field evaluator that together could provide force-feedback to users engaged in modelling drug docking sites of protein molecules. Using the arm, users could orientate the drug and adjust internal twistable bonds in order to find the lowest potential energy of the docked combination.

Augmented reality, in contrast to VR, overlays digital content over the real world, either directly or indirectly. For example, users of see-through head-mounted displays or glasses, such as Google glass, see augmented content directly on the real world, whilst AR applications on smartphones may indirectly show the real world through video or photo displays. AR with see-through displays can provide an alternative to requiring a CAVE environment for sharing data visualisation by enabling a group to see the same visualisation. Using AR, each individual can see the visualisation from their own perspective and the data can even be tailored to their own specific expertise or requirements. Such technology can facilitate collaboration and multidisciplinary work in science. Because there is no requirement for the users to be located within the same room, AR can act as supporting technology for teleconferences or Web meetings to facilitate understanding of the scientific data and visualisations under discussion. Using these technologies for collaborative working can also be extended into the laboratory. For example, to provide access to certain equipment it is possible to make use of a remote laboratory. The interface to the remote laboratory can be 'augmented virtuality', where a computer-based simulation is overlain with a data stream from the real apparatus, or alternatively provide more of an AR environment by providing a video-feed augmented with data or digital objects that can be manipulated to remotely control the equipment.[32]

Although we tend to think of AR overlaying graphical content over the real world, the AR content can also be sound, video, and other data, such as GPS information. For example, audio manufacturer Bose has developed glasses called Bose Frames that use machine-generated sound rather than graphics for AR applications.[33] The product is still early in development at the time of writing, but Bose suggest applications could include music, gaming, fitness, travel, and communications, for example, providing audio to describe what the user is looking at based upon their location and head orientation. In a scientific context, AR technologies together with AI could be used to overlay previously collected data or other information on what we are viewing, or to help us to locate a specific location or object for study. We will examine more examples of how AR is used for scientific applications and how these link to the scientific record below.

Augmented reality is becoming more commonplace, particularly for use within smartphone applications. For example, the camera on a smartphone is often used to show computer-generated information superimposed on the real world. Examples include overlaying characters in games onto the user's environment in applications such as Niantic's Pokemon Go; by providing translations of text being viewed by the camera with translations applications including Google's Translate; or by rendering animations on top of faces that move with movements of the face, often used in video chat applications like Facebook's Messenger app. In many AR applications, data is being taken from the real world in order to provide useful functions. For example, Google's Lens can not only recognise and translate text, but also recognises objects and provides useful information in context. For example, using a photograph or live-view of the real world, Lens can identify objects and places and provide contextual information such as overlaying map directions for landmarks, identifying plants or animals and providing information about them, or identifying objects and finding similar items to buy on shopping sites. Considerations for any application based on smartphones are battery life and available processing power that can limit their capabilities, but both of these are increasing all the time. We will discuss some more uses of smartphones in science later in this chapter.

Augmented reality appears to be a more developed and perhaps a more suitable technology than VR for science applications beyond data visualisation. AR can be particularly useful for educational and exploratory purposes. For example, some medical students use the Microsoft HoloLens to interact and learn about functions of the body through an AR 3D projection, rather than working with cadavers.[34] AR applications in teaching can also make use of the interactive nature of AR. For example, in geology, an augmented reality sandbox was used to teach students principles of topography and surface process.[35] Users can shape topographic surfaces using sand in a sandbox, onto which is projected an elevation colour map, topographic contours, and simulated water in real time. As the shape and depth of the sand changes, the projected features change to reflect the new conditions. Students can build shapes to reflect real environments and observe how these features affect water flow dynamics in that environment. For chemical and biological applications, a number of different AR applications have been developed to enable users to explore molecular structures, including some that make use of data from protein data banks (PDB).[36,37] Currently, these applications for visualising molecular structures in 3D appear to need to convert the 2D to 3D representations in advance, but with advances in AI technology, it should be possible to perform such tasks on the fly.

An example of augmented reality in the field comes from the 'Virtual Vouchers' project, which developed an Electronic Field Guide (EFG) to support botanists identifying species during fieldwork.[38] The system makes use of a head-mounted display, a video camera, and a sensor array for tracking head movements. Users can access high quality images and additional information about physical reference 'vouchers', or type specimens, without the user needing to acquire the fragile physical voucher specimens from a museum collection. These 'virtual vouchers' enable the users to verify if a species found in the field matches an existing species or if it could be a new species. Another field-based example comes from the geological studies of boreholes, with an application that makes use of AR, maps, and a database of borehole data to visualise both borehole locations and their associated data in the field.[39]

It may not seem like these different technologies have much to do with recording science, but in fact there are a number of ways that they can interact with the scientific record. These technologies, AI, in particular, can make use of the contents of the scientific record as mentioned, and they can form part of technologies that can intelligently identify scientific activities and events, as we shall discuss in some examples later in this chapter. Scientific data, and its associated context from both the published and unpublished scientific record, are needed for AI to perform tasks that cannot be directly programmed. AI systems need to be taught using example data and the nature of that data needs to be understood before it can be used to develop these systems. New data produced by AI systems can be fed back into the system itself, but can also be used to help inform other intelligent scientific assistants. The use of AI and associated technologies such as VR and AR change the way that researchers work. Not only are researchers relying on the knowledge that they themselves have acquired from background reading and prior experience, but they may also be relying on knowledge provided by an intelligent system. For example, in the case of the Google AR microscope used for predicting cancer diagnosis, the way that the researcher works is likely to change. Although they are using the same analytical processes as before, it is inevitable that they become more likely to focus on the areas highlighted by the AI system in the pathology samples and may become reliant on the system to identify suspicious areas with mutated cells. The use of such a system may also change what the researcher records. For example, in the AR tool for botanists, the system automatically captures data about the specimen under investigation, including the selected identification. If the system is capturing this information on behalf of the user, they are less likely to be capturing details in a more manual way, such as drawing a sketch of the specimen and noting down the features that prompted the decision about species identification. These are not necessarily problems, but they are likely to have an impact on the record that is captured. Assumptions may be made about what the system is recording or that less information now needs to be recorded. It could be argued that the machine-generated elements of the research should be captured as a part of the record. For example, the augmented image of the pathology slides should form part of the record, along with the considered view of the researcher who examined the slide. It could also be argued that the underlying data used to train the software using machine learning, the algorithms used to generate the predicted diagnosis, and the results of the pathology tests all form a part of the scientific record. Expert systems typically provided reporting functions that enabled the user to be able to ask how the system came to a particular conclusion or why it only asked certain questions. It would be useful to understand how the system made a particular diagnosis or identification and to capture that information as part of the record so that it can be referred to again. With AI systems, it may actually be quite a complicated question to ask, and even though logs could be generated from the software, they would not necessarily be in a comprehensible or usable format.

These technologies can also be used to help communicate the scientific record. Although the published scientific record is mostly in written format, it can often be easier to understand scientific concepts in graphical form. AR and VR are dominated by graphical representation and can be used to enhance the presentation of scientific data through visualisations, as discussed, but also to enhance data within the published record. For example, 3D chemical and molecular structures are represented as 2D flat images within publications. However, for complicated structures such as proteins, it would be useful to be able to not only view the structures in three dimensions, but also be able to manipulate them so that they could be explored and the structure verified. A suggested solution to this problem is to include QR codes within publications next to the 2D images, so that a user can scan the code with their smartphones to display a 3D model that can be manipulated in context with the published materials using AR.[40] Although technically relatively simplistic, using similar technologies as described in the other VR and AR molecular modelling applications described previously, it does represent a step on the road to more sophisticated experiences for interacting with the published record in the future.

Open science and citizen science also have roles to play with these technologies. Without the release of open data, the datasets available for machine learning would be significantly reduced. It is no surprise that many of the VR and AR applications that do exist are making use of data from scientific repositories such as the Protein Data Banks and other biomolecular repositories. In citizen science projects, such as those hosted on the Zooniverse portal, citizen scientists are utilised to classify scientific data that can then be used to train artificial intelligence systems.[41] The technologies themselves can support both the scientific community and the public by providing new ways to disseminate scientific knowledge in a more meaningful and engaging manner.

12.3 The Internet of Things and Invisible Computing

The main premise of the 'Internet of Things' (IoT), is that not just computers are connected, but also sensors, embedded computers, and everyday objects. Sensors are key to providing the 'intelligence' of the IoT, bringing context awareness to digital tools and enabling them to be able to sense and present the information to users or to make use of the knowledge to perform automatic actions.[42] Sensors can measure a variety of physical properties, including strain and pressure, position, direction, distance, motion, speed, orientation, light, radiation, colour, temperature, solids, liquids, gases, humidity, flow, sound, images, and electricity. With computers, sensors, and other devices all connected together, they can not only connect and share data, but also interact. For example, a common use of IoT is in home automation. A 'smart' house may contain sensors that determine whether to turn the heating or lights on based on the presence of people in the house or the proximity of people returning home, for example, based on GPS location. The temperature of the heating and colour of the lights can be controlled by applications on a mobile phone. Sensors do not have to be directly connected to the Internet and there are various ways of connecting devices, sensors, and computers together at a local level, using protocols such as Bluetooth, Zigbee, and Ant+. In addition, radio-frequency identification (RFID) tags are often used to facilitate the tracking of objects in a local environment using radio waves.

Within the laboratory, more common IoT technologies include connected incubators, freezers, instruments, and other sensors for monitoring the environment. Instruments can be monitored and controlled, enabling them to be observed and controlled remotely, have the data collected and uploaded to Electronic Laboratory Notebooks (ELNs), and the usage and workflow of instruments to be assessed.[43] Equipment with sensors can also be used to determine whether they are working correctly. For example, if freezers containing important samples fail, research of significant importance and value may be lost. Using sensors or smart devices can help to provide advance warning that something is going wrong, and provide the opportunity to fix the problem, or at least save the samples. In the field, most sensors and devices are geo-enabled, and can be connected to the Internet *via* a private networking using radio, commonly used for sensor networks, or individual devices may have their own cellular data connections.[44] Sensors may even be able to connect directly through a network if they are located as part of a building, such as a smart house or many modern buildings including schools, hospitals, factories, and offices. In contrast to 'data logger' systems, data can be uploaded in real time, rather than manually uploaded to a computer. Such real-time data collection through the Internet not only provides early warning systems for natural hazards, such as floods and earthquakes, but also makes maintenance of field-based sensors more efficient.

Both the price and size of sensors and embedded microprocessors are being reduced all the time, making the development of novel devices for capturing data for scientific purposes easier and cheaper. The more devices being used to collect data, and the more different kinds of data that are being collected, the more important it is to have digital tools for storing, managing, and making sense of the data. Although we have discussed some of the issues around 'big data' specifically, even relatively small projects, such as environmental data collection within a laboratory, are going to have large quantities of data generated over time. Although some instruments may generate a single reading per experiment, other instruments, for example, those involved in flow chemistry, may produce many readings per second and thousands of readings over the course of an experiment. Some data can be discarded relatively quickly, for example, it may be unlikely that data from temperature sensors in a refrigeration unit will be needed for the long term, but temperature within the laboratory may be important to keep for longer to help assess the reason for differences between experimental results. Associating data produced from different devices and through different methods is a challenge, although timestamping is often the most effective method. Ideally, digital tools for capturing and viewing the information are automatically able to bring together all the relevant data for the benefit of the user and present it in a usable format.

For certain types of information, it can be beneficial to provide the information to the user in a format that enables them to be informed about changes in the data as they happen. An Ambient Intelligence (AmI) system uses connected sensors and devices to collect data, and based on both real-time and historical data, takes decisions that can benefit the users of a particular environment.[42] AmI is part of a move towards 'invisible computing', where software is able to make inferences from collected data to make decisions. For example, if a satellite navigation system in a car can send data as well as receive it, then information about the GPS location of the vehicle and the speed of the vehicle can be used to infer information about the local traffic conditions. The more vehicles that provide data from sensors, the more accurate the inference can become. For other users heading towards a problem area, the satellite navigation systems can be sent a message to suggest alternative routes to avoid areas with traffic jams. Artificial intelligence is commonly used in conjunction with such systems, so that they can learn from historical data that has been captured to make the appropriate inferences based upon the live data. The advantage with these technologies is that the user is provided with useful information without needing to explicitly ask the computer for it. We receive only the information we need, when we need it, to help inform our decisions. Ambient technologies are also useful for providing information about a topic in a subtle way, for example, changing colours of an ambient orb, projected ripples on the ceiling, or 'active wallpaper' where projected light moves more rapidly to indicate more activity or the changing nature of the data values.[45] Ambient devices have been used to communicate information about a variety of subjects, for example, communicating about the weather, stock prices, the activity and presence of users on a social networking system, the temperature of the room, throughput of an instrument, and so on. Rather than providing an onscreen dashboard with numbers, ambient devices can present the information in the background, requiring less attention from the user, but changes in the data used can still be noticed by the user.

An example of where an attempt has been made to create an 'invisible computing' environment in the laboratory comes from the LabScape project.[46,47] The project describes the work of cell biologists in a laboratory where they are required to record every step of the experiment in detail, but do not work entirely with a predefined protocol. Paper notebooks and mobile devices cannot be brought into the laboratory. The experiments may take a long time, so researchers work on multiple experiments at the same time, but also multiple individuals with different expertise work across the same experiments. There are therefore challenges in ensuring that accurate, timely, and thorough records are captured, and that all the appropriate information can be brought together in the record for a single experiment. The proposed solution for LabScape was to automate the recording of the experiment based on the actions of the researchers and to provide data as they needed it to help inform the next step of the experiment. Sensors are used to capture information about the actions of the researchers automatically, but also to provide them with the information they need at the correct time and location based upon the samples they are using and the actions they are performing. The framework for organising the data for the experiment was based upon the rough flow diagrams that researchers were observed to sketch of the steps of the experiment before going into the lab. The LabScape system requires that users create a plan of their experiment before entering the lab, with the steps in the experiment represented as a flow graph. The flow graph captures the appropriate information from the various sensors deployed in the laboratory, so at the end of the experiment, all of the information, no matter by whom or how it was collected, would be available from this single representation. In this particular laboratory, researchers go to a different bench to perform different tasks, and it is therefore possible to track what task in the flow graph the user is doing based on their location using infrared tags. Barcodes are used on containers, together with readers on the benches, to track individual samples. Tablet computers with touch interfaces are provided at each bench, allowing users to enter and view data, with the appropriate flow graph and next steps displayed for the sample most recently scanned at the bench. Some equipment also contained controls and sensors to allow users to modify settings or capture changes automatically. For example, pipettes were monitored by a camera that could be used to calculate exactly how much sample had been dispensed and which samples were present in which of the wells in a microarray tray, with the information projected onto the table-top for the benefit of the researcher. The system described would be particularly effective in laboratories where standard procedures are followed, a template for each sample can be completed and specific tasks are carried out at specific benches. Challenges with such a system come from experiments where there is no established protocol, if changes are required to the experiment protocols in response to an observation during the experiment, or if equipment and procedures are used in different areas of the laboratory each time. Designing an 'invisible computing' system for use in the lab may be easier for more repeatable activities, but similar systems that are more flexible have also been developed, and we will discuss some of those later in the chapter.

The capture of all of the data in one representation, in this case the flow graph, is valuable in terms of being able to locate the information again and for reuse. However, it is less clear whether the solution overcomes the problem of minimal documentation for experiments. The authors suggest that an abstract representation of the protocol steps and results is sufficient for understanding and communicating laboratory results, with users putting screenshots of the flow graph into their notebooks as a representation of the steps of the experiment instead of writing notes.[46] There is always a risk with automatic data capture that the record can be diminished if researchers adopt a 'minimal' approach to recording. Defining a plan for the experiment up front is valuable for providing guidance during the experiment and for capturing the intended protocol, but the steps taken in the experiment may vary from the intended protocol for a number of reasons, and the changes, together with the reasons, need to be recorded. There are ways to extend such a system to enable users to add these changes, and also to capture other valuable observations as they go along, but it is important to recognise that such information is important to capture in one way or another. Although changes to protocols could potentially be captured automatically, as the user moves from one bench to another or uses a different piece of sensor-connected equipment, the rationale cannot be captured automatically.

One of the most important sensors in IoT is the digital camera, which is capable of capturing images and video in a variety of resolutions and angle of view. For example, wide-angle lenses, even up to 360 degrees, can enable cameras to capture a wide view of an environment. Cameras can be used stand-alone, or they can be added to standard scientific equipment such as microscopes, telescopes, and binoculars to capture still images or videos of scientific phenomena in the laboratory and field. Capturing images digitally, separately or in conjunction with sketches, can enhance the scientific record. The capture of video, in particular, enables the 'replay' of scientific activities. For example, the 'Ami' project makes use of a variety of cameras, coupled with other sensors, to facilitate the monitoring and collection of experimental data in order to determine the cause of unexpected results in chemistry experiments.[48] Using Ami, 'Ancillary data' that would not normally be collected can be consulted after the experiment if an unexpected result was observed. Data was collected to monitor the parameters of the reaction vessel itself to determine if the conditions of the experiment varied unexpectedly. Video was also captured from two different angles in order to determine whether the experiment had failed because the wrong chemicals were used, reaction components were not added, whether some other action was mistimed, or for some other reason. One camera provided a close-up view of the reaction, whilst the second provided a wide-angled view of the whole fume cupboard to record all activity in the area, triggered by an infrared movement sensor. All the equipment and materials used in the experiments were tagged with RFID tags, and the researcher's lab coat has an RFID reader built into the sleeve, so that the system can register all equipment and materials used by the researcher into a timestamped log. The user is also able to add observations using a microphone or keyboard. The system enables the researcher to access a timeline for all experiments, containing the records of the sensor logs and videos, and to replay elements of the experiment if required. It is not clear how 'smart' the Ami system is, in terms of being able to recognise the activities of chemists, so again, like the LabScape project, it is unclear whether enough information is being captured in the record to replace the researcher's notes. However, the Ami project was seeking to provide a mechanism for troubleshooting and providing the video together with some additional data about what was done when, enables the user to go back and check if they had missed a step in the process or whether the temperature of the reaction was too high or too low. Ideally, such a system would be able to associate snippets of the video with particular actions and events to provide a 'holistic' overview of the experiment and to highlight unexpected events. For example, did another researcher approach the fume cupboard whilst the reaction was not being observed? Developments in AI could well lead to events being identifiable within video and sensor logs that could automatically be tagged within the record to provide additional insight into what happened during the experiment. It may be more of a challenge to capture the record if the researcher needs to change location during their work because the cameras and sensors are associated with a specific fume cupboard, but as the LabScape project demonstrated, providing a mechanism to identify the researcher and the materials and equipment could overcome this problem. The researcher themselves could also wear a 'life-logging' camera to capture all their own actions for later review, providing another viewpoint on the experiment that can be captured in the digital record. The Ami system enables users to add additional input into the record through a keyboard or microphone as mentioned, acknowledging that automated capture of actions and materials does not capture all of the pertinent information for the record.

12.4 Wearables

The Ami project included an 'intelligent lab coat' with a built-in RFID reader, whilst the LabScape system used small infrared tags to detect a researcher's proximity to the bench. These devices could be considered 'wearables', or smart technologies that can be worn on the body, containing sensors or other devices to give them some element of intelligence. Wearable devices are usually incorporated into clothing or worn as accessories such as watches, wrist bands, and glasses, but they can also include implants such as medical devices and microchips. There are two primary types of wearable device in extensive use, the 'smart watch' and similar activity trackers, and head-mounted displays such as Google Glass and Microsoft HoloLens, previously mentioned. Wearables often contain sensors and are usually connected or connectable in some way. For example, smart watches may contain a variety of different sensors, for example, GPS, pedometers, gyroscopes, accelerometers, and heart-rate monitors, and have the capability to connect *via* Wi-Fi or even cellular data. A key element of wearables is that they are 'invisible' in terms of interfering with the user's activities. The sensors collect and accumulate the data, and they may provide feedback that changes our behaviour, but we do not need to cause the action directly. A primary use of wearable devices is in the healthcare and fitness markets where they can be used to monitor the health and exercise activities of wearers. Such devices may have even more sophisticated sensors, enabling not only vital signs, motor activity, and sleep patterns to be monitored, but even social interactions and correlates of mental health.[49] There are numerous examples of wearable devices being used in healthcare for patient monitoring and other healthcare uses, from health devices for tracking activities at home, to the tracking and monitoring of patients within emergency departments using RFID, ultrasound, and sensor information.[50]

So far, there have been limited use of wearables for scientific applications, although both smart watches and head-mounted units have been utilised. As well as providing sensing capabilities, wearables can provide access to information without needing to leave the laboratory or operate a computer. For example, the Apple Watch has been used as a wearable ELN, providing access to notes and visualising data, but also as a substitute for common bench equipment, such as timers, calculators, and alarms.[51] In addition to the examples already mentioned, wearable technologies have been used in the laboratory environment to support researchers following experiment protocols and capturing information about the experiment as it is performed. For example, Scholl and Van Laerhoven describe using wearable devices to overcome the difficulties of using computers to capture the details of experiments within a life sciences laboratory.[52] The authors describe the use of Google Glass to display experiment protocols and provide access to previously recorded information. Google Glass also provides a way to capture the experiment with video, including observations spoken by the user. A smart watch system with an RFID reader is used to record the motions of the researcher. The motion captured by the smart watch can automatically identify the action that the researcher is carrying out, such as pipetting, and enables a log of actions to be recorded. As with the other lab-based examples discussed previously, RFID tags are used to label samples and materials, and are identified and logged by a wearable RFID reader when the researcher handles them. These captured actions and logged materials provide memory cues that can help the researcher to understand and locate particular events in the experiment record at a later time. As with some of the other examples, it is unclear whether the captured sensor logs and videos now replace the previous written record, although the authors do make it clear that this system is more suitable for reproducing an existing protocol, rather than attempting to document a new protocol. The head-mounted camera can provide a useful 'researcher's eye view' of the experiment, and the automatic capture of activities through wrist movements is an interesting area where further developments are probably needed in order to be able to automatically capture a greater range of actions. The combination of tagged equipment, as used in the LabScape and Ami projects, might help to link movements with actions more effectively. Again, if the automatically created logs replace the paper record, it is essential to ask if anything important is being lost and is the record being created in a format that can be reused by others?

Smart devices can also be used to create scientific instruments in their own right, often at relatively low cost. Such devices not only provide new opportunities for capturing data, but they are also important in opening up science in developing countries.[53] Microscopy and imaging are essential tools in science and enabling the integration of such tools into smartphones is valuable for applications in remote diagnostics, mobile health, and environment monitoring. Smartphones, often with special readers or medical attachments, can be used to test for a variety of organisms and materials, and be used for diagnosis and analysis. For example, devices for use with smartphones exist for blood analysis and cytometry, immunoassays, electrochemical biosensing, detection of bacteria and viruses, colorimetric detection, sensing of allergens, analysis of tumour samples, ultrasound imaging, monitoring of electrocardiogram rhythms, and many more.[53,54] Google Glass has also been used to create a mobile optical reader and in conjunction with a leaf holder and a multi-spectral illuminator, it can be used to measure chlorophyll for plant health much faster than direct measurement using traditional chemical extraction.[55] Smart devices can not only capture the data using such applications, but also provide ways of communicating the data and linking it to the rest of the record. For example, for the diagnosis of disease, applications could upload the data to the patient record, and for environmental monitoring, the data could be uploaded to records of individual localities to examine changes over time or in relation to local events.

12.5 Natural Interfaces

The primary ways that we interact with computers are still *via* the screen, keyboard, and mouse. These are effective interfaces for viewing and inputting information when we are working in a static location and at a desk, but not necessarily the easiest way to work with when moving around or working in the field or laboratory. Scientists who primarily work on *in silico* experimentation and with simulations and modelling, may spend all of their time in an office, but the majority of scientists are spending at least some of the time in laboratory or field where computers with monitors, keyboards, and mice are not very practical. Although laptops help to reduce the footprint of the computer and the various peripherals, they still may not be that practical for getting access to information on the fly and recording notes in particular. As mentioned previously, this can often lead to the situation where researchers capture the records of their activities at a later time and in a different location to where the research was actually carried out, bringing with it the potential for forgetting to record pertinent information. We also know that paper notebooks are considered more practical for use in both the laboratory and the field, and also that they are considered by researchers to be more flexible and easier to use than computer-based systems. One potential way to overcome these disadvantages of traditional interfaces for computers is to make use of more natural interfaces for both presenting and inputting information.

There are numerous possible ways to interact with computers, and most of us will be familiar with both older and more recent technologies, although often the first incarnations of these technologies may be much older than you might imagine. In the earliest days of computing, punched cards were used as the mechanism for providing the program code and data used to drive the computers, although punched cards were originally developed during the eighteenth century to control the patterns woven on looms, and later used to automate them. Punched cards were typically replaced by punched tape in the 1950s and 1960s. Early keyboards were essentially typewriters and allowed the input of textual and numeric information, but the first peripheral to enable the direction manipulation of graphical objects on the screen was actually a pen device. The light pen was used to touch the screen of a cathode-ray tube (CRT) monitor, with its position calculated by changes in brightness of nearby screen pixels. The pen could be used to draw, and also to move and manipulate objects, in much the same way that the mouse is used now. An interesting video clip from 1967 shows a demonstration of the light pen in use, with the presenter manipulating and creating 3D models on the screen.[56] The clip from

Tomorrow's World, a former BBC televisions series showcasing new developments in science and technology, also provides an explanation of how the light pen works. There were some problems with the light pen, for example, the detection of the light pen location was not very precise, and because it needed to be held up to the screen during use, the arms of users would quickly become fatigued, developing so-called 'gorilla arm'.[57] The mouse was developed in 1965 as a cheap replacement for light pens without the same problems of imprecision and need to hold the arm up to the screen.

Pens can provide a more natural interface for writing and drawing than using a keyboard and mouse, and this is reflected in the perceived ease of use of paper notebooks. There are a number of technology solutions that enable the capture of text and images using pen-like devices. Digital pens, such as the Anoto pen, allow the user to both write and draw in ink, but also capture the information in digital format in the background. The data may be transmitted to a computer or stored on the device in order to be uploaded at a later time, a more practical solution for working in the field. There are a few drawbacks to digital pens. They are still quite expensive to buy and replace if they break or are lost, and they typically rely on surfaces with specific patterns on them in order to capture the changing position of the pen, limiting the surfaces on which notes can be captured. The majority of smartphones and tablets have drawing or notetaking applications where notes and sketches can be captured directly using a stylus. A problem with the majority of these devices, currently at least, is a lack of precision of drawing and resolution, so it can be difficult to capture writing neatly and naturally. Graphics tablets can be used as an alternative to a mouse by using a pen on a separate surface to capture digital drawings and text, although they are most widely used for computer art and graphics, rather than as note-taking devices. Graphics tablets are currently still better than mobile tablets in terms of the accuracy and resolution of content capture for handwritten and hand drawn materials. However, graphics tablets are not very practical for use in either the laboratory or field because of their need to be connected to a computer, and they usually require their own power connection or a powered stylus to function. There are two styles of graphics tablets, the cheaper alternative is as a peripheral device where the user draws on a surface and the marks appear on the main screen, or where the graphics tablet itself is a screen and marks can be drawn over the content displayed on the screen. Technology for tablets is improving all the time; and some of the high-end graphics tablets currently available provide an experience with the same feel and accuracy as paper. Some manufacturers of these high-end graphics tablets have developed solutions for mobile platforms, for example, the Wacom MobileStudio Pro.[58] These technologies provide all the benefits of portability of mobile devices with the accuracy of capture in the high-end graphics tablets. Although these systems are currently very expensive and aimed primarily at the digital arts and design market, as the technology improves, it is likely that ordinary mobile technologies will begin to benefit from these advances. It could be argued that when these 'feels like paper' mobile technologies become ubiquitous, it will be harder for researchers to resist the call of digital tools for supporting all aspects of their research.

As described in Chapter 9, there are a number of hybrid electronic notebooks that have been developed to make use of digital pens or graphics tablets to capture writing and drawings as images. Hybrid systems can be relatively smart or dumb. For example, if notes are captured using a digital pen together with software that is able to recognise characters and therefore extract meaning from the notes being written, then the system is relatively smart. Information that is captured may be viewable as an image, but there will also be a textual version that can be mined and searched by the user and other software systems when the files are stored. Particularly smart systems may be context aware and may be able to tell whether you have written text, drawn a reaction scheme, or sketched a daisy. On the other hand, if the digital pen and software only capture an image of what is written without the capability to interpret the image or input, then the system is relatively 'dumb'. I use the term 'dumb' in the same way as 'dumb terminal', a term especially used in the 1960s and 1970s to describe video terminals (typically a display and keyboard) used to connect multiple users to a main computer, but with limited processing power of its own. Hybrid systems, described in Chapter 9, are smarter, recognising basic mark-up and enabling some interactivity between handwritten notes and sketches, as well as providing ways to capture metadata.[59,60] The key is being able to extract the textual content and make sense of content that has been captured in order to provide additional value to users, and to ensure that the record is available for reuse in a usable and meaningful fashion in the future.

The pen interface is included in this section because writing and drawing are certainly more natural than using a mouse and a keyboard. However, there are many other more natural, and potentially unconventional, interfaces for interacting with computers as shown in Figure 12.2. The examples show interfaces both for receiving machine-generated information and for producing information that can be used to control the computer. Vision is the most widely used and familiar interface for receiving information from the computer, usually through the use of monitors and screen displays. As mentioned previously, vision is also an important, although not exclusive, part of virtual and augmented reality. Eye-tracking and gaze-tracking can be used as an input mechanism to the computer, although it has primarily been used to select objects on a screen to control the computer for users with physical and motor disabilities such as paralysis, missing limbs, and cerebral palsy. Eye-tracking and gaze-tracking, along with head-tracking, are also used within virtual reality environments to determine where the user is looking. Although not directly useful for recording information, these could allow the user to interact with a computer hands free to make selections and control the computer to display particular information. The user could

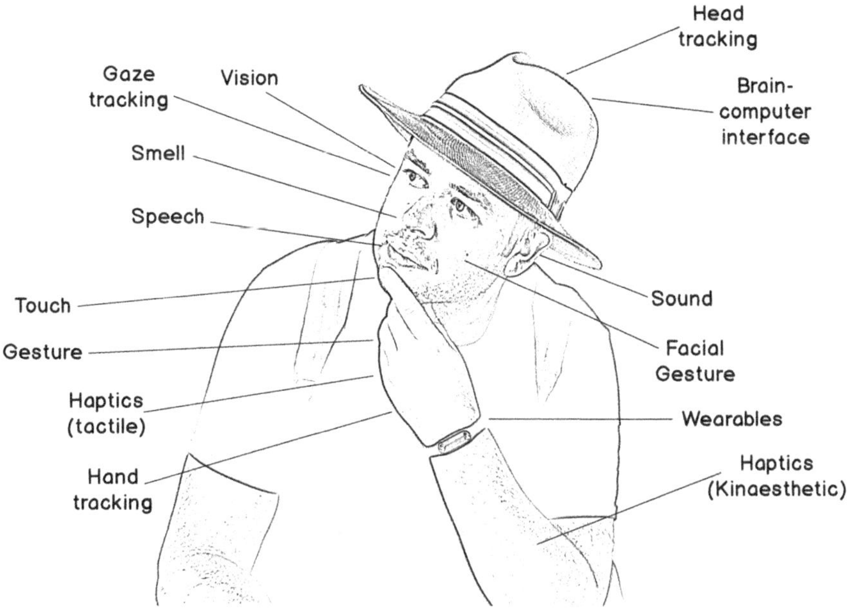

Figure 12.2 Examples of natural interfaces for information input and output between a human and a computer system.

also be detected to be looking at something other than the computer. For example, in a set-up such as the Ami fume cupboard, where the various components of the experiment and location of the researcher is known, head-tracking could be used to determine whether the user would have noticed an event that occurred during the experiment, such as a change in the state of the reaction, based upon where they were looking.

Brain-computer interfaces are a technology related to eye and gaze-tracking systems and are used for similar purposes. Brain-computer interfaces are typically electroencephalogram (EEG)-based interfaces that most frequently are used to provide alphabetic writing for people with serious disabilities. The EEG signal can be acquired from the brain by using sensors inside or outside of the scalp, with outside being much less invasive though the use of wearable devices.[61] Systems have also been developed that enable users with damage to the spinal cord, muscles, or nerves to help them control robotic prosthetic limbs and other systems or devices such as a computer cursor and a television.[62] Brain-computer interfaces may actually be faster than eye-tracking based spellers at around 12 words per minute compared with five to ten words per minute.[63] There are a number of conferences and hackathon events dedicated to researching new ideas in brain-computer interfaces and researchers have developed systems for controlling robot arms, remote control cars, controlling computers, playing games, recording memories, and capturing emotional state.[64,65] In theory, such technologies in the future could allow a user of the system to interact with a computer to display information or to record information at will without needing to write anything down, speak, or touch the computer. Novel information that the user themselves are not even aware of could be captured, such as changes in their emotional state or subconscious attention or connections. However, such applications are probably a long way into the future, and although it is possible to envisage ways in which gaze-tracking and brain-computer interface technology could be used within an environment where scientific activity is being carried out, for able-bodied users, there are probably more natural ways to provide input to the computer.

After vision, sound is probably the most important sense for receiving information from the computer. Sound typically comes from speakers, although headphones can also be useful in both noisy and quiet environments. Sound is commonly used as a way of communicating information in computer systems and other digital tools, for example, sound is used for warning signals, medical monitoring instruments, sonar, Geiger counters, clocks, and is also a useful way to provide feedback during an activity. For example, sound can be used as a warning that a sensor has been activated, such as a trap in the field or a thermometer in the laboratory, or by providing a 'heartbeat' to indicate that an instrument is working correctly. Speech can be an alternative way to receive information from the computer rather than reading from the screen and can be useful when carrying out activities because it does not require the user to be looking in a particular direction to receive the information. Speech has in the last few years become a reasonably reliable method of providing input to the computer. Virtual assistants, such as Amazon's Alexa, Apple's Siri, and Microsoft's Cortana all are able to interpret human speech and respond to commands with synthesised speech, providing a natural way to interact with a computer or other device. Such devices can be used to retrieve information from the computer, particularly small amounts of information such as a description or value for a particular item, for example, details of the weather or to look up the meaning of a word. Voice assistants can also be asked to perform tasks such as creating appointments, ringing telephone numbers, and sending messages.

At the current time, the range of commands and interactions available for voice assistants is relatively restricted, although with Amazon's Alexa, it is possible to create your own syntax for commands and link them to Web services to generate novel actions. For example, the Talk2Lab system developed at the University of Southampton, makes use of Amazon's Alexa as part of a smart laboratory.[66] The laboratory uses a variety of sensors which monitor elements of the environment in a physics laser laboratory using X-ray radiation for imaging and spectroscopy. The sensors include temperature, water flow, power consumption, and a number of network-enabled cameras. An Amazon Echo Dot smart speaker device allows the researchers in the laboratory to query the real-time values from any of the sensors using voice commands to Alexa. Users can also ask questions about the status of instruments and the cameras can be used in conjunction with voice commands to help with remote alignment of lasers. There are also potential ways to automate the capture of experiment information in such an environment. For example, a system could enable the user to ask the virtual assistant to save photos or videos from the cameras or data from instruments into an electronic notebook or data file. Data from different sources could be configured to be captured in a single location with a voice command from the user. For example, rather than manually having to capture or collate data from every sensor and every camera, asking the virtual assistant to 'save environment' at a significant point could command the system to automatically extract all the information and save it in the electronic notebook. Alternatively, all the data for the entire experiment could be recorded by asking the assistant to 'record imaging experiment'. A challenge comes with managing multiple users using an experimental set-up or room at the same time. In Talk2Lab, this problem is overcome by adding the name of the user into the command and configuring the assistant to save information to the correct notebook based on the user's name.

Virtual assistants like Alexa are particularly effective at saving this kind of data automatically, but less effective for having 'a conversation' with the user that might be better for actually recording an experiment in a natural way. We can effectively capture the data from the instruments and sensors, and take a photograph, but it has been harder to ask the assistant to record a stream of words as you narrate your experiment activity as you would within the paper notebook. This is because the virtual assistant works with a 'command-response' or 'question-response' style of interaction and is listening for commands during any speech. This is probably not an insurmountable obstacle, and there is already at the time of writing, an update that allows the recording of short notes. The ability to be able to capture the record of the experiment hands free by using a microphone is a feature requested by many researchers for electronic notebooks and has also been incorporated into some of the electronic notebooks currently available. The major challenge with speech captured as audio is the same as text captured as images: the audio needs to be converted into textual form in order for the information to be captured in a way that can be searched and easily reused. Sound files are also typically quite large, although they can be converted into smaller compressed formats if long-term storage is required. Using appropriate AI software or Web services, voice can be converted to text on the fly or processed later. A process of verifying that the translation from speech to text is correct is likely to be needed, but with developments in AI, accuracy of the software translation is likely to improve significantly over time, so long as the audio quality is sufficient.

Speech is a natural way to converse with others, but some find it difficult to talk in order to narrate their activities and find it unnatural to talk to a device, although most overcome the initial awkwardness quite quickly. Competing sound in the environment can be a challenge for recording sound or interacting with a virtual assistant. Virtual assistants can struggle to understand spoken words in a noisy environment, and loud environments may also impact on the quality of recorded sound. This is particularly a problem for interacting with a static device such as a microphone attached to a computer, smartphone, or an Echo Dot. In a quiet or multiperson environment, it may also be inappropriate to be shouting or speaking loudly in order to be heard by a device. A potential solution to both these problems is to make use of a Bluetooth or wireless microphone system that can capture spoken words effectively whilst cutting out background sound. Headset microphones can be relatively unobtrusive whilst enabling the user to move around freely and continuing to work hands free.

As previously mentioned, wearables such as Google Glass can be used to capture both visual information and audio. Although primarily used for voice commands, the audio can be used to capture speech and be processed in the background to create textual notes in digital format. For example, in some health centres, doctors use Google Glass during consultations, enabling them to give full attention to the patient and ask pertinent questions, whilst the captured conversation is converted into digital notes in the background.[67] Not only does the system improve the experience for the patients, but also significantly reduces the amount of time the doctors spend typing up notes, doubling the amount of time they spend with patients.

Another method of providing input into computer systems, whilst being able to move freely, is through the use of gestures. There are a variety of ways of capturing and using gestures for control and input. Most commonly, we think of 'touch gestures' that we use with smartphones and tablet devices, although we may not actually be thinking about them as we use them. Touch interfaces enable us to interact with objects on the screen. We 'tap' to press or select a control or link on the screen in much the same way that we use a mouse click. We drag objects around the screen, we slide fingers up and down to scroll documents, we make things larger or smaller with pinching, we swipe to change pages, and so on. Some of these actions are more natural and intuitive than others. As mentioned in Chapter 8, on many smart devices these touch controls have become complex and confusing. It may not be obvious what effect different touch gestures may have without the user explicitly learning the meaning, and the same gestures may have different behaviours between different applications. Examples of touch gestures that may be confusing include double taps, long taps, multi-fingered taps and drags, and even multiple fingers and thumb moving in different directions as used on the Apple multi-touch trackpad.[68]

There are other ways to capture gestures from users that make use of physical movement without the need to touch a surface of the computer. A number of prototypes of scientific applications have made use of devices such as the Nintendo Wii remote controller (Wiimote) and Microsoft's Kinect to capture gestures made by users to control computer systems. The Wiimote controller contains a number of different sensors, including accelerometers and gyroscopes, that enable its position and orientation to be deduced, along with an infrared optical sensor, which is used to calculate the position of the Wimote relative to a 'sensor bar' using triangulation. The Kinect system requires no controller and instead used a combination of camera, motion detection sensors, a depth sensor, and a microphone, enabling the system to capture motion, facial recognition, and speech recognition. An example of a scientific application using both of these systems is used to provide ways for users to navigate 3D geographical maps using gestures.[69] In the Wiimote version, the orientation of the remote is used to control whether the view point moves forward, back, left, right, and moves the altitude up and down to zoom into the map. In the Kinect version, users hold their arms in different positions to control a virtual paper plane on the screen. For example, to move forward, the user can put their arms out forwards, or to move upwards, they can put their arms over their heads. In the Kinect example, the gestures are quite large, but gestures with the Kinect can also be subtler and capture detailed hand movements, enabling more complex interaction and capabilities. For example, applications using the Kinect include translation of sign language and enabling surgeons to view and manipulate medical images on the screen whilst working in the operating theatre, an environment where they are unable to touch computer peripherals for hygiene reasons.[70,71] Using AI, some software and camera systems are able to accurately recognise facial gestures.[72] There are other systems that can be used for motion tracking. For example, in animation, reflective markers and cameras are commonly used to capture motion, but there are also body sensor systems, such as Notch sensors, that enable capture of 3D motion to a smart device.[73] Gesture-based interfaces have been used for the development of interfaces to virtual laboratories, for example, by using the Kinect to detect hand movements and voice commands, users can perform experiments by grasping virtual laboratory equipment and mixing virtual materials, with the application providing a simulation of the reactions and their effects.[74]

Gesture capture systems could also be used by multiple people in the same environment, for example, to explore and manipulate data in a meeting room or laboratory. They are also useful in VR environments, where there are limited ways for users to interact with the system; gesture recognition can be used as an effective way to interact with the virtual environment. A number of wearable gloves and suits have been developed that track movements and orientations for use in games and virtual reality environments. For example, the VPL DataGlove, which made use of fibre optic cables and sensors to track movements, and the later, cheaper PowerGlove from Mattel that was used as a controller for the Nintendo Entertainment System (NES).[75] The PowerGlove also had a game controller device built into the glove with a variety of buttons for controlling the NES. VPL also developed the Research DataSuit, which was a full-body outfit with sensors that measured the movement of various body parts. Modern equivalents of the gloves are still available for use in VR environments and for other applications, including the SpaceGlove for VR games, CyberGlove used to translate British Sign Language into English, and Talking glove used to convert American Sign Language into synthesised speech.[76]

Gesture-based systems like those described above could be used in a similar way to virtual assistants and voice control, providing a way to interact with digital tools without speech, so users can interact with the system hands free without disturbing others. As with the other technologies described, there are currently a number of challenges with gesture capture. For some systems, such as the Kinect, the user will need to be within a certain range of the computer or sensing device, and variations in lighting, camera angle, and distance from the device can impact on the ability of the system to recognise the gestures. There are also likely to be variations from gesture-to-gesture and person-to-person that need to be handled by the system, in the same way that voice recognition systems need to be able to deal with differences in voice, pronunciations, and accent. Gestures may also be difficult to learn and there are cultural differences in meaning. For example, simple gestures such as head nodding and shaking are not universal in their meaning and there is therefore a risk that the intended meaning is misinterpreted by a system.[77–79]

An extension to touch and gesture-based interfaces are haptic interfaces. These interfaces provide feedback in the form of mechanical signals, which provide physical information about virtual objects enabling users to understand certain properties of the virtual object. For example, we discussed the use of haptic interfaces when we talked about project GROPE, where users received force feedback from the controls to guide their manipulation of molecules, and also from the surgical instruments used in the VR surgical training where users could feel the amount of pressure required to use the instruments and manipulate human tissue. There are two primary forms of haptic interfaces, tactile and kinaesthetic. Tactile refers to our sense of touch, whilst kinaesthetic refers to our awareness of our body position and movement. The two examples provided above are kinaesthetic, where the user applies some force to the interface device in order to provide input to the computer to control the VR environment, whilst the system in turn provides feedback to the user through simulating the force required to make the movement. Another commonly used example to explain kinaesthetic interfaces is the steering wheel accessory for some racing games. With the wheel, the user can turn it left or right to a greater extent and the game can in turn provide force feedback to reflect the feel of driving the car. Other examples include medical simulators to train students in tasks such as palpation using a virtual finger and needle insertion, and the haptic exploration of computationally modelled fluids.[80] Kinaesthetic devices are also often used as mechanisms to provide feedback for the user when remotely operating equipment or driving robotic devices. For example, the system can give feedback to the user, such as the appropriate resistance if they have reached the limit at how far some part of the equipment can turn or if the robot has encountered an obstacle. There are also force-feedback gloves that can be used to simulate the grasping of objects within VR.[81] Tactile interfaces on the other hand, deal with more delicate sensations such as pressure, vibrations, texture, temperature, softness, wetness, edges, and friction. Tactile interfaces can provide ways for users to experience the properties of surfaces, for example, the topography of a surface could be modelled using a tactile interface. Haptic interfaces can also enable humans to provide input into the computer in mechanical form. Within science, the major applications of haptic interfaces are currently in data visualisation and exploration, particularly within VR environments or the remote control of instruments.

A common method of providing physical representations of digital objects in VR environments is to use actual physical objects in the form of 3D-printed models. Although this is a relatively cheap method of providing a physical representation to use in a VR environment, objects need to be printed in advance, and rapid prototyping is still relatively slow. Other kinds of physical interfaces can be used to represent virtual information or to provide input into the computer. For example, the 'Geographic Illuminating Clay' interface is used in landscape design and analysis by simulating dynamic forces by projecting computational representations directly on to a model landscape made of clay.[82] The model can be manipulated and the projections updated, similar to the augmented reality sandbox for geographical features described previously, but the landscape also provides the input and output for GIS simulations.

Physical interfaces in the real world are often in the form of buttons and controls that are more natural to use than a keyboard and mouse. For example, a dial control could be provided to change the volume of sound from a computer system, as a more familiar control than pressing keys on the keyboard, and buttons can be used to operate instruments. Objects in the physical world can also be tracked and their positions recorded as input into a system. Within a VR system, changing the orientation of a physical model could change the orientation of a virtual representation of the same model, or changing the orientation of a gloved hand could be represented by a virtual model of a hand in the VR environment. The Sensetable interface uses electromagnets to track the positions and orientations of multiple wireless objects on a table-top display surface.[83] The objects, known as 'pucks', also have dials and state modifiers so that they can be used to provide additional input, and the system includes a projector over the table-top that allows data bound to the different objects to be projected onto the table. One of the applications developed for Sensetable was a chemistry application for exploring chemical reactions. The pucks on the table-top are mapped to atoms and molecules, and reactions occur when the molecules are brought into close proximity, showing the resulting products of the reactions. Dials on the pucks can be used to modify the electrical charge of the atoms or molecules. The system can be used for other kinds of data visualisation, such as attaching data of interest to different pucks and moving them around or changing their status in order to see the effects. For example, attaching data about rabbit and fox populations to different pucks and the dials can be used to change the population size of one to see the effect on the other. The simulation is updated as the user moves the pucks or changes their status. Such systems could be valuable for use in data visualisation and may be more natural than manipulating objects in VR environments, for example. There is also the potential to capture such visualisations and explorations within the record. The changes in both position and values for each 'puck' and the resulting effects can be captured. However, without the explanation as to why the researcher changed the positions or values in a particular way cannot be known, although it is likely that they will be influenced by the observed effects. Providing a way to input the rationale for the explorations and what it means to the researcher would be incredibly useful in the record, as both memory and explanation for the benefit of others. Input by voice may be valuable for enabling the researcher using the system to provide a commentary, but a touch table-top could perhaps also enable the capture of written notes and annotations around the position of the pucks.

There are other ways in which dealing with physical objects can be more natural than using keyboards and mice, or virtual objects. For example, when we interact with a paper notebook, we make use of the physical pen as described above. There are also, however, some affordances of paper itself as an interface that could be considered in the design of digital interfaces. For example, the developers of the Smart Tea ELN system, described in Chapter 9, found that chemists spent a large amount of time flipping through their paper notebooks to locate previous experiments and then holding the current experiment and previous experiment pages with their fingers, and rolling backwards and forwards between them so they could compare the notes.[84,85] Even though on tablet devices you can create a 'notebook' with 'flippable' pages, it would be difficult to create an interface where you could easily flip between two experiments to view them at the same time. The Smart Tea team attempted to emulate this digitally by using a drop-down 'palette' within the ELN showing a list of experiments and the steps of the current experiment. A physical interface solution to the problem could be more effective making use of the affordances of paper for flipping. For example, a potential physical solution to the problem could be the use of 'electronic paper' or a PyzoFlex interface. PyzoFlex is a sensing device based on a pyro- and piezoelectric sensor matrix printed on a flexible substrate.[86] The device is available in the form of a transparent tablet cover onto which data can be displayed and written information captured. The flexibility enables it to be peeled back and forwards to reveal information in a very natural way, for example, showing different layers in maps, providing filters for information, or potentially to show the current experiment, whilst the underlying tablet screen shows selected pages from the notebook or related background material. Using AI, it could even be possible to highlight areas of similarity between the two experiments so that the user can easily see differences in quantities, materials, steps between the experiments, or results captured in order to facilitate the activity of 'comparing two experiments'.

Immersive data visualisation in virtual environments and smart environments, such as smart laboratories and smart meeting rooms, often make use of many of these different natural interfaces to support users in their tasks. Using interaction technologies such as pens, touch, and gesture controls provides more natural interfaces to work with data, but also for facilitating collaboration and enabling people to work together better to understand data and research.[87] Immersive technologies also provide ways for collaborators to carry out research together, by exploring and manipulating simulations and carrying out novel *in silico* experiments.[88] It could be argued that capturing the processes of how research decisions were made, how conclusions were reached from data, and how future directions of research were discussed in meetings should be a part of the research record. Such events are important for understanding research paths and can provide a memory for the reason for actions or decisions, but also insight for others who were not present during the activities.

12.6 Intelligent Scientific Assistants and Smart Environments

The majority of the example uses of these technologies within the laboratory make use of similar technologies for similar tasks, suggesting that there is a certain amount of consistency of requirements for lab-based scientists, even though their ways of working may be quite different. The advantage of networked systems making use of relatively low-cost sensors, such as those described, is that they can be implemented incrementally, so the advances of the different technologies can be gained as the opportunity and budget arises. In the commercial market, these various different technologies are starting to be brought together in the form of virtual assistants such as Alexa and Siri, but also in smart robotic devices. For example, Anki's Vector robot, shown in Figure 12.3, is a cross between a virtual assistant and robotic cyber-pet coupled with a variety of sensors and AI software. Vector is internet-connected and voice activated, so can be used to ask questions and be given commands in the same way as the voice assistants. Vector is also able to do a variety of things autonomously, for example, it can move around, plan paths, avoid obstacles and edges, and interact with objects to move them, or even pick them up and move them around. It also knows when it is running low on power and can return automatically to its 'home' to charge. Vector has a video camera, enabling it to take pictures on request, but the AI also provides face and emotion recognition enabling the robot to identify the presence of humans and pets, but also specifically to remember and name faces it recognises.

Figure 12.3 Anki Vector robot, intelligent scientific assistant of the future?

Vector will also be programmable, enabling users to extend the capability and write their own interactions. An early Python SDK is available at the time of writing, and the predecessor to Vector, called Cosmo, also has a block-programming interface that opens up the creation of applications to non-programmers.[89] Vector has enough sophisticated technology that something quite like it could be just a few steps away from being a truly intelligent assistant in the laboratory. The main missing functionality of Vector as an assistant in the ways previously described is a way to display complex data, although there is a small screen that can display text, numbers, graphics, and animations. Given the examples discussed, a robotic assistant that can move around could take photos, videos, control devices, act as an RFID reader, capture audio commentary, and provide information about the status of equipment or the experiment, remind the user about the next step in the procedure, and so on. That Vector is programmed with a little personality may also help with the concept of personal assistant, but equally it would be easy to ignore the technology as just a novelty unless it provided real value to the scientist. Within the laboratory, if we want to encourage users to capture the whole of their experiments in digital form, rather than a mix of digital and paper, then there is certainly scope for an intelligent assistant that can be operated hands-free to help.

In the field, the alternative robotic assistant may take the form of autonomous drones or unmanned aerial vehicles (UAVs). Drones can perform a number of useful functions for fieldwork, such as collecting high quality images from the air, remote sensing applications, transporting field supplies, and providing communication networks. As an example, drones have been utilised by field biologists to monitor and protect endangered animals, for example, protecting rhinos and elephants from poachers and monitoring tigers, whales, and orangutans in various locations across the globe.[90,91] Drones can come in a variety of scales and coupled with sophisticated cameras and sensors can navigate autonomously in both human and natural environments. At a less high-tech level, many of the various 'smart' digital devices and IoT elements such as RFID tags, sensors, and cameras that were used in the smart laboratory examples, along with digital pens, certainly have potential value in field studies and are easier to use than computers in most field environments.

Not only laboratories can be made 'smart' with sensors, digital devices, and natural interfaces. Meeting rooms can also be furnished with a variety of technologies to make them 'smart spaces' that can be used to both access the scientific record and record the discussions that contribute to it. A 'smart space' is usually considered a place where people come together to collaborate, share knowledge, and engage in activities such as exploring ideas, solving problems, and making decisions, although the term is also often applied to IoT enabled environments, such as smart houses, smart cars, and smart buildings such as airports and hospitals.[92] Smart spaces for collaboration may be physical locations such as research laboratories, meeting rooms, museums, galleries, and classrooms, but they can also be virtual spaces where people can come together in an online environment whilst physically remote. Many organisations have 'smart conference rooms' with technology such as smart tables and walls for the display of media and data, and various teleconferencing or videoconferencing facilities to enable the participation of remote team members. Interactive whiteboards may be used to capture in digital format drawings or points recorded as part of the meeting. In both physical and virtual meeting environments, technologies may be used to take a recording of the meeting for later review by participants or others who were unable to attend the meeting. Recordings may include audio and video of presenters and participants. For teleconferences and videoconferences, capturing audio and video is relatively straightforward, enabling all contributions and contributors to be recorded, but it can be much harder in physical spaces to make this recording for a number of reasons. Participants may be unwilling to give permission for audio to be recorded, and unless microphones are placed in front of each participant, the sound quality may be poor. Capturing a video record of the meeting can be better than an audio recording, because non-verbal communication can often be as influential as words spoken.[93] However, video recording is even more intrusive and for a full record would ideally require video footage of each participant, although some systems make use of robotic cameras that can capture the individual speaking.

Smart environments, both physical and online, have the potential to support collaborative working, particularly with heterogeneous participants, for example, where the members of the group are from different background disciplines or where participants do not share the same first language. Technologies can provide each participant with information based on their needs, such as the examples provided with the augmented reality systems, where a different perspective on shared data being visualised can be provided based on the users' specific knowledge or area of interest. There are also AI technologies that can enable real-time translation of speech and text into different languages, enabling participants to converse in the language they are most comfortable with. For example, Skype Translator which claims to translate speech in ten languages and Google's Translate app claims to translate speech in 32 different languages.[94,95] The main challenge with attempting to capture the record of meetings using technology is the same as we discussed with speech and video as natural interfaces for recording, it is necessary to be able to navigate the recording and extract the information of importance. In meetings, the task of capturing the salient information may fall to a participant or dedicated note-taker, but there are potentially problems with points being missed or the information recorded having an element of editorial bias.[93] A technology-based system can overcome these problems by capturing the whole record of the meeting exactly and then converting the captured speech to a textual format. Automatically identifying the important information is a challenge, but word analysis and sentiment analysis can potentially be used to identify topics of conversation, decisions, actions, and areas of agreement and disagreement that can be used to navigate the content of the meeting. The record of the meeting could be linked to other aspects of the record, such as data and research records discussed and viewed during the meeting. New research records created subsequently could also be linked to the original meeting where they were originally discussed or proposed.

12.7 Conclusion

In this chapter, we have discussed a number of areas of technology that can support scientists in the creation, capture, and reuse of research and data. Although many of these technologies are yet to be widely used, and many of the examples presented are prototype applications, the capabilities of AI to drive more capable and useful digital tools in the future cannot be underestimated. The miniaturisation of technology and the ability to connect devices together is extremely valuable in creating new instruments that can function in a variety of locations and contexts to capture novel data, including data about the activities of researchers. The data captured by such systems can be further used to develop new smart and autonomous systems.

A major challenge with any technology is usability, as we have discussed before. If the technology is difficult to use or the user cannot see the benefit of using it, then they will avoid it. If the technology takes too much time to learn, diverts the user from their preferred workflow, and does not provide the promised benefits, then users will fail to use it, and in turn the technology will be wasted. Many of the technologies currently used in smart meeting rooms, for example, can be difficult to use and often requires an expert or 'wizard' to be present in order to deal with set-up and to facilitate use during meetings.[96] Without the wizard, the technology cannot be used, and consequently opportunities to utilise and reuse data or to capture important information are lost. The move towards 'invisible technologies' and more natural interfaces can help with some of these challenges, but the smart technologies themselves have to be easy to use and set-up. A real understanding of what researchers need and how to compliment and support the activities of researchers is required in order to develop genuinely useful digital tools for facilitating science and capturing the research record. Once we have data in digital form, we can search for it, repurpose it, use it to make other technology smarter, and share it. The biggest challenge is creating digital tools that work as well as paper notebooks as a personal memory. Not just capturing the information inside our heads, but ensuring that the right information is collected, without distortion or parts missing, and in a format that can be used by machines to provide value both for us as individual researchers and also for the rest of the scientific community.

References

1. J. Gemmell, G. Bell and R. Lueder, *Commun. ACM*, 2006, **49**(1), 88.
2. M. Ware and M. Mabe, *The STM Report: An Overview of Scientific and Scholarly Journal Publishing*, International Association of Scientific, Technical and Medical Publishers, The Hague, 2015.
3. National Academies of Sciences, Engineering, and Medicine, *Fostering Integrity in Research*, National Academies Press, Washington, 2017.
4. Available at https://home.cern/science/accelerators/large-hadron-collider (accessed Dec 2018).
5. D. Clery, *Science*, 2009, **324**(5932), 1247.
6. Available at https://panstarrs.stsci.edu (accessed Dec 2018).
7. Available at https://www.lsst.org/about (accessed Dec 2018).
8. Available at https://www.genome.gov/12011238/an-overview-of-the-human-genome-project/ (accessed Dec 2018).
9. Available at http://www.earthscope.org (accessed Dec 2018).
10. Available at https://oceanobservatories.org (accessed Dec 2018).
11. B. G. Sutherland and E. A. Feigenbaum, *Mach. Intell.*, 1969, **4**, 209.
12. E. H. Shortliffe, *Computer-based Medical Consultations: MYCIN*, Elsevier, Oxford, 1976.
13. M. I. Jordan and T. M. Mitchell, *Science*, 2015, **349**(6245), 255.
14. Y. LeCun, Y. Bengio and G. Hinton, *Nature*, 2015, **521**, 436.

15. T. Ciodaro, D. Deva, J. M. de Seixas and D. Damazio, *J. Phys. Conf. Ser.*, 2012, **368**, 012030.
16. M. Helmstaedter, K. L. Briggman, S. C. Turaga, V. Jain, H. S. Seung and W. Denk, *Nature*, 2013, **500**, 168.
17. H. Y. Xiong, B. Alipanahi, L. J. Lee, H. Bretschneider, D. Merico, R. K. Yuen, Y. Hua, S. Gueroussov, H. S. Najafabadi, T. R. Hughes, Q. Morris, Y. Barash, A. R. Krainer, N. Jojic, S. W. Scherer, B. J. Blencowe and B. J. Frey, *Science*, 2015, **347**(6218), 1254806.
18. J. Ma, R. P. Sheridan, A. Liaw, G. E. Dahl and V. Svetnik, *J. Chem. Inf. Model.*, 2015, **55**(2), 263.
19. L. Zhang, J. Tan, D. Han and H. Zhu, *Drug Discovery Today*, 2017, **22**(11), 1680.
20. Y. Gil, M. Greaves, J. Hendler and H. Hirsh, *Science*, 2014, **346**(6206), 171.
21. S. M. Leach, H. Tipney, W. Feng, W. A. Baumgartner, P. Kasliwal, R. P. Schuyler, T. Williams, R. A. Spritz and L. Hunter, *PLoS Comput. Biol.*, 2009, **5**(3), e1000215.
22. R. Aragon, S. J. Bailey, S. Poon, K. Runge and R. C. Thomas, *J. Phys. Conf. Ser.*, 2008, **125**(1), 012091.
23. P. C. Chen, K. Gadepalli, R. MacDonald, Y. Liu, K. Nagpal, T. Kohlberger, J. Dean, G. S. Corrado, J. D. Hipp and M. C. Stumpe, 2018, Available at https://arxiv.org/abs/1812.00825.
24. P. Milgram, H. Takemura, A. Utsumi and F. Kishino, in *Proceedings of SPIE: Telemanipulator and Telepresence Technologies*, 1995, **2351**, p. 282.
25. B. Frischeretal, in *Virtual Reality in Archaeology*, ed. J. A. Barcelo, M. Forte and D. Sanders, BAR International Series 843, Oxford, 2000, p. 155.
26. S. Bryson, *Commun. ACM*, 1996, **39**(5), 62.
27. A. Huang and H. Lin, *Comput. Geosci.*, 1999, **25**, 1167.
28. M. Guthold, M. R. Falvo, W. G. Matthews, S. Paulson, S. Washburn, D. A. Erie, R. Superfine, F. P. Brooks and R. M. Taylor, *IEEE/ASME Trans. Mechatronics*, 2000, **5**(2), 189.
29. R. Mintz, S. Litvak and Y. Yair, *J. Comput. Math. Sci. Teach.*, 2001, **20**(3), 293.
30. M. A. Spicer and M. L. Apuzzo, *Neurosurgery*, 2003, **52**(3), 489.
31. F. P. Brooks, M. Ouh-Young, J. J. Batter and P. J. Kilpatrick, *Comput. Graph.*, 1990, **24**(4), 177.
32. C. Onime and O. Abiona, *Int. J. Commun. Netw. Syst. Sci.*, 2016, **9**(4), 67.
33. Available at https://developer.bose.com/content/tuning-augmented-reality (accessed Dec 2018).
34. D. Metcalf, S. T. Milliard, M. Gomez and M. Schwartz, *IEEE Pulse*, 2016, 7(5), 35.
35. T. L. Woods, S. Reed, S. Hsi, J. A. Woods and M. R. Woods, *J. Geosci. Edu.*, 2016, **64**, 199.
36. P. Safadel and D. White, *TechTrends*, 2018, Available at https://link.springer.com/article/10.1007/s11528-018-0343-0.

37. D. Behmke, D. Kerven, R. Lutz, J. Paredes, R. Pennington, E. Brannock, M. Deiters, J. Rose and K. Stevens, *Proc. Interdiscip. STEM Teach. Learn. Conf.*, 2018, **2**, 3.
38. S. White, S. Feiner and J. Kopylec, in *IEEE Symposium on 3D User Interfaces*, ed. W. Stürzlinger, Y. Kitamura and S. Coquillart, IEEE, New York, 2006, p. 133.
39. S. Lee, J. Suh and H. Park, *Comput. Geosci.*, 2015, **76**, 41.
40. P. Wolle, M. P. Müller and D. Rauh, *ACS Chem. Biol.*, 2018, **13**, 49.
41. Available at https://www.zooniverse.org (accessed Oct 2018).
42. D. J. Cook, J. C. Augusto and V. R. Jakkula, *Pervasive Mob. Comput.*, 2009, **5**, 277.
43. J. M. Perkel, *Nature*, 2017, **542**, 125.
44. I. Hart and K. Martinez, *Earth Space Sci.*, 2015, **2**(5), 194.
45. C. Wisneski, H. Ishii, A. Dahley, M. Gorbet, S. Brave, B. Ullmer and P. Yarin, in *Cooperative Buildings: Integrating Information, Organization, and Architecture: Lecture Notes in Computer Science 1370*, ed. N. A. Streitz, S. Konomi, H. J. Burkhardt, Springer, Heidelberg, 1998, p. 22.
46. L. Arnstein, C. Hung, R. Franza, Q. H. Zhou, G. Borriello, S. Consolvo and J. Su, *IEEE Pervasive Computing*, 2002, **1**(3), 13.
47. G. Borriello, *Phys. Eng. Sci.*, 2008, **366**(1881), 3669.
48. B. J Brooks, A. L. Thorn, M. Smith, P. Matthews, S. Chen, B. O'Steen, S. E. Adams, J. A. Townsend and P. Murray-Rust, *J. Cheminf.*, 2011, **3**, 45.
49. A. Pentland, *Computer*, 2004, **37**, 42.
50. M. C. O'Connor, *RFID Journal*, 2006, Available at https://www.rfidjournal.com/articles/view?2199.
51. S. Guerrero, G. Dujardin, A. Cabrera-Andrade, C. Paz-y-Miño, A. Indacochea, M. Inglés-Ferrándiz, H. P. Nadimpalli, N. Collu, Y. Dublanche, I. De Mingo and D. Camargo, *PLoS ONE*, 2016, **11**(8), 0160428.
52. P. M. Scholl and K. Van Laerhoven, in *Proceedings of the 2014 ACM International Joint Conference on Pervasive and Ubiquitous Computing*, ACM, New York, 2014, p. 1381.
53. A. Ozcan, *Lab Chip*, 2014, **14**(17), 3187.
54. S. K. Vashist, O. Mudanyali, E. M. Schneider, R. Zengerle and A. Ozcan, *Anal. Bioanal. Chem.*, 2014, **406**(14), 3263.
55. B. Cortazar, H. C. Koydemir, D. Tseng, S. Fenga and A. Ozcan, *Lab Chip*, 2015, **15**(7), 1708.
56. Available at https://www.youtube.com/watch?v=jRIeLZ-_Y2Y (accessed Dec 2018).
57. J. D. Hincapié-Ramos, X. Guo, P. Moghadasian and P. Irani, in *Proceedings of the SIGCHI Conference on Human Factors in Computing Systems*, ACM, New York, 2014, p. 1063.
58. Available at https://www.wacom.com/en-us/products/pen-computers/wacom-mobilestudio-pro (accessed Dec 2018).

59. P. Brandl, C. Richter and M. Haller, in *Proceedings of the 28th International Conference on Human Factors in Computing Systems*, ACM, New York, 2010, p. 599.
60. W. E. Mackay and G. Pothier, *ERCIM News*, 2001, **46**, 52.
61. N. N. Chu, *IEEE Consum. Electron. Mag.*, 2017, **6**(2), 31.
62. L. R. Hochberg, M. D. Serruya, G. M. Friehs, J. A. Mukand, M. Saleh, A. H. Caplan, A. Branner, D. Chen, R. D. Penn and J. P. Donoghue, *Nature*, 2006, **442**, 164.
63. X. Chen, Y. Wang, M. Nakanishi, X. Gao, T. Jung and S. Gao, *PNAS*, 2015, **112**(44), E6058–E6067.
64. E. Yin, Z. Zhou, J. Jiang, Y. Yu and D. Hu, *IEEE Trans. Biomed. Eng.*, 2015, **62**(6), 1447.
65. Available at https://brainihack.org (accessed Dec 2018).
66. N. J. Knight, S. Kanza, D. Cruickshank, W. S. Brocklesby, J. G. Frey, *IEEE Internet Things J.*, submitted.
67. Available at https://blog.x.company/a-new-chapter-for-glass-c7875d40bf24 (accessed Dec 2018).
68. Available at https://support.apple.com/en-us/HT204895 (accessed Dec 2018).
69. R. Francese, I. Passero and G. Tortora in *Proceedings of the International Working Conference on Advanced Visual Interfaces*, ACM, New York, 2012, p. 116.
70. X. Chai, G. Li, Y. Lin, Z. Xu, Y. Tang, X. Chen and M. Zhou in *Proceedings of the IEEE International Conference on Automatic Face and Gesture Recognition*, IEEE, New York, 2013.
71. Available at https://www.gestsure.com (accessed Dec 2018).
72. M. Z. Uddin, M. M. Hassan, A. Almogren, M. Zuair, G. Fortino and J. Torresen, *Comput. Electr. Eng.*, 2017, **63**, 114.
73. Available at https://wearnotch.com (accessed Dec 2018).
74. P. Jagodziński and R. Wolski, *J. Sci. Educ. Technol.*, 2015, **24**, 16.
75. Available at https://www.microsoft.com/buxtoncollection/detail.aspx?id=20 (accessed Dec 2018).
76. D. J. Sturman and D. Zeltzer, *IEEE Comput. Graph. Appl.*, 1994, **14**(1), 30.
77. D. A. Norman, *Interactions*, 2010, **17**(3), 6.
78. D. Särg and K. Jokinen, in *Proceedings of the 2nd European and the 5th Nordic Symposium on Multimodal Communication*, Linköping University Electronic Press, 2015, p. 87.
79. E. Andonova and H. A. Taylor, *Cogn. Process.*, 2012, **13**(1), 79.
80. J. McCormack, J. C. Roberts, B. Bach, C. D. Freitas, T. Itoh, C. Hurter and K. Marriott, in *Immersive Analytics. Lecture Notes in Computer Science*, ed. K. Marriott, F. Schreiber, T. Dwyer, K. Klein, N. H. Riche, T. Itoh, W. Stuerzlinger and B. H. Thomas, Springer, Cham, 2018, vol. 11190, p. 57.
81. V. Hayward, O. R. Astley, M. Cruz-Hernandez, D. Grant and G. Robles-De-La-Torre, *Sens. Rev.*, 2004, **24**(1), 16.

82. H. Ishii, C. Ratti, B. Piper, Y. Wang, A. Biderman and E. Ben-Joseph, *BT Technol. J.*, 2004, **22**(4), 287.
83. J. Patten, H. Ishii, J. Hines and G. Pangaro, Duquesne, in *Proceedings of the SIGCHI Conference on Human Factors in Computing Systems*, ACM, New York, 2001, p. 253.
84. H. Hughes, D. Mills, D. De Roure, J. G. Frey, L. Moreau, M. C. Schraefel, G. Smith and E. Zaluska, *Org. Biomol. Chem.*, 2004, **2**, 3284.
85. R. Cooke and M. C. Schraefel, 2004, Available at https://eprints.soton.ac.uk/259251/.
86. M. Haller, in *Proceedings of the 8th International Conference on Human Systems Interaction (HSI)*, IEEE, New York, 2015, p. 15.
87. T. Chandler, M. Cordeil, T. Czauderna, T. Dwyer, J. Glowacki, C. Goncu, M. Klapperstueck, K. Klein, K. Marriott, F. Schreiber and E. Wilson, in *Big Data Visual Analytics (BDVA)*, ed. U. Engelke, J. Heinrich, T. Bednarz, K. Klein and Q. V. Nguyen, IEEE, New York, 2015, p. 73.
88. T. Czauderna, J. Haga, J. Kim, M. Klapperstück, K. Klein, T. Kuhlen, S. Oeltze-Jafra, B. Sommer and F. Schreiber, in *Immersive Analytics. Lecture Notes in Computer Science*, ed. K. Marriott, F. Schreiber, T. Dwyer, K. Klein, N. H. Riche, T. Itoh, W. Stuerzlinger and B. H. Thomas, Springer, Cham, 2018, vol. 11190, p. 289.
89. Available at https://developer.anki.com/blog/news/the-vector-sdk-hits-alpha/ (accessed Dec 2018).
90. R. Schiffman, *Science*, 2014, **344**(6183), 459.
91. D. Floreano and R. J. Wood, *Nature*, 2015, **521**, 460.
92. J. Frey, C. Bird and C. Willoughby, in *Innovation through Knowledge Transfer 2012. Smart Innovation, Systems and Technologies*, ed. R. Howlett, B. Gabrys, K. Musial-Gabrys and J. Roach, Springer, Heidelberg, 2013, vol. 18, p. 19.
93. J. Frey, C. Bird and C. Willoughby, in *Innovation through Knowledge Transfer 2012. Smart Innovation, Systems and Technologies*, ed. R. Howlett, B. Gabrys, K. Musial-Gabrys and J. Roach, Springer, Heidelberg, 2013, vol. 18, p. 11.
94. Available at https://www.skype.com/en/features/skype-translator/ (accessed Dec 2018).
95. Available at https://itunes.apple.com/us/app/google-translate/id414706506?mt=8 (accessed Dec 2018).
96. M. Back, G. Golovchinsky, J. Boreczky, P. Qvarfordt, L. Denoue, T. Dunnigan, J. Doherty, B. van Melle and G. Filby, *UbiComp*, 2006, Available at http://www.fxpal.com/publications/the-use-project-designing-smart-spaces-for-real-people/ (accessed Dec 2018).

CHAPTER 13

Telling the Story

13.1 Introduction

In this book, we have explored a variety of topics related to recording science, and in particular looked at the transformation in science from paper-based to digital recording. There are many aspects of science that are changing, from the nature of the science under investigation to the makeup of the teams conducting the research, such as 'big science', and increasing multidisciplinary and global teams, and even citizen scientists. Along with these changes are growing drives to make both the results of research and the products of research more accessible to a wider audience, facilitating not only reuse and reproducibility of the science, but also quality improvements in both the record and the science itself.

In this chapter, we will focus on some of the unresolved questions that may change the way that science is recorded in the future. Although the issues may seem unrelated initially, they may in fact converge when we think about the potential role of digital tools in the future. First, we will take a look at what belongs in the scientific record and expand upon our definition from Chapter 1. We will consider the nature of Electronic Laboratory Notebooks (ELNs) and discuss whether there is a need for the development of digital research management systems to support the needs of researchers. Next, we will examine the arguments for and against paper, and consider whether the move to paperless technology is essential for progress in science. Finally, we will consider storytelling and serendipity in science and different ways that stories about research and researchers might be told using digital tools and other methods.

Recording Science in the Digital Era: From Paper to Electronic Notebooks and Other Digital Tools
By Cerys Willoughby

Published by the Royal Society of Chemistry, www.rsc.org

13.2 The Nature of the Record

The scientific record has a number of different functions, as we discussed in Chapters 2 and 3. The published record provides the basis upon which the scientific community as a whole is able to build upon scientific knowledge, 'standing on the shoulders of giants' in the words of Isaac Newton. However, the published record does not stand alone, and the unpublished record, especially the careful notes we take in our notebooks, provide the evidence to help us when we need to prove our claims and demonstrate our priority in novel research or to refute accusations of scientific misconduct. The notes we make act as both an 'aide-memoire' to remind us what we did and as an institutional memory so that others can both see and repeat what we did. However, the scientific record is comprised of more than just these two parts.

Shankar describes the scientific record as being somewhere between 'data' and the 'formal paper'.[1] A paper must stand on its own to describe research, but as mentioned in previous chapters, the format of the formal paper does not provide the space or scope for the author to provide the full details of the methodology of the research, nor the complete data used to arrive at the results discussed within. In Chapter 1, we discussed the scientific record as being generated by a process of 'inscription', with the day-to-day details of scientific activities recorded primarily in paper notebooks and the results of those activities published in academic journals. However, we also produce other inscriptions as a result of 'doing science'. Raw data is inscribed by hand or recorded by instruments. Intermediate documents are often produced as part way steps towards formally published materials, such as research reports, presentations, and drafts of papers. These items, known as the 'grey literature', may be collected by institutional libraries or repositories, but are not usually published formally, and yet they arguably belong as part of the scientific record. There are a number of other information artefacts that might be considered to belong to the scientific record because of the contribution that they make to the story of the research. In Chapter 5, we defined 'research data' as all the information collected as part of scientific activities contributing to the production and validation of research findings, including methodologies, protocols, workflows, survey materials, models, algorithms, and code, as well as photographs, maps, specimens, and other physical artefacts. Even living specimens, or 'biorecords', can be considered to be living information objects that form a part of the record, for example, cell lines, bacteria, and genetically engineered laboratory mice.[2] Models, algorithms, code, software, and related resources used as part of scientific activities can be considered as 'research tools' that are typically needed in order to replicate studies, and therefore also belong within the scientific record, as we have discussed previously. Without these research tools, it is impossible to replicate the research of others, and even with access to them there can be challenges. There are also other information artefacts that

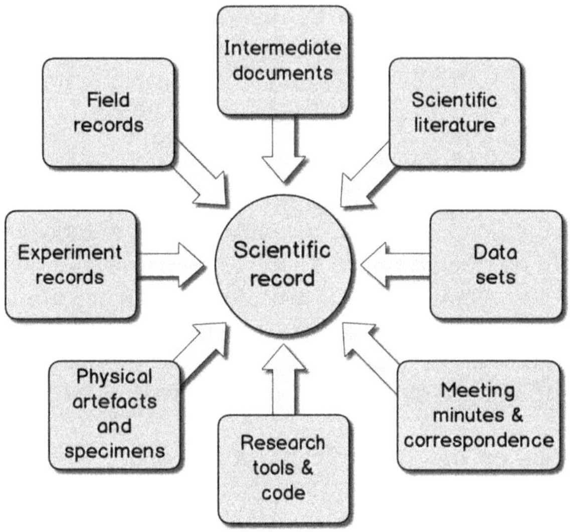

Figure 13.1 Potential components of the scientific record.

might be considered to belong to the scientific record because of the contribution that they make to the story of the research, such as correspondence and meeting minutes as shown in Figure 13.1. Figure 13.1 expands on Figure 1.1 provided in Chapter 1, to include additional information artefacts produced during the scientific process that could be considered to be a part of the scientific record, including physical artefacts, specimens, and research tools.

Before the development of the academic journal, there were three main ways of communicating scientific research: through the writing of books, through the sending of letters, and by meeting in person. Books take a long time to produce, and so the most up-to-date research would be communicated by letter, including information about studies and scientific findings as they were made across the world. Even long after the creation of academic journals and formal publishing, much scientific research was still being communicated and discussed by letter. Communication by letter has not just included descriptions of results, but also ideas about new research, acquisition of materials and equipment, discussion of proposals for funding, details of the planning of the research, providing status reports, and requests for assistance during the research process, as well as providing accounts of findings. Correspondence written by eminent scientists is recognised as having significant importance in the study of the history of science, with particularly important letters even generating substantial revenue when brought to auction.[3,4] The contents of correspondence from eminent scientists, coupled with details from their paper notebooks often provides an insight into the research paths and thoughts of those scientists, and in particular how the interaction and collaboration with different scientists may have influenced their thinking. Correspondence can also fill in gaps in our knowledge if the notebooks are missing or if particular interactions were not recorded in the notebooks. For example, over 9000 letters written between Charles Darwin and his correspondents are available online.[5] These letters provide insight into Darwin's scientific and data collection activities, but also how discussions with friends and colleagues influenced the development of his scientific theories. Letters such as Darwin's have an important place in the history of science for the insight that they can give us by filling in gaps in notebooks, revealing the hidden thoughts, personalities, and interrelationships of scientists, and by situating the science within the context of contemporary dogma.[6]

The history of science is not the only domain within which letters between scientists have had value. They can often provide important background to scientific bad behaviour or accusations of fraud, particularly where the plans for research or activities of the research have not been adequately recorded. For example, letters provide the background, sequence of events, and roles of responsibility in a notoriously unethical clinical study, the Tuskegee Syphilis Study.[7] The study ran for 40 years and deliberately failed to treat African American suffers of syphilis so that the effects of the disease could be studied until death, even though an effective cure for the disease became available during the study. The infrequently published reports on the study were able to obfuscate some of the unethical practices that were going on as part of the study, but various letters and meeting minutes documented the different strategies used to keep the participants sick for the study and to keep them coming to examinations. These documents are certainly an important part of scientific history, although as a record of what should not be done rather than as credible research to be built upon. The question is, should such documents be considered a part of the scientific record?

Capturing of correspondence is evident in some of the more modern efforts to archive the research record of individual researchers, as well as for eminent scientists from history. In the UK, preservation of the records of individual scientists has been undertaken since a pilot by the Royal Society and Royal Commission on Historical Manuscripts to archive the work of distinguished and recently deceased members of the society in the 1960s, and also include an exercise to archive the materials of scientists contributing to the Human Genome Project.[8] In these archives, artefacts include both digital and physical correspondence relating to topics such as published papers, discussions about the research, invitations to conferences, correspondence about computer systems, grant proposals, job vacancies, and applications. Whether the artefact is physical or digital, the correspondence relates to the research and is deemed worthy of being archived for the benefit of future understanding of science.

In the digital era, most communication in science, outside of formal publishing and conferences, is done by e-mail, telephone, in meetings, and to a smaller extent through the use of social media. E-mail correspondence is very similar to letters, although perhaps less personal and generally more mundane. However, the content of e-mails can be just as important to the record for highlighting poor scientific behaviour or for refuting it. A prime example of this comes from the 'climategate' scandal, where around 1000 e-mails and documents were stolen and subsequently released from the Climatic Research Unit (CRU) at the University of East Anglia (UEA) in 2009, including private correspondence from the head of the unit.[9] The materials released were used by the climate change sceptics community to accuse the CRU of hiding data with 'tricks' and manipulating the peer review system, whilst the content of the e-mails also demonstrated a lack of willingness to share data. Many of the e-mails were, in the author's own words, 'awful', with blunt refusals to share data in response to Freedom of Information (FOI) requests and rudeness about other scientists and their papers.[10] Ultimately, the accusations levelled at the CRU were found to be false, although it was noted that their unwillingness to share data probably escalated the problems. In fact, the released e-mails were only a small percentage of all the e-mails and highlighted frustrations with being inundated with FOI requests, where a response to each request was followed by more and more requests; in consequence, releasing the earlier e-mails in the chain after the initial theft and release may have helped to dispel the suspicions.[10,11] The scandal does raise certain questions about what belongs in the record. Many of the unflattering e-mails would have been standard fair for discussions in person, and had they not been recorded digitally through e-mail, they would never have been public knowledge. Animosity between researchers is nothing new in science, and there are numerous examples from history, where not only letters but even entire symposia have been used to disparage competitors.[12–15] The use of informal language within the climategate e-mails reflects the informal nature of the medium. In particular, the use of the phrases 'Mick's Nature trick' and 'hide the decline' were sufficient to communicate to other scientists within the unit the methodology to use to represent the data, but suggests a different, more nefarious, meaning to a non-expert audience.[10] Stolen e-mails also revealed that scientists were also actively encouraged to delete certain emails pre-emptively in order to avoid sharing. Although perhaps understandable from the perspective of avoiding the irritation of intrusive FOIs, the suggestion is nonetheless concerning. Should scientists be deleting materials that might be considered a part of the record? Do scientists need to keep everything? If something is deleted there is unlikely to be a record of when and why. Ultimately, decisions need to be made about what needs to be kept and what can be disposed of, as we have mentioned when we discussed data management in Chapter 5, but it may be impossible to know what materials will become important or valuable at a later time. Are discussions about research planning, peer review comments, comments about other publications, research proposals, applications for funding, and discussions about candidates for job applications a part of the record? These are the kind of materials that are being archived along with the traditional elements of the record, such as paper notebooks and presentations, for eminent scientists as mentioned above. However, it is unlikely that scientists are deliberately archiving these materials outside of their own e-mail inboxes. If these materials are considered part of the record, then there is a major challenge in providing additional context to translate them or describe them for a different audience. It could be argued that only 'formal' elements of scientific inscription should be considered to be a part of the record; notebooks although private and personal, are meant to be read, whereas e-mails and letters are not. However, they can become important and significant providing both evidence and insight into scientific activities in their own right.

Correspondence by letter or e-mail is perhaps not where the most important discussions happen in science. Discussions about research and decisions about the future directions of research are typically made in meetings, such as regular team meetings or one-to-one sessions between supervisors and researchers. In fact, team meetings are not just important for discussions of the research, but are also a key setting for the generation of new ideas, proposals for new experiments, and the development of hypotheses and interpretations, because they provide the opportunity for the asking and answering of questions and the rationalising of thoughts.[16] Meetings are clearly important for the generation and development of scientific ideas, and therefore such discussions have significant relevance to the scientific record. It is common for some notes to be captured from meetings in notebooks, but this is not universal, and any such notes are likely to be incomplete and biased towards the interests of the individual researcher, rather than an accurate and complete record of the discussion. As mentioned in Chapter 12, there are possible strategies for capturing complete records of meetings, such as using dedicated note-takers or capture through audio recording or video, but such activities are likely to be rare in the scientific environment. Digital tools have a potential role here in facilitating the capture, storage, and dissemination of such records, but there would first need to be a recognition of the value of meeting minutes as part of the record. Recordings of meetings can be accepted as normal, and online meetings are often recorded for the benefit of absent members of the team, but recordings of 'in person' meetings are still relatively unusual. It is, as mentioned previously, important that the actual content of such recordings can be extracted in a textual form in order that they are easily accessible if a point or decision needs to be checked. It could be argued that if meetings were captured, a conversion to a textual format would only be needed if there was a need to query something about the meeting, but finding the correct meeting to reference would still be much easier if the content could be searched; and text has a much smaller footprint than audio or video files.

The exposure of discussions and decision-making processes within science from correspondence and meetings may have a potentially interesting role in Open Science. It is already possible to see more public discussions of ideas on social media sites, blogs, and wikis. On such sites, users can explore the influence of individuals on the discussions and see how the research developed and what decisions were made. Conferences are often 'live-tweeted', so that people not attending can hear about the presentations and contribute their own thoughts to the discussion. Where collaboration or community engagement is part of the objective for a project or a particular team then it makes sense for these discussions to be out in the open. Recordings of online or tele-meetings facilitate the inclusion of members of the team in different locations or who could not be present, and they also provide a way to share the activities of the project with a wider audience. For such Open Science projects, open discussion is desirable and discussions taking place on public platforms are recognised as not being private, which is, of course, quite different to public release of private correspondence and meeting minutes. That private correspondence could become public in the future is perhaps a consideration for researchers, to both avoid embarrassment, but also to consider what might constitute valuable correspondence to retain.

One of the elements of correspondence that has been included in archived records of researchers are discussions about published papers and also reviews made as part of the peer review process. The current peer review process is under scrutiny for a variety of reasons, not least that the process has been ineffective in identifying cases of scientific fraud, fake papers generated by computer algorithms, and even faked reviews.[17,18] In the interests of transparency, a number of publishers have begun to publish the reviews from the peer review process alongside the published papers, and recently *Nature* has started a trial to name reviewers on the reviewed paper, if it is published, in order to acknowledge the contribution of reviewers and in recognition of their importance in shaping scientific papers.[19] In effect, these originally private reviews have become part of the published record by being published alongside the papers they address.

The various elements of the scientific record need to be kept safe and accessible for those who need to access them, at whichever point in the scientific process they need to be used. The published record, in the form of journal articles and conference proceedings, is already well managed by the many scholarly publishers, but the unpublished record, comprising the details and the data of science, is troublingly distributed across a variety of media and systems.

13.3 The Nature of Electronic Laboratory Notebooks

In Chapters 4 and 6, we took a look at a number of examples of electronic notebooks and ELNs in particular. There are advantages and disadvantages to different kinds of electronic notebooks, and as discussed, many ELNs are targeted towards a specific audience in a particular discipline. For example, many ELN vendors specifically target an industrial or commercial audience and often contain specialised modules for the benefit an individual disciplines such as chemistry, biology, life sciences, and pharmaceuticals. However, there is great diversity and many ELNs also target both industrial and academic audiences and provide discipline-specific features for more than one field. We discussed 'generic' electronic notebooks, such as Microsoft's OneNote and Evernote, that provide basic capabilities for content capture and organisation and are relatively simple for anyone to use. There are also a number of ELNs that provide minimal functionalities beyond these generic electronic notebooks that can be used by any discipline and audience, but many others have vast capabilities in comparison, often reflected in their complexity. Beyond discipline specific modules, there are numerous capabilities that are important to many scientists, including digital signing and witnessing, security and access control, versioning, and data storage for example. Many other ELNs provide advanced functionality, enabling the control of instruments, connecting with data repositories, managing samples, connecting with other systems such as Laboratory Information Systems (LIMS), and project and task management, and reporting. The market of ELNs is in fact so diverse it can be difficult to define exactly what an ELN is and what it should do.

In Figure 13.2, some of the capabilities an ELN can have are shown, together with access to systems locally and remotely. At the very least, an ELN is expected to provide tools for note-taking, and even in the most basic electronic notebooks there are usually ways to add drawings and links to other data. Some of the features that are usually present in ELNs that may not be within generic electronic notebooks include the ability to be able to

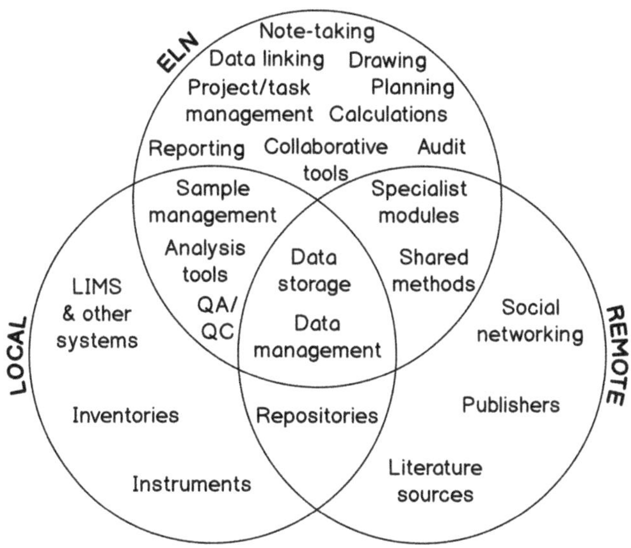

Figure 13.2 ELN capabilities and connections.

perform calculations, tools for collaboration, and tools that enable planning of experiments such as protocol templates and experiment plans. Beyond that, the other capabilities shown in Figure 13.2 are less common, although many are more common within ELNs targeted at industrial audiences, such as task management, reporting, and auditing. Some of the capabilities that could be found within ELNs are often present and already available as tools running externally to the ELN. For example, tools may already be used locally for managing samples, performing analysis on data, and tools for Quality Assurance (QA) and Quality Control (QC). Specialist modules can be available within an ELN, but it is also possible for ELNs to connect to remote systems in order to access specialist information such as dictionaries, databases, and discipline-specific tools. Methods can be shared internally through functions within the ELN, or it may be possible to import or access methods remotely through the ELN. ELNs can also connect to other local systems such as LIMS, inventories, and instruments, as well as remote systems such as other data sources and social networks.

One of the big dilemmas with ELNs is where the data should be stored. Although the majority of ELNs have storage space and enable the upload of certain files into notebook records, not all have the capacity or capabilities to hold large quantities of data or to automatically import data directly from instruments. Links can be made to data stored in local or remote locations, or data can be stored centrally within the same system. There are advantages to having the data stored together with the notes: when they are stored together there is less chance of them becoming separated or disconnected and makes it more likely that the files can be opened *in situ*, but this is not always possible. There are also questions about when and if content needs to be moved from the ELN into a different system for sharing or for archive. Although the use of digital files is highest after creation and then drops off over time, easy access is likely to remain useful for a longer period of time, and in fact may have patterns of both use and non-use over time.[20] For example, during an initial project, files may be used intensively, and the use may drop off after completion of the project, but may pick back up again if a similar project begins or if a new researcher attempts to duplicate the work. In Open Notebook Science, files are perhaps likely to be accessed by others more often than the original author as time progresses. In either case, sensible decisions need to be made about how long files should remain active, and when they can be archived. As mentioned previously, many ELNs provide some options for exporting the content of the ELN for sharing or for importing into another system, but most often this is in PDF format, or sometimes XML. These formats may be considered suitable for archive, and it may be possible for ELNs to directly connect to specified repositories, locally or remotely to enable automatic deposition.

The expectations that users have about ELNs and what they will do with them, do not necessarily match up with the capabilities that they provide. As previously mentioned, to some researchers, an electronic notebook means simply a digital pen and the associated software that is used to produce text files, rather than software within which notes are captured directly, but the majority have expectations beyond this. For example, Kanza *et al.* provide an extensive list of 'desired features' for ELNs including less common features, such as 'post-it notes' and 'to-do lists', voice searches, linking to reference managers, paper notebook and digital pen integration, text recognition, audio capture, 'get involved' pages, and bulletin boards.[21] As discussed in Chapters 7 and 8, the working practices of scientists can help inform the capabilities that are most appropriate, but it is also important to consider how needs might change over time.

Many ELNs attempt to incorporate numerous capabilities with the aim of facilitating the paperless lab and providing a single tool that can do the majority of tasks that might be required in the laboratory. A potential future area of development may be the creation of platforms for scientific research, for example, the concept of a 'a research management platform' or ELN-type tool that can do everything. There are a variety of ways to build such platforms, but from the user's perspective, such a tool would enable them to perform all the tasks related to their scientific research within the same tool; such as planning experiments, collecting data, recording the experiment, analysing the data, and creating reports for publication, and even submitting to publishers and depositing data into repositories. Such a tool might also contain project management elements, so that the profiles of users and their interests could be accessed within the same tool, notifications could be sent to collaborators, and searches could be performed across the whole system to find materials, experiments, and people of interest. Such a tool might also include, or at least integrate with, systems for managing instruments and materials, such as inventories and sample tracking systems. There are pros and cons to such an approach. On the one hand, there is a certain amount of desire from users to be able to do everything within a single familiar interface. On the other hand, the more capabilities a tool has, the more complex the tool is likely to be. It is also likely that the implemented capabilities will be less well developed than those in a specialist tool. It is difficult for a software development organisation to be expert across all disciplines and to be able to dedicate as much effort to every element of the software. Often, the specialist tool will be developed by experts within that area who can focus on creating a good quality experience for that single function, compared with the tool that does 20 different things. Those ELNs that provide extra extensions and specialist modules are close to providing a platform like experience, by being able to have third-party experts develop the specialist functionality on their behalf.

There is no right answer to what an ELN is and what capabilities it should have. As the diverse market shows, different organisations and different teams have different needs, and therefore there is 'no one size fits all' answer. Those solutions that are tailored to a specific audience are ideal for those organisations that match them, but for many scientists, their research does not fit neatly into a box, and therefore the more generic options are likely to be more appropriate. What is important is for ELN developers to understand their audiences and to provide functions that can provide value for customers without overwhelming them with unnecessary complexity. As discussed in Chapter 8, designing for usability is extremely important to enable users to get the best out of software and to encourage them to both use and recommend the software to others. Ease of use and anxiety about the complexity of ELNs is one of the most significant barriers to ELN adoption. As mentioned previously, ELNs are an area of technology where there is a great interest and many potential new users to recruit. There is a lot of potential to develop new and valuable tools to support scientists in their research and to enable them to create a high-quality record that can be reused by themselves and others.

13.4 Paper *versus* Digital

For many researchers, the tool of choice for capturing their research is still the trusty paper notebook, even though the majority are using computers and digital tools for other aspects of their scientific practice. As mentioned in Chapter 1, notebooks can be a reflection of the minimum information that a researcher was taught to record or mere scribbles to remember some information just for a moment, but more often, the notebooks are much more than that and are recognised as special for the precious information held within. As we discussed in Chapter 6, there are various benefits of the medium of paper for capturing notes, such as portability, ease of use, and familiarity; however, it is likely that our attachment to our paper notebooks goes beyond these affordances. Although we are more likely to become attached to physical objects, compared to digital objects, digital objects can be treasured if they possess certain qualities, such as being self-made, augmented, changed through use over time, or had been the focus of engagement for a long time, qualities that apply to paper notebooks, but could also potentially apply to electronic notebooks.[22] It has been suggested that digital note-taking loses some of the personal imprint of the author and obscures their personalities and emotions.[23] The handwritten notes and hand-drawn images within the paper notebook add not just personality and life to the narrative on the page, but can also provide a connection to the author. Handwriting is unique and provides a connection to the individual that a digital font cannot, although it is possible to create a unique font based on an individual's handwriting that could perhaps recapture this unique and personal element of paper-based notes.

There are also a variety of ways in which the digital capture of the record may impact the quality of the record as we have previously discussed. The designs of interfaces can have a substantial impact on what information gets recorded, as discussed in Chapter 9. With increasing quantities of data now being captured in digital form, there is also the chance that the focus will become on the capture of data, to the detriment of the capture of important context information about the data. As mentioned in Chapter 6, some authors have observed that their note-taking behaviour has changed as a result of switching from paper to digital, and others lament that they no longer seem to write down their observations and feelings like they once did, focused instead on the collection of data rather than capturing the personal experience.[23] As described in Chapter 9, when participants in the Lego Rocket Cars study were given a table to complete to collect data from the experiment as part of the study, they were so focused on data collection that they were distracted from recording information that actually described the experiment and from reflecting on the results. In Chapter 12, we looked at a number of systems where technology such as cameras, radio-frequency identification (RFID) tags, alternative computer interfaces, and wearables were being used to capture elements of the scientific record automatically. A concern raised about such systems is whether they might deter users from writing full notes to describe the experiment, and whether what is captured automatically is enough information to replace what the researcher would have recorded without them. There is a significant risk that not enough or inappropriate information may be recorded because of an over-reliance on the system and assumptions about what the system is recording or how much additional information is needed. Although a representation of the experiment, such as the various flow-chart representations described in Chapter 12, may serve as a useful 'memory of events' for the researcher after the experiment, it is likely that after some time when the researcher's memory has faded, the representations alone will not be enough to enable the researcher to recall the detail of the experiment. It is also debateable whether such representations provide enough information for another researcher to be able to understand and duplicate the work.

Basic representations such as flow charts can suffer from similar problems as those we see with the replication of methods included in scholarly publications, in that they provide a high-level overview of experimental procedures, but do not provide sufficient details for another researcher to duplicate the work. A similar problem can also occur when researchers rely on screenshots of workflows to describe an automated process; the image alone provides no information about the settings and code underneath and although provides a high level overview, does not allow another researcher to reconstruct the workflow or understand the exact process represented.[24] However, using visual representations such as flow charts and workflows can facilitate understanding of the experiment or flow of activities, when enough information is captured alongside them and made available to describe the details of the process. The danger is in developing systems where a screenshot of a workflow or even a high-level list of activities is considered information enough, and the actual detail fails to be recorded. If the detailed information is later needed, but fails to be recorded, there is no way to go back and retrieve it after the event. The problem of focusing on the data and failing to capture the context may be even more acute for new researchers who have yet to develop a note-taking style, and are therefore perhaps less aware of the value of capturing the personal experience and the detailed process of the research, or 'showing their working'.[25]

There are other advantages of note-taking that might be lost if the nature of note-taking in science changes. For example, note-taking is considered a way of helping authors to organise their thoughts and come up with new ideas.[26] The use of narrative whilst note-taking is considered to capture the thought process of the author, and that if the time to properly record work is not taken, then opportunities for generating new ideas and solutions may be missed.[27] That note-taking provides the time and space to reflect on the work is a crucial issue. Having the opportunity to reflect, to consider the outcomes, to record the problems and to develop solutions, and to design the next stage of the research is essential to generate the ideas that leads to the path of innovation or progression. Whether reflection takes place in a paper notebook or on a computer screen probably does not matter, but it is most important that the reflection takes place. If limited notes are taken, or if the design of an interface doesn't support it, then there is a chance that the process of reflection is hindered, and opportunities will be lost. Creativity may also be lost or stifled under such circumstances. Within some environments, the purpose of recording is just to capture the data and no more. For example, in a test laboratory, the routine may be that a sample moves between tests and the results of the test are recorded, with no further observation or comment, because these are not required. However, in the majority of scientific environments, more is required, because the results of an activity have a bearing on the future course of the research and the interpretation of the findings. As Simonton says, 'Creativity in science does not result from a static, repetitive, and well-delineated process as found in assembly-line manufacturing. Rather creative output emerges out of a dynamic, incessantly changing, and ill-defined process' (p. 78).[28] If you always follow the same recipe, you may lose creativity, and it is possible that restricted formats for capturing notes could stifle creativity by forcing the collection of some information, whilst reducing the capture of free form thought. As discussed in Chapter 9, the interfaces that we use to capture research notes have a significant impact on what is captured and failing to ask the right questions can result in the loss of information, particularly the personal experience. It may be that the blank page encourages reflection, where completing a form-style template does not.

The flexibility and familiarity of paper notebooks are probably their biggest benefits. They enable researchers to record whatever they want however they want to. Nevertheless, as discussed in Chapter 6, there are drawbacks to paper, with the primary problem being access. It is only really practical for one person to use a notebook at a time making it difficult for others to even know what it contains, let alone reproduce or validate the research within. Records that are carefully preserved and made accessible from within repositories, archives, or other software systems, such as ELNs, have the potential to impact and shape future scientific activity.[29] However, records locked away in paper notebooks are likely to be inaccessible and lost from scientific memory. Although researchers tend to focus on the potential negative sides to sharing their data such as being scooped in publication, misuse of the data, or having a competitor pass the data off as their own, there are in fact benefits to sharing. Open practices and digital tools can improve both the discoverability and visibility of a researcher's work, increasing citations, but also facilitating a reciprocal access to novel data, software resources, and making new connections.[30] Open projects are also more likely to lead to a competitive advantage for funding as funders continue to mandate open sharing in publicly funded projects.

Despite the many benefits and capabilities of ELNs, researchers are reluctant to give up their paper notebooks. There is anxiety and resistance to overcome with ELNs, and it is worth stressing that a large number of sceptical users of ELNs do like and value them once they have tried them for a while. In many environments, especially the laboratory, they can effectively remove the need for paper notebooks and provide immense value at the same time. However, it is unlikely that replacing paper notebooks everywhere is either practical or desirable. There are undoubtedly many situations and environments in which paper notebooks are more practical to use than digital tools. Technologies such as touch screens and electronic paper that can be written on can arguably provide a bridge between paper and desktop ELNs, although they need to have sufficient accuracy and resolution, as discussed in Chapter 12. There are an increasing number of laptops that are becoming hybrid between laptop and tablet, that provide not only a smaller, more portable format, and the ability to write on the screen, but also the capability to run the other scientific tools that users might want to use in their work, which are less likely to be available or fully featured on a tablet device. Technology is expensive, and users do not want to buy multiple tools to use in different locations, making it more likely that many researchers will move away from tablets towards more capable but more slim-line laptops in the future.

Some would argue that we need to find a way to make digital tools that work with paper rather than trying to replace paper.[21] As previously mentioned, we could scan or take photos of the paper notebook and include them in ELNs or other digital tools, as some users already do. However, this is a very manual process and the files uploaded are next to useless in terms of mining data, being able to search the contents, sharing with others, and for providing any additional value for the user. If paper notebooks must be used, then a better approach would be to adopt digital pens for use with them. In that way, the author can capture the notes in the way that they desire, but the actual information recorded can also be captured and utilised in digital format. Digital pens are expensive and perhaps not as robust as they need to be in certain environments, but they do at least provide a workable compromise for ease of use and capturing the record in digital form. As with ELNs, a culture change is needed to adopt digital pens, but at least the change in working practices is much smaller. Digital tools need to be provided that can make good use of the information captured in such a way that value can be provided for the user.

There is a risk that when notes are captured in digital form that the researcher will feel inhibited because of the knowledge that digitally captured notes are more likely to be shared than those only captured on paper. However, it seems likely that the technology element of a digital pen is more likely to blend into the background and become invisible in comparison with taking notes on a computer using typing rather than writing. Using the pen, the user can capture the notes in the way they want to, without having to work around the computer. For example, capturing chemical names and materials on the computer in just text form can be difficult, and different researchers tend to use different strategies to overcome this, such as using formal chemical names, common names, abbreviations, or the molecular formula. On paper, however, they may also use one of these strategies or may be much more likely to use a drawing of the molecular structure to represent the materials in their experiment, something which cannot be done on a computer without a specialist tool, and even with a specialist tool, it is still easier on paper. That the researcher can keep their paper notebook and share the notes in digital format may also be considered a benefit.

If we record our research only on paper, then it cannot easily be shared, and opportunities are lost for not only the wider scientific community, but also for the author of the notebook. Information captured within the paper notebook cannot be mined, and therefore cannot be used for search, cannot be used for linking together information, and cannot be used to provide added value that can be provided by digital tools, as we will discuss in the next section. It must also be remembered that ELNs are not simply replacements for paper, they provide a myriad of additional functions that can support researchers with their work, from the humble spell-checker to advanced discipline-specific tools. Although data sharing and open notebook science are relatively new, with the ubiquity of the Web and increasingly open access to resources, any part of the record that is not shared will not be found, used, and subsequently built upon. If the scientific record provides the narrative of scientific discovery and technological development, then the loss of any part of the record leaves the narrative incomplete and the players missing.

13.5 Storytelling and Serendipity

The majority of the public think that researchers should be telling their own stories about their research, and it is likely that more of this personal communication about science will be expected in the future, perhaps as part of the drive towards more open science or through a desire for greater public engagement. As discussed in Chapter 11, blogs can be an effective way for researchers to tell their stories and share their research to a wider audience.[31] Narrative and storytelling do not have a particularly good reputation within science, but for effective communication with others outside of an area of expertise, they can become much more important.[32] Stories can help readers or listeners to understand, remember, and connect with the story and the person telling it.[33] When communicating with experts, it is enough to detail the work and let the science do the talking, but more is needed to help non-experts to develop an emotional connection with the story.[34] Narrative is sometimes considered as just a step-by-step account of an event, whereas the story typically has a plot, recognisable characters, and ultimately a resolution.[33] The classic 'heroes journey' structure to popular stories are quite suited to tales of scientific research, with the 'hero' researcher overcoming adversity or making use of unexpected results to come up with a new piece of the puzzle towards their ultimate goal. Such a format provides the opportunity to talk about problems and mistakes, as well as the triumphs of research. The public are particularly interested in human interest stories, and much of the history of science plays to this interest with stories of the scientists as characters overcoming adversity, making great discoveries, carrying out bizarre experiments, overcoming the orthodoxy, and engaging in deep rivalries.[35] As previously mentioned, journal articles never represent the full story of the research, and valuable information such as the dead-ends and failed experiments do not get recorded, and yet these can provide an important insight into how the research developed, as well as helping others avoid the same mistakes. We often learn more from things that go wrong than when everything happens the way we expect it to.

Narrative in textual form is one way to tell a story, but it is also possible to tell stories using other formats. Even a single image, a diagram, or photograph, together with an appropriate short description, can convey a whole story, much more than the description or the image alone.[36] Collections of photographs with descriptions that may be related or unrelated to each other are a common journalistic format. For example, *Nature* publishes monthly and annual 'best science pictures', combining both fascinating and important images and stories from science.[37,38] A sequence of photographs together with some explanatory text can also be an effective way to communicate a story, for example, by providing a visual representation of change or activities over time. *Nature* again provides an example of the use of sequences of photographs with their visual storytelling blog series 'Nature India Photo Story' telling a variety of science related stories from India, including stories about the research of individual scientists.[39] The US National Oceanic and Atmospheric Association (NOAA) also publish 'photo stories' about their activities, often focused on the stories of the scientists or public engagement activities.[40,41] Another method of using sequences of images to tell stories is through the use of comics. Although some view comics as childish, because they appeal to a younger audience, they can, in fact, be a way to communicate sophisticated information effectively to an audience unfamiliar with a subject, both young and old.[42] There have been a number of different 'science comics', developed for communication and educational purposes, with many aimed at adults just as much as children. For example, a series of manga comics explaining a variety of scientific concepts have been produced by the Solar-Terrestrial Environment Laboratory of Nagoya University, and 'Science Stories' and 'Science Stories 2' are books containing short but informative comic-based stories about real researchers and their work.[43–45] An introductory comic about synthetic biology has even been published in *Nature*.[46] There are also science- and mathematics-based graphic novels.[47] Many of the science comics use fictional elements to provide entertainment, but also to convey concepts in an understandable way. There have been numerous studies that demonstrate that comics provide an effective way to both educate and enthuse readers about science, especially those audiences with less knowledge and less interest in science, but without having a negative effect on the those who already have an interest in science.[47–49] Communicating science in comic-based form can improve both enjoyment of science and attitudes towards science. Comics do not have to be sophisticated, but most are likely to be developed as a collaboration between scientists and a graphic designer of some kind. Digital tools do exist to help design comics, although they can be quite complex, but even simple drawing tools and 'stick people' can be used to convey the message, as seen in many of the images in this book.

Maps as images can also provide effective ways to tell stories, especially for activities or stories that visualise changes over time and place, and can be particularly useful for talking about research in the field, such as environmental studies, archaeology, ecology, and earth sciences. An example tool for telling stories using both images and maps is Story Maps.[50] Story Maps lets users craft their own stories with a variety of different templates and allows them to add their own images, videos, sounds, and associated commentary. There is a focus on map-based stories, but not exclusively. For example, 'London's lost river' from the Museum of London Archaeology (MOLA) tells the story of the research into the lost rivers of London through photographs, paintings and drawings, videos, cross sections, and maps.[51] The maps have overlays to explain parts of the story such as excavation locations, borehole sites, topography, and projected courses of rivers. The story is also used to link to the associated publications. Another example, is the 'Sounds of the Wild West' where photographs, maps, and sound files are used to explore the wildlife and environments of locations in North America.[52] A Story Map has also been used to create a poignant tribute to an employee of Esri and one of the designers of Story Maps, with the story told with photographs, screengrabs, and memories by colleagues.[53] The diversity of community Story Maps demonstrates how a few different formats illustrated by images and maps can be very effective at telling a story without the need for vast quantities of text. Such tools could be valuable for helping researchers to create more engaging material for a wider audience, in addition to drier formal papers.

Outside of actively creating stories about the research and scientists to share, there are ways that digital tools can help us to explore the scientific record and the research paths and interactions of individual scientists. Digital tools, with the right information, can allow us to generate visualisations or even textual representations, that can allow users to see connections between people, activities, digital artefacts, and changes over time. Providing these connections in a visual form enables users to be able to gain insights and be reminded of the nature of those connections over time. For example, when Viégas *et al.* created visualisations based on e-mail archives, showing relationships between both people and dates, the visualisations prompted the recall of particular events that users wanted to share.[54] Clusters displayed show interesting relationships or connections for specific projects; over time, different people come and go whilst some are more constant. In fact, when users have used the visualisation to identify important events or periods, they desire to highlight them for future use. If something as mundane as an e-mail archive when visualised can provide a spark to memory or prompt users to want to capture and highlight the connections, then using visualisations could have enormous value when applied to the scientific record.

Time is an important aspect of the scientific record. Dates are usually the one consistent element recorded as a part of individual entry in the notebook, and when scientists seek out previous research, time is frequently the method that is used to locate it by flicking backwards to the time period of that particular research.[21,55] For any information that we can capture in a system, we can produce a timeline describing that information, and such timelines can represent a variety of scales. For example, Figure 13.3 shows a timeline feature from the LabTrove ELN showing entries in the notebook, but with a graphical representation of activity over time, with increasing scale from a period of days at the top, to months, and years at the bottom. This view not only provides a way to access the individual entries, but also shows the pattern of activities in the notebook over time. The owner of the notebook can quickly navigate to time periods of activity to explore the entries of interest.

Although LabTrove does not record the length of experiments, in a system that did capture this information, the timeline could show the overlap between different experiments, when the experiments are carried out in parallel for example. Any metadata that is captured about the experiments or within each entry can be used to filter the content of the timeline. In the example in Figure 13.3, the entries relate to different researchers, and metadata has been used to capture that, enabling the user of the timeline to display the activity in the notebook from different researchers. Any other metadata fields that have been added to the entry can be used in the same way to filter the content on a topic of interest.

Although we tend to think of timeline representations as being linear with events displayed along a line, they do not have to be. For example, the Timeline feature in Facebook displays posts on a user's wall in either a

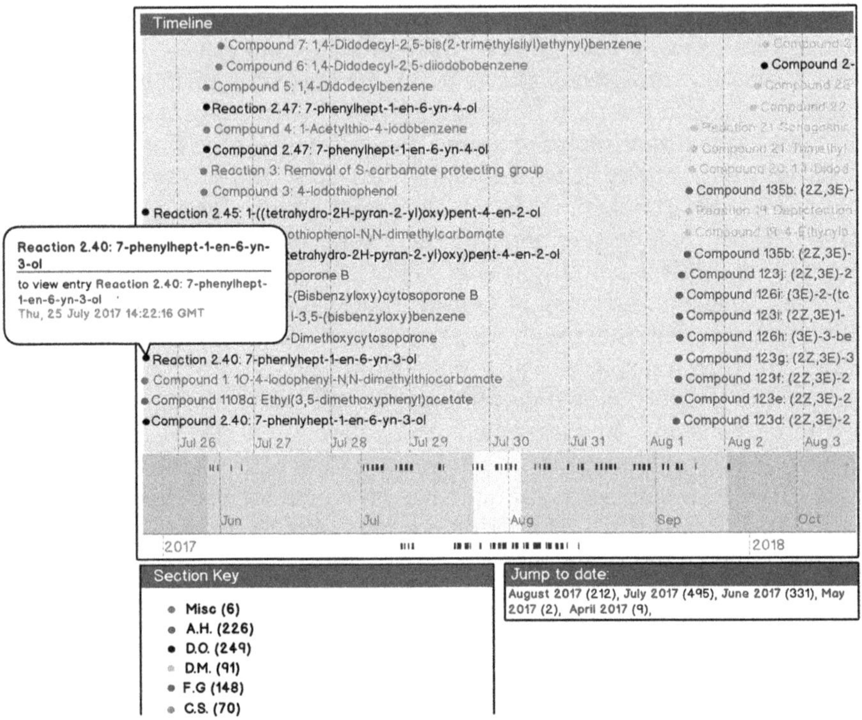

Figure 13.3 Mock-up of the LabTrove timeline showing the activity in a notebook over different timescales ranging from a few days to a whole year in the same view.

standard newsfeed list style or alternatively a grid style. The grid style, shown in Figure 13.4, enables a lot of information to be displayed in a relatively compact form, but also makes it easy to pick out and find individual posts from a particular month. Metadata is also used to provide additional information, presented in graphical form, such as who created the post, the kind of post (such as a photograph, text, or link), the number and type of reactions the post has received, in addition to providing extra descriptive information. The user is also able to filter the content based on metadata such as whether the posts were created by themselves or by others, and whether the user is tagged in the posts. Although not provided, a vast amount of other metadata about the posts could be used to filter the contents. In an ELN, images attached to notebook entries or other images, such as reaction schemes, for example, could be used to create a similar representation that can provide a way to both locate relevant information and to tell the user something about the story of the research over time.

The majority of timelines in use are relatively simplistic, but they could be quite sophisticated, enabling users to visualise the research trail and any element involved within it. For an individual experiment, we could view a timeline of updates made to the experiment, for example, measurements,

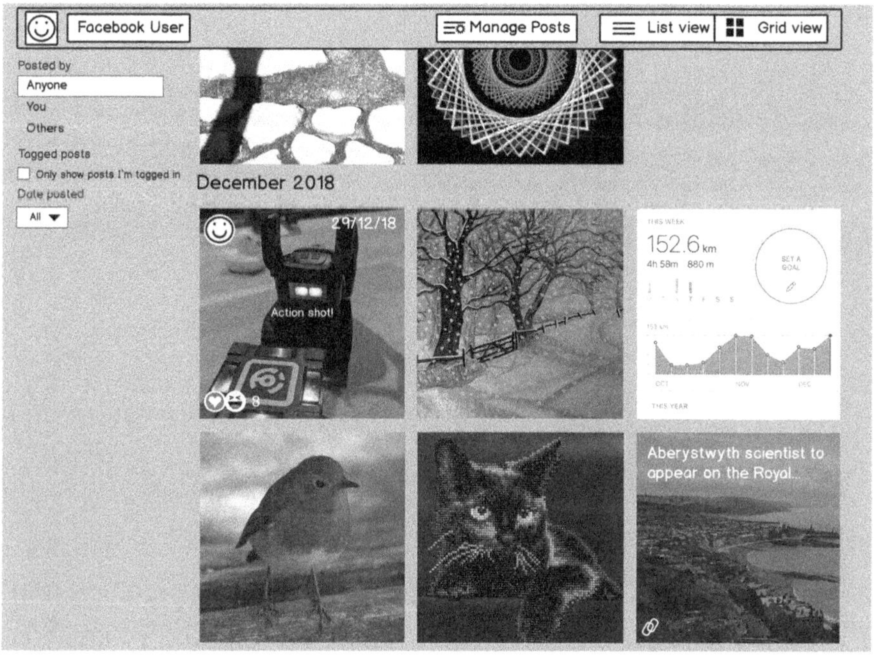

Figure 13.4 A mock-up of the Facebook timeline grid view.

observations, or information from an analysis running as the experiment proceeds. The changing conditions and data points recorded could be plotted on a timeline, as a graph or even as an animation enabling the experiment to be 'replayed'. If the experiment is one of a sequence, a repeat, or just a related experiment, they could be plotted together with others and compared and analysed to find new patterns or examined to explain differences between them or their results. Over longer periods of time, we could plot changes in the topics of the research activities recorded, for example, for a chemistry project, we might be able to plot changes in the reactions and materials used. A more complex, but perhaps more visually appealing idea could be to visualise the changing chemical structure of the target compound over time.

Timelines and other visualisations are not just limited to experiments and results data, but also to projects and people. For instance, relationships between different people are typically examined through their connections, for example, connections to other researchers, institution, projects, materials and instruments in common, and through shared topics of interests. A common way of visualising relationships between people is to use a network diagram where people are represented by nodes and connections between them are represented by links. Figure 13.5 shows an example of a network diagram, in this case, showing the relationships between people on my friends list on Facebook. The clusters produced by the visualisation encourage exploration. It can be seen that many of my friends know each

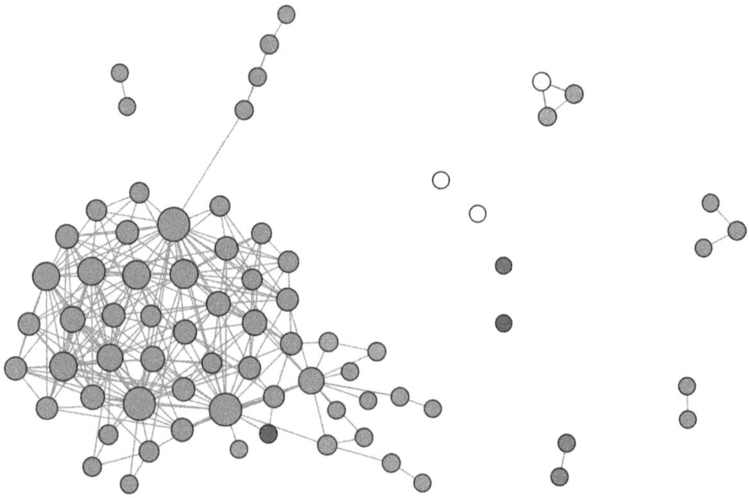

Figure 13.5 Example of a network diagram based on connections in Facebook.

other, but there are individuals who do not know anyone else, others who know each other but no-one else, and groups linked by only one other friend in my list. The visualisation highlights relationships from different activities, but it is also interesting to find people from different places who unexpectedly know each other.

There are many other kinds of visualisation that can be utilised to tell stories and to communicate information, and as mentioned in Chapter 12, visualisation is very important in science for exploring data, identifying patterns, and making new connections. Of course, the ability to create any kind of visualisation requires that we have managed to capture the appropriate information within the system to generate the visualisation. Smart software can help to generate and represent the story of the research. The career of an individual could be visualised on a timeline or through animations, visualising their 'trail of research' by showing how the topics and elements of their experiments change and mutate over time. Their projects, institutions, and research connections could also be visualised to provide a unique tool for viewing the research paths and career evolution of an individual. Relationships between individuals, projects, and research activities can tell the story of individual projects, but also whole teams, and with enough information, whole disciplines. As mentioned in Chapter 2, the questions 'who', 'what', 'where', and 'when' can make effective cues to help us remember information, but coupled with visualisations, they can also provide effective tools for telling stories. The key is capturing the information and developing tools that understand the information in order to be able to turn that information into something useful.

Visualisations can help to tell stories, but as mentioned, they can also help to facilitate new understandings and discovery. Serendipity can be thought of as fortunate accidental discoveries; under the right circumstances, we can find something interesting when we are not looking for it. For example, clusters of connections in a network visualisation can reveal hidden connections between data that we were not aware of. There are many accidental discoveries that have gone on to facilitate important developments in science, for example, Alexander Fleming's discovery of penicillin, Jocelyn Bell's discovery of pulsars, Roy Plunkett's discovery of Teflon, and William Röntgen's discovery of X-rays. There are a number of ways that systems can be designed to help facilitate serendipity, by helping users to access information that is of value to them and potentially to provide new knowledge and help to make new connections along the way.[56] Providing visualisations is one way to facilitate serendipity by providing new ways to look at information. For example, representing search results in a visual way rather than a list can help to encourage curiosity and exploration. For example, tags clouds, although less common now than they were, can potentially lead to more exploratory behaviour. They can also provide additional information for the user, such as the popularity of terms, as shown in Figure 13.6.

Making discoveries requires appropriate context in order for the connections to be made. Context comes from both one's own knowledge and also from the knowledge provided by the record. Ensuring that a comprehensive and inclusive scientific record is captured within digital tools is a good way of acquiring such context. A challenge within science is finding content and connections outside of the knowledge that you already have. Browsing in a physical library can enable you to come across materials that you did not already know existed, but in the digital realm, you often need to know what you are looking for in order to find it, based upon keyword searches which can limit what you find. The use of different terminologies across domains means there may be very relevant content out there, but it can be difficult to find if you do not already know it exists. Many systems provide help in the form of 'recommendations' that can overcome some of these limitations. For example, many of the scholarly journal publishers provide

Figure 13.6 Example tag cloud based on the most commonly used words in Chapter 11.

a number of different ways of accessing related content when you view a particular article, such as links to other articles that have been referenced, links to articles that cite the work, and recommendations for related articles based upon keywords or co-authors. The same kinds of recommendations can also be used in other systems. For example, most social networks offer recommendations for connections to people you might know, people of interest to follow, or people who are connected to people you know. Blogging platforms often provide recommendations for posts to read or bloggers to follow. However, these systems rely on context to generate the recommendations. If there is no underlying context because the system does not know very much about us, then the recommendations suggested are not likely to be very useful or accurate.

An example of where access to appropriate context can influence the relative usefulness of recommendations comes from researcher networks and reference managers such as ResearchGate and Mendeley. These networks provide recommendations for both journal articles to read and also researchers to connect with, based upon work you have previously read, cited, or researchers you have co-authored with, and research topics of interest. The designs of the systems are quite similar, with both recommending potential articles of interest supplemented with information about how many people are reading them and the number of citations if appropriate. Both also provide suggested people to follow together with a short explanation about why you might want to follow them, for example, they follow the same people you already do, they are followed by people you know, or they are popular within a discipline of interest. ResearchGate relies on the information based on your own articles, whereas Mendeley can use the whole library of articles contained within. Personally, I find that Mendeley is much better at guessing and recommending appropriate articles I might be interested in reading. I believe the reason for this is because Mendeley typically has much more information to work from in generating the recommendations. For example, my Mendeley library contains thousands of papers, whereas ResearchGate relies on a much more restricted base for recommendations, with a much smaller number of 'research items' that the system is aware of. The Mendeley library contains a much broader base for recommendation, with much more diversity in terms of topics and authors of articles compared to the works of a single author. The information that ResearchGate knows about an individual is more limited because researchers tend to author work with a more limited range of co-authors and publish on a more limited range of topics than they actually read. Both platforms enable the user to provide more information about themselves and their interests, but if the user chooses to create a minimalist profile, then less helpful recommendations can be made. Social networking tools are very reliant on the information that can be acquired about users automatically, or the information that users provide to them either directly or indirectly. A user's commitment or interest in a particular platform will determine how much information they are willing to share, and consequently impact on how much value the system can provide for them.

Electronic Laboratory Notebooks have the opportunity to capture much more information from the user directly, compared with systems such as researcher networks and reference managers which have to extract their knowledge about users automatically. As discussed in Chapters 9 and 10, a lot of information about the experiments recorded in ELNs can be captured by asking researchers to provide user-defined metadata and from information that can be mined from the notes recorded. Key to this, is that the relevant information is provided by users. As we have discussed previously, if the system makes use of the information to provide value for the user, they are more likely to add it. ELNs can make use of metadata or mined information to provide advanced search facilities, to enable grouping of entries based on common properties, and to generate visualisation of elements of the experiments. For example, a visualisation could be created to show yield or other measurement of an experiment over time, or a comparison of conditions and results in a sequence of experiments. As an aside, in terms of direct storytelling, the notes taken by users are the most important. Not only do researchers need to thoroughly and accurately record the 'how' and 'what' of the experiment, but as mentioned previously, we also need to capture the 'why'. Why was this experiment or study undertaken? Why did we change what we initially planned to do? Why were the results different than we expected? What we did and what happened are incredibly important, but reflecting on the 'why' is fundamental to telling the story, and each is necessary to help others to understand and reproduce the research. Perhaps researchers could be encouraged to consider their note-taking as a way of telling their stories, including both failures and triumphs; each entry as a step on the road to reaching a particular scientific goal. Notebooks containing reflection and narrative are likely to be more effective for telling the story both during the research and at the end.

Electronic Laboratory Notebooks collect information about the scientific activities captured within, but it is less common for them to collect information about the people and their relationships, beyond determining accessing and security. ELNs with collaborative tools may be able to derive enough information about relationships to create networks of interaction, for example, connections between researchers working within the same notebook or relationships based upon comments or reviews within the notebooks. Recommendations could be generated about researchers working on similar materials, techniques, or instruments, and recommendations for notebooks or individual experiments to view based on these elements. For ELNs with many users, such recommendations could help researchers to find others with similar interests and skills. More sophisticated ELNs that include project and task management capabilities are likely to have access to additional information on which recommendations based upon relationships could be derived, but all have the potential to be able generate visualisations and recommendations around the elements of the research. For example, timelines of experiments, how the development of materials has changed over time, and recommendations for related experiments to view. Such capabilities are likely to be particularly useful within Open Science, where such tools could provide the opportunity for storytelling of the science within and also functions for the discovery of related work, tools, and potential collaborators. However, at the present time, there is no one coherent system that can connect researchers, research, data, and research tools, although individual systems may partially fulfil some of these roles.

With different parts of the scientific record and information about researchers stored in a diverse array of different digital tools, it is difficult for researchers to keep a track of all the different elements of their research such as contacts, data, methods, tools, and published research. There are services that take away some of the need for tracking. For example, for published papers, services such as ORCID, Mendeley, ResearchGate, Google Scholar, and ImpactStory frequently trawl the Web and collate published works on behalf of the author. Depsy is an example of a service that allows users to track their contributions to code projects.[57] As discussed in Chapter 11, there are many kinds of digital tools and networks available for collating information about researchers and the published literature, but also data, tools, blogs, methods, and a small number collating information about open notebooks. All this information can be used to create opportunities for visualising the record and creating serendipity in research, but it is not in any way joined-up. The most hidden aspect of the record is the information recorded in the notebooks and in ELNs, and this is arguably the information that is the most valuable and the most important to capture well. Making connections and discovering similar work whilst research is being carried out is probably the best time to do so.

There must be a potentially vast untapped universe of yet to be developed digital tools that can make use of all this information, assuming that the appropriate standards and Application Programming Interfaces (APIs) can be developed to facilitate such integration. Such tools may even help to encourage researchers to share their stories, from the bird's eye view of research paths, to the narrative and reflection that provides depth to the story of the research.

References

1. K. Shankar, *J. Doc.*, 2009, **65**(1), 151.
2. A. Acker, *Arch. Sci.*, 2015, **15**(1), 1.
3. Available at http://www.sothebys.com/en/auctions/ecatalogue/2017/history-of-science-technology-n09686/lot.32.html (accessed Feb 2019).
4. Available at https://www.sothebys.com/buy/a63edcf7-3bf9-4ba5-82cc-9aaeb71f8aef/lots/96ae535d-5044-4a3f-84b9-bfc6b6ebe2b0 (accessed Feb 2019).

5. Available at https://www.darwinproject.ac.uk/letters/darwins-letters-timeline (accessed Feb 2019).
6. A. Abbott, *Nature*, 2018, **561**(7724), 441.
7. A. M. Brandt, *Hastings Cent. Rep.*, 1978, **8**(6), 21.
8. J. Shaw, *Stud. His. Philos. Biol. Biomed. Sci.*, 2016, **55**, 61.
9. D. Adam, *Nature*, 2010, **468**, 362.
10. House of Commons Science and Technology Committee, *The Disclosure of Climate Data from the Climatic Research Unit at the University of East Anglia*, 2010.
11. O. Heffernan, *Nature*, 2009, **460**, 787.
12. F. J. Fine, *Nature*, 1977, **266**, 19.
13. M. Romano and R. L. Cifello, *Science*, 2015, **350**(6263), 915.
14. R.M. Clary, *Geol. Soc. London Spec. Publ.*, 2016, **442**, 189.
15. C. N. Johnson, *J. His. Biol.*, 2018, **52**(1), 45–85.
16. K. N. Dunbar, in *Conceptual Structures and Processes: Emergence, Discovery, and Change*, ed. T. B. Ward, S. M. Smith and J. Vaid, American Psychological Association Press, Washington, 1997, p. 461.
17. R. Van Noorden, *Nature*, 2014, Available at https://www.nature.com/news/publishers-withdraw-more-than-120-gibberish-papers-1.14763 (accessed Feb 2019).
18. C. Ferguson, A. Marcus and I. Oransky, *Nature*, 2014, **515**(7528), 480.
19. Nature Communications, 2018, Available at https://www.nature.com/articles/s41467-018-07336-1 (accessed Feb 2019).
20. K. Shankar, *Arch. Iss.*, 1999, **24**(1), 21.
21. S. Kanza, C. Willoughby, N. Gibbins, R. Whitby, J. G. Frey, J. Erjavec, K. Zupančič, M Hren and K. Kovač, *J. Cheminf.*, 2017, **9**, 31.
22. C. Golsteijn, E. van den Hoven, D. Frohlich and A. Sellen, in *Proceedings of the Designing Interactive Systems Conference*, ACM, New York, 2012, p. 655.
23. J. L. Reveal, in *Field Notes on Science and Nature*, ed. M. R. Canfield, Harvard University Press, Cambridge, 2011, p. 187.
24. C. Willoughby and J. G. Frey, *Int. J. Digit. Curation*, 2017, **12**(1), 72.
25. C. L. Bird, C. Willoughby and J. G. Frey, *Chem. Sci. Rev.*, 2013, **42**(20), 8157.
26. A. Eisenberg, *J. Chem. Educ.*, 1982, **59**, 1045.
27. M. R. Canfield, in *Field Notes on Science and Nature*, ed. M. R. Canfield, Harvard University Press, Cambridge, 2011, pp. 1–18.
28. D. K. Simonton, *Creativity in Science: Chance, Logic, Genius, and Zeitgeist.*, Cambridge University Press, Cambridge, 2004.
29. J. A. Bastian, in *Research in the Archival Multiverse*, ed. A. J. Gilliland, S. McKemmish and A. J. Lau, Monash University Publishing, Victoria, 2016, p. 269.
30. E. C. McKiernan, P. E. Bourne, C. T. Brown, S. Buck, A. Kenall, J. Lin, D. McDougall, B. A. Nosek, K. Ram, C. K. Soderberg, J. R. Spies, K. Thaney, A Updegrove, K. H. Wool and T. Yarkoni, *eLife*, 2016, **5**, e16800.
31. A. Gerber, in *Opening Science*, ed. S. Bartling and S. Friesike, Springer, Cham, 2014, p. 73.

32. M. F. Dahlstrom, *PNAS*, 2014, **111**(4), 13614.
33. S. J. ElShafie, *Integr. Comp. Biol.*, 2018, **6**(1), 1213.
34. S. Martinez-Conde and S. L. Macknik, *PNAS*, 2017, **114**(31), 8127.
35. H. Chang, *Notes Rec.: R. Soc. J. His. Sci.*, 2017, **71**(1), 20160042.
36. Available at https://www.nature.com/articles/d41586-018-07684-4 (accessed Feb 2019).
37. M. Tatalovic, *Nature*, 2018, **564**, 318.
38. D. Castelvecchi, 2018, Available at https://www.nature.com/articles/d41586-018-07599-0 (accessed Feb 2019).
39. Available at http://blogs.nature.com/indigenus/category/nature-india-photo-story (accessed Feb 2019).
40. Available at https://www.noaa.gov/education/photos-images/photo-story-science-meets-education-where-river-meets-sea (accessed Feb 2019).
41. Available at https://www.noaa.gov/education/photos-images/photo-story-aquarium-partners-engage-public-in-creating-healthy-ecosystems (accessed Feb 2019).
42. Available at http://www.asor.org/anetoday/2018/09/Comics-Narrative-and-Archaeology (accessed Feb 2019).
43. Available at http://www.stelab.nagoya-u.ac.jp/ste-www1/doce/outreach.html (accessed Feb 2019).
44. Available at http://www.elliott-design.com/science.html (accessed Feb 2019).
45. Available at http://www.elliott-design.com/science2.html (accessed Feb 2019).
46. D. Endy, I. Deese and C. Wadey, *Nature*, 2005, **438**(7067).
47. A. N. Spiegel, J. McQuillan, P. Halpin, C. Matuk and J. Diamond, *Res. Sci. Edu.*, 2013, **43**(6), 2309–2326.
48. S. Lin, H. Lin, L. Lee and L. D. Yore, *Int. J. Sci. Edu. Part B*, 2015, **5**(3), 276.
49. J. Hosler and K. B. Boomer, *CBE Life Sci. Edu.*, 2011, **10**(3), 309.
50. Available at http://storymaps.arcgis.com/en/ (accessed Feb 2019).
51. Available at http://molarchaeology.maps.arcgis.com/apps/Cascade/index.html?appid=6b00daa1acac4df7a2fcde06104bac1a (accessed Feb 2019).
52. Available at https://storymaps.esri.com/stories/2018/sounds-of-the-wild-west/index.html (accessed Feb 2019).
53. Available at https://story.maps.arcgis.com/apps/Cascade/index.html?appid=f2e8448fef064238ace4f324ffc16fde (accessed Feb 2019).
54. F. B. Viegas, D. Boyd, D. H. Nguyen, J. Potter and J. Donath, in *Proceedings of the 37th Annual Hawaii International Conference on System Sciences*, IEEE Computer Society, Los Alamitos, 2004, 1265287.
55. R. Cooke and M. C. Schraefel, 2004, Available at https://eprints.soton.ac.uk/259251/.
56. S. Makri, A. Blandford, M. Woods, S. Sharples and D. Maxwell, *J. Assoc. Inf. Sci. Technol.*, 2014, **65**(11), 2179.
57. Available at http://depsy.org (accessed Feb 2019).

Subject Index